THE STORY OF THE DINOSAURS IN 25 DISCOVERIES

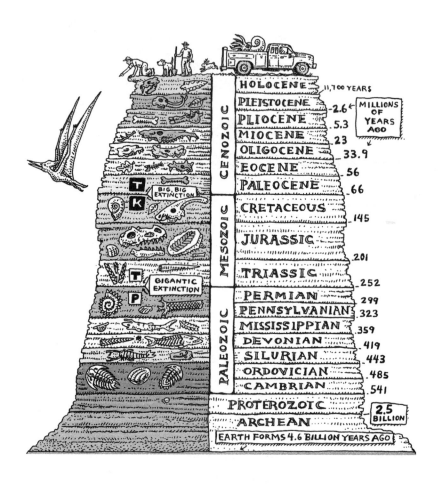

HOLOCENE ..... 11,700 YEARS

PLEISTOCENE 2.6

MILLIONS OF YEARS AGO

PLIOCENE 5.3

MIOCENE 23

OLIGOCENE 33.9

EOCENE 56

PALEOCENE 66

CENOZOIC

CRETACEOUS 145

JURASSIC 201

TRIASSIC 252

MESOZOIC

T K — BIG, BIG EXTINCTION

PERMIAN 299

PENNSYLVANIAN 323

MISSISSIPPIAN 359

DEVONIAN 419

SILURIAN 443

ORDOVICIAN 485

CAMBRIAN 541

PALEOZOIC

T P — GIGANTIC EXTINCTION

PROTEROZOIC

ARCHEAN

2.5 BILLION

EARTH FORMS 4.6 BILLION YEARS AGO

THE STORY OF THE DINOSAURS

in 25 DISCOVERIES

AMAZING FOSSILS AND THE PEOPLE
WHO FOUND THEM

DONALD R. PROTHERO

COLUMBIA UNIVERSITY PRESS   NEW YORK

COLUMBIA UNIVERSITY PRESS

Publishers Since 1893
New York   Chichester, West Sussex

cup.columbia.edu
Copyright © 2019 Donald R. Prothero
All rights reserved

Library of Congress Cataloging-in-Publication Data
Names: Prothero, Donald R., author.
Title: The story of the dinosaurs in 25 discoveries : amazing fossils
    and the people who found them / Donald R. Prothero.
Description: New York : Columbia University Press, [2019] |
    Includes bibliographical references and index.
Identifiers: LCCN 2018044694 | ISBN 9780231186025 (cloth : alk. paper) |
    ISBN 9780231546461 (e-book)
Subjects: LCSH: Paleontology—History. | Paleontologists. | Dinosaurs.
Classification: LCC QE705.A1 P76 2019 | DDC 567.9—dc23
LC record available at https://lccn.loc.gov/2018044694

Cover design: Julia Kushnirsky
Cover image: Trudy Nicholson
Frontispiece: Courtesy of Roy Troll

• ◉ •

THIS BOOK IS DEDICATED TO THE GENERATIONS OF
GREAT PALEONTOLOGISTS OF THE AMERICAN MUSEUM
OF NATURAL HISTORY.

HENRY FAIRFIELD OSBORN
WALTER GRANGER
BARNUM BROWN
WILLIAM DILLER MATTHEW
WILLIAM KING GREGORY
JACOB WORTMAN
ELMER RIGGS
GEORGE GAYLORD SIMPSON
EDWIN H. COLBERT
BOBB SCHAEFFER
MALCOLM C. MCKENNA
EUGENE GAFFNEY

THEY ESTABLISHED VERTEBRATE PALEONTOLOGY IN THE UNITED
STATES; THEIR DISCOVERIES PUT DINOSAUR PALEONTOLOGY
ON A FIRM FOUNDATION; AND THEIR TALENT FOR MAKING
DINOSAURS EXCITING BROUGHT ABOUT "DINOMANIA." THEY
WERE MY INTELLECTUAL MENTORS AND FORBEARS WHEN I WAS
TRAINED THERE FROM 1976 TO 1982.

CONTENTS

PREFACE

I am one of those kids who got hooked on dinosaurs at age four and never grew up. I knew I wanted to be a paleontologist as soon as I understood what the word meant. However, when I was a child in the 1950s and early 1960s, there was almost no dinosaur merchandise to buy—just a handful of kids' books and few plastic toys. One or two amateurish movies and TV shows had a few seconds of stop-motion Claymation dinosaur effects, advanced CG dinosaur movies certainly didn't exist in the 1950s or 1960s, and there were only a few sci-fi movies with crude stop-motion animation. Moreover, my classmates all through school did not seem interested in dinosaurs, and I was considered somewhat of a freak for knowing all those complex names. It was so unusual that they had me lecture the sixth graders about dinosaurs when I was only in fourth grade.

Today that has all changed. "Dinomania" has been growing since at least the late 1970s and 1980s, and it's difficult to pinpoint the specific cause. A huge media onslaught of dinosaurs exists today, and a gigantic array of dinosaur merchandise is available. Perhaps it was the effect of the "Dinosaur Renaissance" (chapter 17), which brought dinosaurs to life on more and more TV specials, that gradually got people interested. Certainly, by the time Michael Crichton's novel *Jurassic Park* hit the best-seller lists and the movie version blew away the box office records in 1993, a whole new generation of kids was fired up, most of whom are now just entering the profession of vertebrate paleontology. Whatever the reason, the norm these days is for most kids to go through a "dino phase" from ages 6 to 10; after that most of them gradually lose interest. Their hormones kick in, and they focus on the opposite sex, sports, video games, and other high school obsessions.

This huge upsurge in interest has been a mixed blessing for vertebrate paleontology. We are certainly more popular and respected then when I was trying to make it into the profession, but the job market is still horrendous. With so many eager students now, only one in one hundred will land a job as a professional research paleontologist in a museum or university. Not only are jobs lacking, but there is little or no grant funding for all but a fortunate few. Most vertebrate paleontologists struggle to get field funds or even to pay for simple supplies for camping or prep work.

No jobs, no money, few chances to find funds to do much—it's not a pretty picture. But the profession chugs along, with hundreds of graduate students spending 10 years of their lives in college with no prospect of a job at the end of the line, and hundreds of scientific papers are published each year. A few of these discoveries are mentioned in the science media if they concern dinosaurs or other charismatic extinct beasts such as mammoths or saber-toothed cats. But as my graduate advisor, Malcolm McKenna, put it, "there are two types of people in paleontology: the rich and dedicated, and the poor and dedicated." Nobody goes into the field to get rich, that's for sure! A few of the leading members of the profession—Edward Drinker Cope, Henry Fairfield Osborn, Childs Frick, Wann Langston (and Malcolm McKenna)—came from wealthy families, but most do not. You either have money already and can afford to practice an esoteric profession that doesn't find oil or coal or make big money, or you do it because you love paleontology and wouldn't consider any other career despite the bad odds and poor pay.

I persisted through high school, college, and six years of graduate school at the American Museum of Natural History and Columbia University, the best place in the world to get the right background and connections in vertebrate paleontology at that time. Most graduate programs had terrible track records for placing their students, but nearly all of my classmates at the Columbia/ American Museum program succeeded in getting good jobs. We had the best collection in the world to work from, and we were trained by the leading paleontologists in the world, particularly my advisor, Malcolm McKenna.

As an undergraduate working with Mike Woodburne at the University of California-Riverside, I realized there were far more opportunities to do many different kinds of interesting projects with fossil mammals than there were with dinosaurs. This remains true today, even though the number of new dinosaur discoveries has skyrocketed. In addition, a lot of people are trying to study dinosaurs, so the field is very crowded. This is a big contrast with the available research opportunities in fossil mammals, with thousands of specimens available for study if you know where to look. I have done a few projects with dinosaurs here and there and tried to keep up with the incredible pace of new discoveries and new ideas, so I feel on top of most of the major developments over the past 50 years. I also started teaching a dinosaur course at the university level for the first time, which has brought me up to date much faster than anything else could.

After the great reception for my books *The Story of Life in 25 Fossils* and *The Story of the Earth in 25 Rocks*, I thought it was about time to do a dinosaur book using the same format. Each chapter focuses on a particular discovery or genus or specimen of scientific and historical significance, and it weaves a story around it of the interesting people who found it, what they thought, and a summary of what we now think about this or that dinosaur. Paleontology is replete with colorful characters and great human interest stories, from the eccentric William Buckland, to the nasty feud between Gideon Mantell and Richard Owen, to the Bone Wars between Edward Drinker Cope and Othniel Charles Marsh, to the aristocratic Henry Fairfield Osborn or the humble Walter Granger, or to the legendary collectors José Bonaparte and Barnum Brown, or the wild gay Transylvanian Baron Franz Nopcsa, who tried to become king of Albania.

The book is organized into four sections. The first section, "In the Beginning," talks about the first discoveries of dinosaurs in England and elsewhere (*Megalosaurus, Iguanodon, Cetiosaurus,* and *Hadrosaurus*) and describes the earliest ancestors of the dinosaurs such as *Eoraptor*. The second, "The Long-Necked Giants," reviews the major types of sauropods, the largest land animals that ever lived. The third section, "Red in Tooth and Claw," surveys the predatory theropods from their earliest forms to their most bizarre forms, including birds and the weird herbivorous therizinosaurs and ornithomimids. The final section, "Horns and Spikes and Armor and Duck Beaks," looks at the herbivorous ornithischians and their wide variety of armor, spikes, horns, frills, and crests.

I hope you enjoy this grand tour through the dinosaurs and come to know some of the interesting people who found them. Their stories and the (sometimes horrible) conditions they endured in pursuit of this knowledge are sure to amaze you.

ACKNOWLEDGMENTS

This book idea was suggested by my supportive editor at Columbia University Press, Patrick Fitzgerald. I thank him and Kathryn Jorge at Columbia University Press, along with Ben Kolstad at Cenveo, for all their hard work on the book. Most of the illustrations were generously provided by my colleagues who are acknowledged as appropriate in the captions. I thank Drs. Thomas Holtz, Darren Naish, and Hans-Dieter Sues for providing a careful scientific review of the entire book; and Daniel J. Chure, William Hammer, and John Foster for reading certain chapters in areas of their expertise. I particularly thank my youngest son, Gabriel, who is even crazier about dinosaurs at his young age than I ever was; he gave me helpful feedback on certain chapters. I thank my two other sons, Erik and Zachary, for their love and support on this project as I disappeared in front of my computer for one and one-half months in winter 2017–2018 to write this book. I especially thank my amazing wife, Dr. Teresa LeVelle, for her encouragement and support on this project, especially when my hands were stiff with carpal tunnel syndrome every night after writing a 4,000- to 5,000-word chapter every day.

THE STORY OF THE DINOSAURS IN 25 DISCOVERIES

PART I

IN THE BEGINNING

Behold now behemoth, which I made with thee; he eateth grass as an ox.

Lo now, his strength is in his loins, and his force is in the navel of his belly.

He moveth his tail like a cedar: the sinews of his stones are wrapped together.

His bones are as strong pieces of brass; his bones are like bars of iron.

He is the chief of the ways of God: he that made him can make his sword to approach unto him.

Surely the mountains bring him forth food, where all the beasts of the field play.

He lieth under the shady trees, in the covert of the reed, and fens.

The shady trees cover him with their shadow; the willows of the brook compass him about.

Behold, he drinketh up a river, and hasteth not: he trusteth that he can draw up Jordan into his mouth.

He taketh it with his eyes: his nose pierceth through snares.

—JOB 40:15-24

MEGALOSAURUS

There were giants in the earth in those days.

—GENESIS 6:4

GIANTS IN THE EARTH

For centuries, people had picked up huge bones in the ground and puzzled over their origins. In some parts of the world, they were thought to be the remains of legendary dragons, sea monsters, or Cyclopes. On the Greek island of Samos, the numerous large bones were thought to be the remains of Amazon warrior women who had died in battle. (We now know they are the remains of elephants, giraffes, antelopes, cattle, hyenas, and other mammals that were abundantly fossilized there.) The red color of the rocks was thought to be bloodstains (the color actually comes from the rusty iron oxides in the rocks). Some people have argued that the striking fossils of *Protoceratops* found in the Gobi Desert of Mongolia were the source of the myth of griffins, which had a lion's head and wings; the frill covering part of the neck reminded the ancients of wings (chapter 24). Large bones of dinosaurs, marine reptiles, and large mammals have been known since prehistoric times, and they have been explained by whatever mythology was prevalent among the peoples who found them.

In Europe, the myths about Earth found in the Bible strongly influenced what people saw in specimens. For example, in 1726 the Swiss scholar Johann Scheuchzer obtained and described a large fossil skeleton

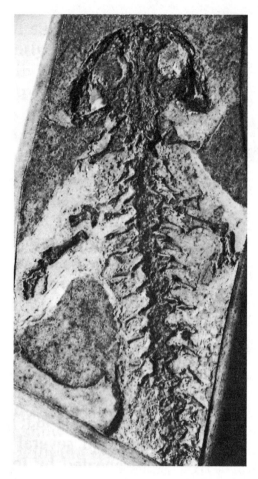

Figure 1.1 ▲

The giant extinct salamander interpreted as *Homo diluvii testis*, "Man, a witness of the Flood," by Johann Scheuchzer in 1726. (Public domain)

of a creature found in Switzerland. His biblical bias was so strong that he thought it was the skeleton of a man who had died in Noah's flood. He even named it *Homo diluvii testis*, "man, a witness of the Flood" (figure 1.1). Many years later the famous anatomist and paleontologist Georges Cuvier showed it was not human at all but a giant salamander, which is a close relative of the living giant salamanders found in China and Japan. Today it seems laughable that Scheuchzer could mistake a salamander fossil for a human skeleton, but comparative anatomy was in its infancy in 1726, and everyone was preconditioned to see giant humans as described in the Bible.

Numerous accounts of giant bones are known in the literature before 1800, but most cannot be tracked down or identified because the fossils have been lost and adequate illustrations, measurements, or locality information are not available to determine what they were talking about. One of the first such fossils that might be identifiable was illustrated in Robert Plot's 1677 book, *Natural History of Oxfordshire* (figure 1.2). Plot was the first professor of chemistry at the University of Oxford, and later he was curator of the Ashmolean Museum of Oxford, established in 1683. He

Figure 1.2 ▲

Robert Plot's fossil of the knee end of a dinosaur thighbone that a later author, Richard Brookes, interpreted as a fossil of giant petrified testicles, "Scrotum humanum." (Public domain)

was interested in all areas of natural history known at the time. His book described the living animals and plants of the Oxfordshire region as well as some of the rock and fossils. Among these was a strange piece of fossil bone that Plot correctly guessed was the end of a thighbone (femur) of a very large creature. This description was unusual for Plot because he did not believe that most fossils were the petrified remains of animals. Instead he imagined that they were formed by crystallization inside rocks.

The specimen had been found in quarries north of Oxford, from a formation we know now is the Middle Jurassic Taynton Limestone. It was much too large to match any animal currently found in England, so Plot guessed that it came from the skeleton of a war elephant used when the Romans conquered Britannia. Later he thought that it might be from a giant human, as mentioned in the Bible. In his words:

> Come we next to such [stones] as concern the . . . Members of the Body: Amongst which, I have one dug out of a quarry in the Parish of Cornwell, and given me by the ingenious Sir Thomas Pennyston, that has exactly the Figure of the lowermost part of the Thigh-Bone of a Man or at least of some other Animal, with capita Femoris inferiora, between which are the anterior . . . and the large posterior Sinus . . . : and a little above the Sinus, where it seems to have been broken off, shewing the marrow within of a shining Spar-like Substance of its true Colour and Figure, in the hollow of the Bone. . . . In Compass near the capita Femoris, just two Foot, and at the top above the Sinus . . . about 15 inches: in weight, though representing so short a part of the Thigh-Bone, almost 20 pounds.

The fossil has since been lost, but it is the first adequately illustrated dinosaur fossil known and is almost certainly from the dinosaur discussed in this chapter, *Megalosaurus*.

This first known dinosaur was almost given a truly inappropriate name. In 1763 Richard Brookes republished Plot's illustration and called the fossil "Scrotum humanum" in a figure caption. Indeed, to someone who doesn't know anatomy well and is conditioned to see every large fossil as a relict of biblical giants, it does look a bit like a pair of huge human scrota and the base of a penis. In addition, the form of Brookes's name suggests the genus and species binomial naming system devised by Carl Linnaeus in 1758 and earlier, not only for animals and plants but also for natural curiosities found in rocks. In 1970, the eccentric British paleontologist Lambert Beverly

Halstead (famous for trying to act out dinosaurs mating when giving presentations at scientific meetings) published an article suggesting that the first named dinosaur was properly called "Scrotum humanum." Later paleontologists asked the International Commission on Zoological Nomenclature (the body that makes the rules about scientific names) to formally suppress the name "Scrotum humanum." The commission ruled that this was unnecessary because the name was only published in a caption, without adequate description or diagnosis, the only specimen was lost, and it was not certain that it was the same as *Megalosaurus*.

Even though the original fossil was lost, specimens continued to be found in Stonesfield that ended up in the Oxford collections. Most were huge but fragmentary, so it was impossible to identify the animal to which they belonged. However, one lower jawbone with several teeth in various states of eruption found in 1797—bought for Oxford by anatomist and physician Sir Christopher Pegge for the then princely sum of 10 shillings 6 pence—was placed in the anatomy collections of Christ Church College in Oxford where Pegge taught. By 1815, there were quite a few bones, and they caught the attention of the legendary naturalist Sir William Buckland (figure 1.3).

Figure 1.3 ▲

Portrait of William Buckland. (Courtesy of Wikimedia Commons)

THE ZOOPHAGE

Buckland was one of the most amazing and colorful figures in the history of science. Born in 1784 in Axminster in Devon in southwestern England, he accompanied his father (the local church rector) on fossil-collecting walks on the coastline at Lyme Regis (later famous for the marine reptiles found by Mary Anning) and developed an early interest in natural history. After attending several schools, he ended up studying mineralogy and chemistry at Corpus Christi College in Oxford, where he would spend most of the rest of his life. By 1813, he succeeded his mentor John Kidd as a reader in mineralogy at Oxford, and he soon became famous for his popular and engaging style of lecturing. Buckland was known for his dramatic delivery and gestures during lectures, sometimes acting out the behavior of the animals he was describing (figure 1.4). According to one story from *The Life and Correspondence of William Buckland, D.D., F.R.S.* (Gordon [1894]),

> He paced like a Franciscan Preacher up and down behind a long show-case, up two steps, in a room in the old Clarendon. He had in his hand a huge hyena's skull. He suddenly dashed down the steps—rushed, skull in hand, at the first undergraduate on the front bench—and shouted, "What rules the world?" The youth, terrified, threw himself against the next back seat, and answered not a word. He rushed then on me, pointing the hyena full in my face—"What rules the world ?" "Haven't an idea," I said. "The stomach, sir," he cried (again mounting his rostrum), "rules the world. The great ones eat the less, and the less the lesser still."

He even gave lectures on horseback. He wore his heavy academic robes at every lecture and scrambled around outcrops on a field excursion in formal clothes.

His eccentricities extended to his household, which was loaded with specimens of fossils, minerals, and animals he had collected. He and his wife, Mary Morland (a talented naturalist and illustrator in her own right), would debate which species of fly was buzzing around the table, a common phenomenon in the early 1800s when both men and women were obsessed with natural history. The entire family (nine children, five of whom survived into adulthood) was recruited to collect natural history specimens. Buckland was so enthusiastic about experiencing animals directly that he claimed he had eaten his way through the animal kingdom. The Bucklands

Figure 1.4 ▲
Woodcut showing Buckland lecturing to spellbound Oxford students, holding up fossils of an ammonite. (Courtesy of Wikimedia Commons)

tried to make a meal from almost every animal they could obtain, a practice known as zoophagy. Mole and bluebottle fly were apparently the most disgusting, but his guests also record him eating panther, crocodile, and mouse. Another account has them eating (and offering to their guests) crisp mice in golden batter, panther chops, rhino pie, trunk of elephant, crocodile for breakfast, sliced porpoise head, horse's tongue, and kangaroo ham. According to Augustus Hare, in his autobiography *The Story of My Life* (1900), "talk of strange relics led to mention of the heart of a French king [possibly Louis XIV] preserved at Nuneham in a silver casket. Dr. Buckland, whilst looking at it, exclaimed, 'I have eaten many strange things, but have never eaten the heart of a king before,' and, before anyone could hinder him, he had gobbled it up, and the precious relic was lost for ever."

Buckland had been looking at the huge bones in the Oxford collection from the Stonesfield Slate Quarry for some time. After the end of the Napoleonic Wars, in 1818, the legendary anatomist and paleontologist Baron Georges Cuvier came to Oxford to see these remarkable huge fossils. Cuvier looked at them closely (especially the jaw with its reptilian teeth, many of which were just erupting) and decided they were the remains of a huge lizard. Buckland and his friend William Conybeare continued to call it the "Huge Lizard," which Conybeare rendered into Greek as *Megalosaurus*, and informally used that name in 1822 in an unpublished article intended for one of Cuvier's volumes. James Parkinson published this unofficial name that same year, but luckily this publication had no standing in the codes of zoology (there was no description or any other indications with the name). Finally, on February 20, 1824, Buckland presented a paper on *Megalosaurus* at the annual meeting of the Geological Society of London. At the same meeting, Conybeare described the first *Plesiosaurus*, found at Lyme Regis by Mary Anning. A few months later Buckland's formal descriptions (with illustrations of the bones done by his wife) appeared in the *Transactions of the Geological Society of London*, giving *Megalosaurus* official status. Buckland did not give it a species name, however, and in 1826 Ferdinand von Ritgen named it *Megalosaurus conybeari*. This name never caught on, and in 1827 Gideon Mantell (chapter 2) named it *Megalosaurus bucklandii*, in honor of the man who described it for science.

Buckland tried to reconstruct what the animal looked like, but he had very little to go on: just a jaw and part of a skull, some hind limb bones, the hip bones, and part of the spinal column (figure 1.5A–B). Because everyone thought it was a huge lizard, nearly all the reconstructions showed it walking on four huge stumpy limbs (figure 1.5B). Analyzing the shape of its hind limbs, Buckland realized that they had been held in an upright vertical posture, so *Megalosaurus* was not reconstructed as a giant sprawling lizard but as something that had a reptilian head and an elephantine body. In his original 1824 description, Buckland followed Cuvier in suggesting that it was 40 feet long and was as heavy as a 7-foot tall elephant. In the printed version of his lecture (influenced by the huge size of Mantell's discoveries described

Figure 1.5 ▶

Megalosaurus: (*A*) the only parts of the dinosaur known, as displayed in the Oxford Museum of Natural History; (*B*) Buckland's illustration of the jaw with the dagger-like teeth; (*C*) early reconstruction of *Megalosaurus* attacking a lizard-like predatory *Iguanodon* (with the horn on its nose). ([*A*] photograph by the author; [*B–C*] Courtesy of Wikimedia Commons)

in chapter 2), Buckland upped the length estimate to 60 to 70 feet. There was really no way to tell because none of the tail bones were found.

For many people, a huge reptile with wicked looking teeth conflicted with their literal interpretation of the Bible, with the lion lying down by the lamb, and all the animals living in a "peaceable kingdom" before Adam's fall. Buckland was a good enough naturalist to reject the ridiculous idea that *Megalosaurus* was a vegetarian with those sharp, slashing teeth (an idea revived by modern-day creationists, who claim that *Tyrannosaurus rex* used its sharp teeth to open coconuts). Instead, Buckland proposed that God had assigned *Megalosaurus* a benign role: getting rid of old and sick animals "to diminish the aggregate amount of animal suffering" (in his words).

By 1842, Richard Owen had united *Megalosaurus* with two other discoveries (*Iguanodon* and *Hylaeosaurus*) to create the taxon Dinosauria, and the public began to appreciate the diversity of huge extinct monsters. Owen recruited the sculptor Benjamin Waterhouse Hawkins to make life-sized models of these amazing creatures for the Great Exhibition of 1851 as part of the incredible Crystal Palace exhibits. These are on display in Crystal Palace Park in Sydenham today where anyone can view them (figure 1.6).

Figure 1.6 ▲

Benjamin Waterhouse Hawkins's concrete sculptures of *Megalosaurus*, as visualized by Richard Owen, now on display in Crystal Palace Park in Sydenham, South London. (Photograph by the author)

Hawkins reconstructed *Megalosaurus* according to Owen's directions as a huge lumbering elephantine predator, and in some illustrations they were shown preying on the larger herbivore *Iguanodon*. Even though the models were grossly inaccurate by modern standards, they were consistent with the then current idea that dinosaurs resembled huge lizards on four limbs. Owen viewed them as "super-reptiles," and the much smaller modern reptiles were the degenerate successors to them. Unfortunately, at that time there were not enough fossils to prove otherwise. These models were among the first to raise the public consciousness of the existence of dinosaurs, and they stirred the first wave of "dinomania."

THE REAL *MEGALOSAURUS*

Since the days of Buckland and Owen, more than 100 specimens of *Megalosaurus* have accumulated at Oxford, but there are still not enough of all the key bones to assemble a complete skeleton (see figure 1.5A). As the first dinosaur named, and especially the first theropod genus named, it became a sort of archetypical theropod genus, and a huge taxonomic "wastebasket." Over the years, many hundreds of additional theropod specimens from all over the world—ranging from Tibet and China and India to North America—have been referred to as *Megalosaurus*. Dozens of species have been named as well. Today most paleontologists reject these additional fragmentary fossils as true specimens of *Megalosaurus bucklandii*, but these mistakes are still common in the dinosaur literature.

So what was *Megalosaurus* really like? The original material had almost no front limbs or tail or most of the spine, and only fragments of the skull, so it was very hard to reconstruct it. In 1859, the tiny theropod dinosaur *Compsognathus* (familiar as part of the inspiration for the pack of tiny "compy" dinosaurs in the second *Jurassic Park* movie) from the Solnhofen Limestone was described, and the first clearly bipedal dinosaur was entered into the scientific literature. This dinosaur was so small that few thought to compare it to *Megalosaurus*.

In 1870 workers in the Summertown Brick Pit just north of Oxford found a nearly complete skeleton of another large theropod dinosaur. It was discovered in the Oxford Clay, Middle Jurassic beds just slightly older than the Taynton Limestone that produced *Megalosaurus*. Due to its strong resemblance to Buckland's species, it was initially assigned to *Megalosaurus bucklandii*, but it eventually was named *Eustreptospondylus*. This skeleton was complete enough to give paleontologists a sense of the proper proportions and posture of *Megalosaurus*. John Phillips, the scientist who first described the *Eustreptospondylus* (then called *Megalosaurus*), soon rearranged the Oxford *Megalosaurus* fossils in the proper bipedal pose, held in position in the display by cardboard sheets. In the 1870s and 1880s, the discovery of complete skeletons of *Allosaurus* in North America further confirmed the bipedal posture and limb configurations of smaller theropod dinosaurs. Today most of Buckland's original fossils are on display at the Oxford Museum of Natural History, with the proper pose (see figure 1.5A).

Many additional types of megalosaurs are now known, many from fairly complete skeletons. With this knowledge, *Megalosaurus* can be reconstructed with a fair degree of confidence (figure 1.7). It was about 7 meters (23 feet) long and weighed about 1.1 metric tonnes. The inhabitants of the environment of the Stonesfield Slate lived with other large theropods, a few sauropods, and the possible pterosaur *Rhamphocephalus* on the shores of the great Jurassic seaways that used to flood most of southern England.

The close relatives of *Megalosaurus* include several other megalosaurids in Europe (such as *Eustreptospondylus* from England, *Wiehenvenator* from Germany, and another fragmentary genus known as *Duriavenator*, from much older Middle Jurassic rocks in Dorset, England). The closest relative of *Megalosaurus* was the almost *T. rex*–sized megalosaur *Torvosaurus* from the Late Jurassic of Portugal—and from the Morrison Formation of Colorado. More distant relatives included the Afrovenatorinae, found mostly in the Jurassic of Europe and Asia, and *Afroventaor* itself, which is from Africa as the name suggests.

Figure 1.7 ▲

Modern reconstruction of *Megalosaurus*. (Courtesy N. Tamura)

The megalosaurs apparently underwent a significant evolutionary radiation in the Middle Jurassic, and they were widespread on all the continents that once made up the supercontinent of Pangea. By the Middle Jurassic, though, they faced competition from allosaurs in North America and from spinosaurs and their relatives in Africa and Europe. The entire megalosaur radiation seems to have vanished at the end of the Jurassic, except for the Cretaceous spinosaurs.

Megalosaurus, the first dinosaur to be named and discovered, opened the door to the real "Jurassic World" of giant reptiles that ruled the planet. But *Megalosaurus* itself was always incomplete, was badly misinterpreted and reconstructed, and was almost saddled with the name "Scrotum humanum."

FOR FURTHER READING

Cadbury, Deborah. *The Dinosaur Hunters: A True Story of Scientific Rivalry and the Discovery of the Prehistoric World*. New York: Harper Collins, 2000.

——. *Terrible Lizard: The First Dinosaur Hunters and the Birth of a New Science*. New York: Henry Holt, 2001.

Colbert, Edwin. *Men and Dinosaurs: The Search in the Field and in the Laboratory*. New York: Dutton, 1968.

Maddox, Brenda. *Reading the Rocks: How Victorian Geologists Discovered the Secret of Life*. New York: Bloomsbury, 2017.

Mayor, Adrienne. *The First Fossil Hunters: Paleontology in Greek and Roman Times*. Princeton, N.J.: Princeton University Press, 2000.

McGowan, Christopher. *The Dragon Seekers: How an Extraordinary Circle of Fossilists Discovered the Dinosaurs and Paved the Way for Darwin*. New York: Perseus, 2001.

Naish, Darren. *The Great Dinosaur Discoveries*. Berkeley: University of California Press, 2009.

Oke, Elizabeth. *The Life and Correspondence of William Buckland, Sometime Dean of Westminster*. New York: Forgotten Books, 2012.

Rudwick, Martin. *Bursting the Limits of Time: Reconstructing Geohistory in the Age of Revolution*. Chicago: University of Chicago Press, 2007.

——. *The Meaning of Fossils: Episodes in the History of Paleontology*. New York: Science History, 1976.

——. *Worlds Before Adam: Reconstructing Geohistory in the Age of Reform*. Chicago: University of Chicago Press, 2010.

Spaulding, David A. E. *Dinosaur Hunters: Eccentric Amateurs and Obsessed Professionals*. Rocklin, Calif.: Prima, 1993.

IGUANODON

A thousand ages underground

His skeleton has lain;

But now his body's big and round,

And he's himself again!

His bones, like Adam's, wrapped in clay,

His ribs of iron stout,

Where is the brute alive today

That dares to turn him out?

Beneath his hide he's got inside

The souls of living men;

Who dare our saurian now deride

With life in him again?

(*Chorus*)

The jolly old beast

Is not deceased

There's life in him again (*ROAR*)

—SONG BY EDWARD FORBES, 1852, SUNG INSIDE THE *IGUANODON* MODEL IN THE "DINNER IN A DINOSAUR"

THE WEALDEN

If you drive southeast from the suburbs of London, you soon leave behind the dense housing and crowded streets and find yourself in the quiet countryside of Sussex (from the Old English "South Saxony"). The landscape is truly pastoral, with quiet country fields full of sheep and horses, and quaint villages here and there that have changed very little in 400 or 500 years or more. The roads were built to deal with one slow-moving single carriage or oxcart at a time and are extremely narrow; modern cars have a challenge squeezing through the tight spots and blind corners at high speed. Don't even think about trying to park!

Thanks to the abundant rainfall typical of all the British Isles, the countryside is green and well watered without visible outcrops except in an occasional roadcut or beach cliff. These exposures are nothing like the famous desert landscapes of the Rocky Mountains that yield most of North America's dinosaurs, nor are they like the spectacular cliffs and buttes of the Gobi Desert or the dry Argentinian Pampas and Andean foothills that are famous for their dinosaurs. Rest assured that dinosaurs are underfoot, but only luck and a rare good exposure provides any chance of collecting them.

Even a good geologist is hard-pressed to recognize the formations beneath the landscape until he or she acquires a more discerning eye. Geologists soon realize that the high ridges (perversely, they are called the "Downs" or "downlands") are held up with the famous Cretaceous Chalk beds, which are exposed along the southern coast of England from the White Cliffs of Dover to the Isle of Wight. The Chalk was the first geologic unit recognized in the area and gave the Cretaceous its name (*creta* is the Latin word for "chalk"). The Chalk is made up of trillions of tiny shells of plankton (especially phytoplankton, or golden-brown algae known as coccolithophorids), which are only a few tens of microns across. Such plankton once bloomed in the shallow warm seas that drowned most of Europe. Similar chalk beds are found across the English Channel in Belgium and northeastern France, in Denmark, and even in Austin, Texas, and western Kansas.

Across Sussex and Kent, the Chalk crops out in a big parabola opening to the east (figure 2.1), which developed due to an upward fold that geologists call a plunging anticline. In the center of the parabola is a lowland area, once heavily forested, known as the Weald (Old English for "woodland" and related to the German word for "forest," *Wald*). The bedrock beneath the Weald is made of Lower Cretaceous rocks that once lay beneath the Chalk but are now exposed as the chalk beds were folded upward and eroded away (figure 2.1B). A dome of layered Lower Cretaceous rocks has been cut by weathering to expose the layers as sandstone ridges and clay valleys. The oldest rocks exposed at the center of the anticline are correlated with the Purbeck Beds of the Upper Jurassic. Above these, the Cretaceous rocks include the Wealden Group of alternating sands and clays: the Ashdown

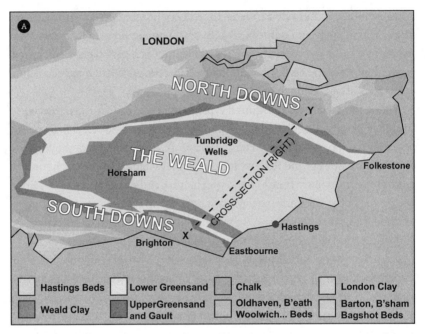

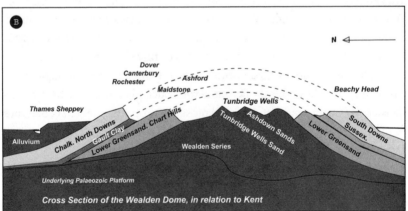

The Wealden anticline and the Jurassic and Cretaceous rocks that make it up: (*A*) geologic map of Sussex and Kent, showing the outcrop pattern with the large parabola of the beds of a plunging anticline, with the older Jurassic and Lower Cretaceous Wealden beds in the middle and the Upper Cretaceous Chalk on the flanks (North and South Downs); (*B*) cross section showing the relations of the strata. (Redrawn from several sources)

Sand Formation, Wadhurst Clay Formation, Tunbridge Wells Sand Formation (collectively known as the Hastings Group), and the Weald Clay. The Wealden Group is overlain by the Lower Greensand and the Gault Formation, consisting of the Gault and the Upper Greensand. These in turn are capped by the Chalk.

The rocks of the central part of the anticline include hard sandstones, and these form hills now called the High Weald. The peripheral areas are mostly softer sandstones and clays and form a gentler rolling landscape, the Low Weald. The Weald-Artois Anticline continues some 64 kilometers (40 miles) further southeastward under the Straits of Dover and includes the Boulonnais region of France.

THE COUNTRY DOCTOR AND THE GIANT TEETH

In the middle of the Lower Cretaceous beds of the Weald lies the tiny village of Cuckfield. Today it is just a quiet country town with Tudor-style white buildings with dark timber beams that date back 500 years or more. In 1822, however, it was the starting point for the first extensive discoveries not only of dinosaurs but of entire dinosaurian faunas. That year Gideon Mantell, a local doctor (figure 2.2) from the nearby town of Lewes, was working in

Figure 2.2 ▲
Portrait of Gideon Mantell. (Courtesy of Wikimedia Commons)

the area, visiting his patients and collecting fossils on the side. Although most people pronounce the name with the accent on the second syllable, in Lewes and throughout Kent and Sussex it is pronounced MAN-tle (like the name for the cloak, or the layer in the earth's interior).

Mantell is an interesting example of how paleontology first blossomed; it was not solely due to the work of the educated scholars of Oxford and Cambridge (such as Buckland) but also due to the people of many walks of life who were interested in natural history. Some were dedicated field collectors with limited education, such as the famous Mary Anning. Her discoveries of marine reptiles and other fossils from Lyme Regis began the entire field of vertebrate paleontology in England. Others were dedicated amateurs with limited specialized education who studied fossils as a hobby. Many, such as Mantell (or James Hutton, the founder of modern geology), were trained as doctors and had as much scientific education as could be obtained in those days.

Mantell was typical of the naturalist doctors of his time. Born in Lewes in 1790, he was one of five children of a humble shoemaker. At an early age, he showed an interest in collecting fossils and understanding the geology of the Weald, discovering ammonites, sea urchins, fish bones, corals, and other typical fossils of the region (especially the Chalk). A family of Methodists, the Mantells were forbidden from attending any of the local schools (which were only open to Anglicans), so young Gideon was tutored by local women who operated their own small schools at home. At age 15, he was apprenticed to the local doctor, James Moore, for five years. Originally his job was cleaning tools and vials and separating and arranging drugs and doing the doctor's billing and bookkeeping, but eventually he learned to make pills himself and to extract teeth from patients. When his father died in 1807, Gideon began to study medical anatomy texts closely and subsequently wrote his own book based on what he had learned. By 1811, he had received a medical education in London and was admitted to the Royal College of Surgeons. There he did an internship as an obstetrician in a women's hospital delivering babies. Once he had finished his London duty, he returned to Lewes and became a partner of Dr. Moore, his former master and mentor. He soon became very busy, typically seeing 50 patients a day, traveling to see each one at home. He workload was staggering. He helped deliver about 300 babies a year. He often stayed up six or seven nights in a row when delivering babies all over Sussex.

Despite this exhausting schedule, Gideon took time when he could to search for fossils and other interesting natural objects on his trips around Sussex. By 1813, he was corresponding with fossil collector James Sowerby about his finds and sent some of his best specimens to Sowerby. A species that Gideon had collected was named after him by Sowerby, *Ammonites mantelli*. By December 1813, Mantell had published his first paper on the fossils of the Lewes area and had been elected as a Fellow of the Linnean Society of London.

During those years of grueling work, Mantell and Dr. Moore dealt with epidemics of cholera, typhoid, and even plague, which killed hundreds of his patients. The daughter of one of those victims was Mary Ann Woodhouse, and she and Gideon fell in love when he was treating her father. They married in 1816 when he was 26 and she was 20 years old (still technically underage). She had to obtain special permission to marry because her father was dead. That same year Gideon set up his own medical practice, and his wife often went with him to help with patients. She also spent time collecting fossils while Mantell was busy with work.

Inspired by Mary Anning's sensational discovery of ichthyosaurs and plesiosaurs at Lyme Regis in Dorset, Mantell became passionately interested in the study of the fossil animals and plants found in his area. By 1819, Mantell had begun acquiring fossils from a quarry near Cuckfield (figure 2.3). These included the first remains of terrestrial and freshwater ecosystems at a time when all the known fossil remains from the Cretaceous of England were marine fossils. As Mantell began to map and understand the geologic sequence of the rocks of the Weald, he called the new beds the Strata of Tilgate Forest, after a historical wooded area. Today that same Tilgate Quarry is covered by a park and soccer field called Whiteman's Green. Standing there today watching kids playing soccer, it is difficult to visualize the steep rock quarry, which has been filled in, but if you stand in the right spot, you can still see the church steeple in the background and determine that you are in the right place. The tiny Cuckfield Museum in town exhibits quite a few small fossils of dinosaurs that were found there.

By 1820, Mantell had acquired some very large bones from the quarry men at Cuckfield. They were as large as those discovered and named *Megalosaurus* by William Buckland at Stonesfield Slate Quarry north of Oxford. In 1822, Gideon was finishing his first book (*The Fossils of South Downs*) and was on the lookout for more fossils. Legend has it that while waiting

Figure 2.3 ▲

The frontispiece of Mantell's 1827 book showing workers in the quarry in the Wealden rocks of the Tilgate Forest, where the first *Iguanodon* fossils were found. (From Gideon Mantell, *The Geology of Sussex*, 1827)

for Gideon to come back from his rounds, Mary Ann found several large teeth in the roadside that looked like nothing anyone had ever seen before (figure 2.4). (Other scholars say that Gideon rarely took his wife along and found the fossils himself, which is what he claims in his own books.) By 1821 Gideon had written and published his next book on the geology of Sussex. It was an immediate success, with 200 subscribers. One of these was none other than King George IV himself. The letter the king wrote at Carlton House Palace read: "His majesty is pleased to command that his name should be placed at the head of the subscription list for four copies."

The king probably heard about Mantell's work when he spent summers at the nearby beach resort town of Brighton. We definitely know how Gideon responded. Now realizing he was onto something important, Mantell showed the teeth to other scientists, but none of them could identify them.

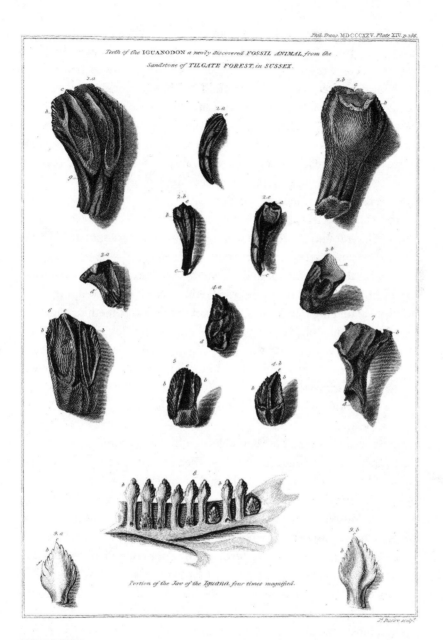

Teeth of the IGUANODON a newly discovered FOSSIL ANIMAL from the
Sandstone of TILGATE FOREST, in SUSSEX.

Portion of the Jaw of the Iguana, four times magnified.

Figure 2.4 ▲

Illustration of *Iguanodon* teeth and the jaw of a modern iguana for comparison. (From Gideon Mantell, "Notice on the Iguanodon, a newly discovered fossil reptile, from the sandstone of Tilgate forest, in Sussex," *Philosophical Transactions of the Royal Society*, no. 115 [1825]:179–186)

Buckland and some others called them teeth of a huge fish or teeth from a large mammal. Others doubted that they were that old, suggesting they might have come from a younger deposit. There were many instances of Ice Age beds deposited on top of the older Mesozoic bedrock that sometimes yielded rhinoceroses, hippos, and mammoths, whose teeth are also large; these were clearly much younger than the ancient Tilgate Forest fossils. No one could imagine that such huge, flat, grinding teeth could have come from a reptile because most reptilian teeth are pointed and cone-shaped for grabbing prey. They are not grinding teeth for eating vegetation—and not even the largest crocodile had teeth this big.

On a trip to Paris, Charles Lyell (author of *Principles of Geology* in 1830, which helped found modern geology) brought some of Mantell's teeth to the famous Baron Georges Cuvier, founder of comparative anatomy. During his weekly salon with friends and admirers, Cuvier examined the teeth and told Lyell that they were rhinoceros teeth. According to Lyell, "the next morning he told me that he was confident that it was something quite different." Unfortunately, Lyell never told Mantell about Cuvier's retraction in Paris, and Mantell felt hurt by having been rejected by the great Cuvier. In 1824, Mantell sent some more teeth to Cuvier, who then admitted he was wrong and that the teeth were from some gigantic herbivorous reptiles. Cuvier even printed a retraction in his *Recherches sur les Ossemens Fossiles* later that year.

Convinced he was right, Mantell looked harder and found more bones and teeth that were unquestionably from the Mesozoic rocks of the Wealden Group. Meanwhile, other people looked among living animals for something with similar teeth. Samuel Stutchbury, then an assistant curator at the Hunterian Museum, discovered that they resembled the teeth of an iguana (which he had just been working on) but were 20 times larger (figure 2.4). From these isolated teeth and a few broken chunks of limb bones, Mantell estimated that the creature must be at least 18 meters (60 feet) long.

Mantell's ideas were finally confirmed in 1825 when publication of the formal description of *Megalosaurus* opened the possibility of gigantic extinct reptiles. Most scholars soon agreed with Mantell that the teeth looked like those of an immense iguana. But what should he call this new creature? In his original presentation Mantell named it "Iguana-saurus," but he then received a letter from William Daniel Conybeare objecting to that name: "Your discovery of the analogy between the Iguana and the fossil teeth is very interesting but the name you propose will hardly do, because it is equally applicable to the recent iguana. Iguanoides or Iguanodon would be better." Of those two names, *Iguanodon* ("iguana tooth") made the most sense, and thus it got its name.

As the years progressed, the Mantells collected enough fossil limb bones to demonstrate that the forelimbs of *Iguanodon* were much shorter than its hind legs. Their best find was a slab from the Lower Greensand found in 1824 that contained a partial skeleton of an iguanodontid (now considered a distinct genus named in his honor, *Mantellisaurus*). This helped improve Mantell's coverage of the bones and enabled him to make a more accurate reconstruction. Newer specimens showed that it was not built like a rhinoceros or elephant, as Sir Richard Owen and others claimed. Even more discoveries showed that fossil vertebrae found in the Tilgate Quarry were from *Iguanodon*. Unfortunately, Mantell also made mistakes. Among the fossils found in Tilgate Quarry was a large spike-like object, which Mantell thought was a nose horn, making *Iguanodon* look even more rhinoceros-like.

But Mantell went further than Buckland or the others: he found not just a single dinosaur but a complete fauna of dinosaurs and other animals that had lived in the area in the Early Cretaceous. Some of the fossils came not from *Iguanodon* but from something with a very different build. In 1833, Mantell gave a partial skeleton the new generic name *Hylaeosaurus*. This name was based on limited material, but *Hylaeosaurus* is now known to be a small nodosaurid ankylosaur (see chapter 21). He found and named two other local dinosaurs, *Pelorosaurus* (an enormous sauropod known from an enormous upper arm bone and a few vertebrae) and *Regnosaurus* (later found to be a stegosaur, only known from a partial jaw), along with fish fossils and crocodiles. Together with his other discoveries, Mantell had found and named four of the first five known dinosaurs and was the world's leading authority on prehistoric reptiles.

Mantell did something even more groundbreaking: he documented not only the fossils but also their geological context and provided the first paleoecological speculation about the environment inhabited by *Iguanodon*, *Hylaeosaurus*, *Pelorosaurus*, *Regnosaurus*, and other Wealden creatures. From the fish fossils, fossilized mud cracks, and even dinosaur tracks, he demonstrated that the beds were of freshwater origin—not marine rocks like the Chalk and much of the Cretaceous sequence in southeastern England.

Over the rest of his life, Mantell wrote more and more about Sussex fossils. Given his busy medical practice, he was amazingly prolific. He published 67 books and 40 scientific papers over his lifetime. In 1833, he moved his medical practice south from Lewes to the fashionable coastal resort town of Brighton, the favorite summer holiday spot for King George IV, and set up his own museum in addition to his medical duties. His medical practice suffered, however, because he spent most of his time collecting fossils and doing geology instead of attending to patients. In his darkest hour, the Brighton town council supported him by turning his house into a museum. There he gave a very successful lecture series titled "The Wonders of Geology, or A Familiar Exposition of Geological Phenomena: Being the Substance of a Course of Lectures Delivered at Brighton." This lecture was published as a book in 1838, which helped his finances somewhat. Unfortunately, Gideon's home museum failed because he neglected to charge admission to people who wanted to see his spectacular collection. Hitting rock bottom financially, he offered his entire collection to the British Museum for £5,000; they lowballed him and paid him only £4,000. There his collection still resides.

By 1839, Mantell had fallen on hard times. His wife left him with their four surviving children, his son emigrated to New Zealand (where he acquired and sent his father the first fossils of extinct giant birds known as moas), and then a year later his daughter died. Seeking new patients, Mantell moved to Clapham Commons in southern London and spent the rest of his life treating patients. In 1841, he had a terrible carriage accident; he fell from his seat, was entangled in the reins, and was dragged along the ground for some distance, severely injuring his back. He suffered ever-increasing pain and severe scoliosis throughout the remainder of his life. He used chloroform and later became addicted to opium to alleviate the extreme pain. (After his death, Richard Owen had Mantell's severely deformed and fused spine dissected and preserved at the Royal College of Surgeons as an example of scoliosis. It was destroyed during the bombing of London in World War II.)

During his last few years, Mantell tried to maintain his medical practice while publishing books about fossils and geology, but he felt abandoned and neglected because others had stolen his thunder and built on his research. Crippled and sickly, he frequently visited the Great Crystal Palace Exhibition in Hyde Park in 1851. Mantell gave a lecture on November 10,

1852, and died of an opium overdose later that night in his home (possibly a deliberate suicide, or possibly an accidental overdose after falling on the stairs in his home).

Today his home on 166 High Street in Lewes is a historic landmark maintained by the National Trust. It is marked by a rectangular blue plaque that reads:

Dr. Gideon A. Mantell, F.R.S.
Surgeon and Geologist
Born in Lewes 1790. Died in London 1852
Lived here
He discovered the fossil bones
Of the prehistoric Iguanodon
In the Sussex Weald.

Tilgate Quarry is covered by an athletic field, but there is a monument at Whiteman's Green where the first *Iguanodon* was discovered, and the tiny town museum in Cuckfield commemorates Mantell and his fossils. Mantell's work was truly pioneering; his 1849 description of *Iguanodon* won the Royal Medal of the Royal Society, the second-highest award in science at that time. More important, Mantell holds a prime place in the history of paleontology: the first to discover a nearly complete dinosaur skeleton (*Megalosaurus* was very incomplete) from a locality we still can locate (the position of the type locality of *Megalosaurus* is unknown), and the first to describe a dinosaur fauna.

The following was written as a memorial to him by a member of the Council of the Clapham in 1853:

> Living in the midst of a most interesting geological district, his quick appreciation could not fail to be struck with its interesting characteristics. As on his professional visits, he rode or drove over the South Downs and Weald of Sussex, he was continually searching for the organic treasures imbedded in the quarries or lying by the roadside, which afforded him an inexhaustible source of delight and instruction; and he thus accumulated materials which eventually enabled him to establish the fresh-water character of the Wealden,—a discovery which alone will hand down his name to the latest posterity as one of the great founders of the science of Geology,—and brought together the

fragments of fossil bones which afterwards gave him the power of building up the skeletons of those gigantic reptiles, the hyleosaurus, iguanodon, pelorosaurus, others, with which he astonished and delighted, not only the public generally, but the scientific world. The number of specimens so collected amounted to upwards of 1,200, and with these he founded the Mantellian Museum, which was visited, while he lived at Lewes, by the most eminent men of the day; among others by Baron Cuvier, and by the Royal Princes. This collection he afterwards removed to Brighton, when he went to reside there, and he made great efforts to have it established in the county from the strata of which it had been gathered, as the nucleus of a local geological museum, but the requisite funds were not forthcoming, and it was ultimately sold to the British Museum.

DINNER IN A DINOSAUR

Meanwhile Mantell's archrival, Richard Owen, was up to his usual tricks. Throughout his career, Owen did everything he could to dismiss or to minimize Mantell's discoveries, even stealing credit for some of them. Mantell was too ill to work on the fossils that were now in the British Museum, so Owen took charge of them, publishing several papers between 1849 and 1884 that were later collected in a huge four-volume *History of British Fossil Reptiles*. In that book, Owen redescribed and reinterpreted Mantell's specimens.

Owen saw an opportunity for great publicity in the upcoming Great Crystal Palace Exhibition and commissioned Benjamin Waterhouse Hawkins to build huge concrete reconstructions of nearly all the prehistoric beasts then known (see chapter 1). These 33 sculptures included Buckland's *Megalosaurus* and Mantell's *Iguanodon* and *Hylaeosaurus*, as well as nearly every other extinct creature: the plesiosaurs and ichthyosaurs that Mary Anning had found; the pterodactyls, mosasaurs, and marine crocodiles; the giant amphibians then known as labyrinthodonts; some of the Eocene mammals reported from the Paris and London basins; and even some Ice Age mammals. The reconstructions of ichthyosaurs, plesiosaurs, the giant ground sloths, and Irish elk were reasonably accurate by today's standards because they were based on nearly complete fossils. But the *Megalosaurus* was rendered as a giant quadrupedal lizard because so little was known of it at that time (see figure 1.6).

Even worse was Owen's version of *Iguanodon*. Although as early as 1849 Mantell had published illustrations of the smaller, delicate front limb bones to show that it could walk on two legs, Owen had it reconstructed in a quadrupedal pose like an elephantine lizard (see figure 1.5 and figure 2.5). Being the largest creatures there, the two *Iguanodon* reconstructions dominated the exhibit (figure 2.5A). They were so large and spectacular that Hawkins organized a famous 1853 New Year's Eve party for Owen and 20 other leading paleontologists with a table set inside the mold of the larger *Iguanodon* (figure 2.5B). Sadly, Mantell had died just a few months earlier, and although his name was commemorated on the walls, he was no longer around to take his rightful place among the other giants of paleontology.

When the Great Exhibition closed, the sculptures and the giant Crystal Palace made of iron and glass were moved from Hyde Park in central London to Sydenham Park in southeastern London, which became known as Crystal Palace Park. They were placed in a corner of the park on a series of islands, and the water protected them from being climbed on by overcurious visitors. Over the years, they suffered much damaged through weathering as the elements broke down the concrete. The Crystal Palace itself burned down in 1936, and by the end of the twentieth century, the dinosaur sculptures were decaying and badly overgrown. In 2002, money was raised to restore them to their original glory, and they remain available for inspection by anyone visiting London's Sydenham Park during daylight hours. They are amazing monuments of where paleontology was more than 160 years ago and provide a benchmark for how much has been learned since then.

DINOSAURS IN THE COAL MINE

Despite Mantell's suggestion that *Iguanodon* might have walked on two legs, Owen's quadrupedal reconstruction dominated for a few more decades. With such incomplete and fragmentary fossils, it was impossible to know what these extinct creatures looked like. All of this changed on February 28, 1878, when two coal miners, Jules Créture and Alphone Blanchard, accidentally hit fossil bones at a depth of 322 meters (1,056 feet) in a coal mine near Bernissart, Belgium. By May, the mine workers began to excavate the skeletons from the depths of the mine under supervision of a technician from the museum in Brussels, and they were brought to the surface. By 1882, there

DINNER IN THE IGUANODON MODEL, AT THE CRYSTAL PALACE, SYDENHAM.

Figure 2.5 ▲

Benjamin Waterhouse Hawkins' reconstruction of *Iguanodon* (guided by Richard Owen) in the Crystal Palace Exhibition, now in Crystal Palace Park in Sydenham: (*A*) the sculptures as they appear today; (*B*) image from the *Illustrated London News* in 1853 of Owen hosting a dinner in the hollow shell of *Iguanodon* (before it was completed) for the leading scientists of the day; the tent pavilion around them has the names of Owen, Mantell, Buckland, and Cuvier displayed prominently. ([*A*] Photo by the author; [*B*] Courtesy of Wikimedia Commons)

were no less that 38 fossil skeletons in the Belgian collections. Most of them were complete, articulated skeletons of adult *Iguanodon*.

Their study fell to the Belgian paleontologist Louis Dollo, later famous for "Dollo's Law," which said that evolution is not reversible. Dollo published a few short notes on the fossils, but it was almost a century later, in 1980, before David Norman finally completed the work and published a lengthy monograph on them. In 1881, the famous herpetologist George Albert Boulenger formally gave them a new species name, *Iguanodon bernissartensis*. The type specimen was mounted on a scaffold with adjustable ropes holding it in place, which was the first time any dinosaur had been mounted in a life-like pose for public display (figure 2.6). It was opened for public viewing in the Palace of Charles of Lorraine in July 1883, and in 1891 the fossils were moved to the Royal Museum of Natural History (now the Royal Belgian Institute of Natural Sciences, where they still stand today). This row of nine huge standing mounts made a very impressive display, and 19 more sit in the museum basement because there was not room to display all of them. Casts of the Belgian specimens were eventually sent to the museums in Cambridge and Oxford, where they are still on display.

Such complete articulated skeletons allowed scientists to put *Iguanodon* together correctly for the first time (figure 2.7). Not only were they capable of bipedal walking, but the front limbs were only of limited use for locomotion. The "nose horn" that Mantell and Owen had put on the snout of the models turned out to be a thumb spike.

Dollo also made some mistakes that have been corrected by more recent work. In the 1970s, paleontologists began to realize that most of the bipedal dinosaurs, from the large predatory theropods to the duck-billed dinosaurs and many others, must have held their bodies horizontal to the ground, balanced on their hip joint, and held their tails out straight behind them as a counterbalance. This was established by dinosaur trackways, which seldom show any tail drag marks, and also by the presence of a network of tendons to hold the tail rigid in many dinosaurs, especially duck-billed dinosaurs and long-necked sauropods. Even the Bernissart specimens were discovered with straight tails and ossified tendons. Yet the bias of nearly all paleontologists of that time was that dinosaurs were stupid, slow, sluggish monsters living in swamps, dragging their heavy bodies and tails around. Nearly all the original Belgian mounts (and their copies in England) posed the skeleton like a kangaroo, leaning back on a curved tail braced against the

Figure 2.6 ◀
(*continued*)

Figure 2.7 ▲
Modern reconstruction of *Iguanodon* as it looked in life, with a long straight tail and spikes on the thumbs rather than on the nose. (Courtesy of N. Tamura)

ground. (Dollo had to break the straight row of tail vertebrae in the mount to make it curve in the kangaroo pose.) Many museums (such as those in Belgium) have remounted some of their *Iguanodon* skeletons in the modern balance-beam stance (figure 2.6B), but others (such as those in Cambridge and Oxford) do not have the money or staff to completely tear down their century-old replicas and remount them in the modern fashion (figure 2.6C). Thus they remain as a relict of an earlier way of thinking about dinosaurs.

Iguanodont material has been found in numerous other places, from the Isle of Wight in England to Mongolia to Africa to South Dakota to Utah. Most of it was originally assigned to the genus *Iguanodon*, which once had dozens of different species, but these finds have since been redescribed and renamed in new genera. The huge herd at Bernissart is not the only place where a large numbers of specimens have been found. In Nehden, in North Rhein-Westphalien in northwestern Germany, a quarry yielded another assemblage killed in a flash flood. At least 15 individuals, from 2 meters (6.6 feet) to 8 meters (26 feet) in length, were preserved there.

Today we know *Iguanodon* was a bulky bipedal herbivore that could occasionally shift to a quadrupedal posture. The Bernissart specimens show that they reached 10 meters (33 feet) to 13 meters (43 feet) as adults and weighed about 3 metric tonnes (3.3 U.S. tons). This range is much smaller than estimates given by Mantell or Owen, who thought *Iguanodon* was the largest creature that had ever lived on land (before much larger dinosaurs were found). Iguanodonts had long narrow skulls with a toothless beak, and those distinctive iguana-like teeth are arranged in dense rows in the jaw.

The long slender forearms had three robust middle fingers to bear the weight when they leaned forward in a quadrupedal posture, plus the thumb spike projecting forward and inward. The function of the thumb spike is debated. They could have been used for defense against predators or fighting other members of their herd, or possibly for foraging for food. Their "pinky" finger was long and flexible and may have been able to manipulate the plant fronds on which they fed. They had powerful hind legs, but their proportions and size do not suggest that they were very fast runners. As mentioned previously, the backbone and tail were stiffened by ossified tendons, which held them rigid in a straight line, but these were often left off the skeletal mounts because they didn't fit when they were posed like kangaroos (although they are preserved intact on the vertebrae of the Brussels mounts).

Thus *Iguanodon* is not only one of the earliest dinosaurs to be discovered but today is also one of the best studied and represented, with dozens of specimens from many continents, including two large quarry samples. We have come a long way from Mantell's or Owen's early notions of a giant saurian pachyderm to the more delicately built biped and quadruped we know today.

FOR FURTHER READING

Cadbury, Deborah. *The Dinosaur Hunters: A True Story of Scientific Rivalry and the Discovery of the Prehistoric World*. New York: Harper Collins, 2000.

——. *Terrible Lizard: The First Dinosaur Hunters and the Birth of a New Science*. New York: Henry Holt, 2001.

Colbert, Edwin. *Men and Dinosaurs: The Search in the Field and in the Laboratory*. New York: Dutton, 1968.

Maddox, Brenda. *Reading the Rocks: How Victorian Geologists Discovered the Secret of Life*. New York: Bloomsbury, 2017.

McGowan, Christopher. *The Dragon Seekers: How an Extraordinary Circle of Fossilists Discovered the Dinosaurs and Paved the Way for Darwin*. New York: Perseus, 2001.

Naish, Darren. *The Great Dinosaur Discoveries*. Berkeley: University of California Press, 2009.

Rudwick, Martin. *Bursting the Limits of Time: Reconstructing Geohistory in the Age of Revolution*. Chicago: University of Chicago Press, 2007.

——. *The Meaning of Fossils: Episodes in the History of Paleontology*. New York: Science History, 1976.

——. *Worlds Before Adam: Reconstructing Geohistory in the Age of Reform*. Chicago: University of Chicago Press, 2010.

Spaulding, David A. E. *Dinosaur Hunters: Eccentric Amateurs and Obsessed Professionals*. Rocklin, Calif.: Prima, 1993.

CETIOSAURUS

In conclusion, it is stated that the vertebrae described in the paper prove the existence of a saurian genus distinct from *Megalosaurus, Steneosaurus, Poikilopleuron, Pleisosaurus*, or any other large extinct reptile, remains of which have been discovered in the oolitic series; that the vertebrae, as well as the bones of the extremities, prove its marine habits; and that the surpassing bulk and the strength of the *Cetiosaurus* were probably assigned to it with carnivorous habitats, that is might keep in check the Crocodilians and Plesiosauri.

—RICHARD OWEN, "A DESCRIPTION OF A PORTION OF THE SKELETON OF *CETIOSAURUS*, A GIGANTIC EXTINCT SAURIAN REPTILE OCCURRING IN THE OOLITIC FORMATIONS OF DIFFERENT PORTIONS OF ENGLAND," *PROCEEDINGS OF THE GEOLOGICAL SOCIETY OF LONDON*, 1842

GIANT BONES

In 1825, an amateur collector named John Kingdon was poking around an old quarry near Chipping North in Oxfordshire when he came across some huge bones. These included some gigantic vertebrae and fragments of limb bones. On June 3, 1825, Kingdon read a brief description of the bones at a meeting of the Geological Society of London and speculated that they might have come from a whale or a huge crocodile. Over the years, other collectors found more bones of the huge creature. A certain Miss Baker donated some from the area of Blisworth, Northampton, and other specimens came from Staple Hill northwest of Woodstock; from Buckingham, Garsington; and some far to the northwest in Yorkshire. Many of these bones went to the Oxford Museum, the closest scholarly institution to the finds, and others ended up in the Scarborough Museum in Yorkshire.

They all came from Middle Jurassic layers known as the Oolites, a distinctive set of limestones made of tiny particles about the size and shape of pellets from a BB gun. In those days, no one knew what produced oolites, but in the 1950s oolites were discovered forming in shallow tropical seas such as those around the Bahamas. When you slice open an oolite and examine it under the microscope, it has a concentric layered structure, like a snowball or an onion. Modern oolites have a form similar to that of tiny lime snowballs. They form when some nucleus, such as a shell fragment, rolls around in fine lime mud particles on the sea bottom due to agitated, back-and-forth currents. As it does so, it builds layer upon layer of coatings of fine lime mud around the nucleus, forming the concentrically layered "snowball" pattern we find inside. The presence of oolitic limestones is a good indicator of shallow warm tropical shoals, with frequent currents and agitated water that rolls particles around.

THE COMPLEX RICHARD OWEN

By 1841, enough specimens had accumulated of the enormous "whale" or "crocodile" that they caught the attention of Richard Owen (figure 3.1), whom we met briefly in chapters 1 and 2. Born in Lancaster in 1804, he was the son of a West Indian merchant; his mother had Huguenot ancestry. He attended schools in Lancaster, then (like Gideon Mantell) he was apprenticed to a local apothecary and surgeon to begin learning the medical trade. In 1824, he spent a year at the University of Edinburgh medical school (where he preceded Charles Darwin by one year). Dissatisfied with the situation at Edinburgh, he finished his medical education at St. Bartholomew's Hospital in London, mentored by the surgeon Dr. John Abernethy.

After earning his medical degree, Owen realized he was not cut out for the life of a country doctor and instead tried to focus on anatomy. Abernethy recommended him for a post at the museum of the Royal College of Surgeons, where he soon organized and cataloged all of their collections, including the recently acquired Hunterian collection of 14,000 specimens of some 3,000 animals, many already dissected. By studying these specimens, he taught himself zoology and comparative anatomy and soon had

Figure 3.1 ▲

Portrait of Richard Owen. (Courtesy of Wikimedia Commons)

an encyclopedic knowledge of anatomy second to none. Indeed, later in life he was nicknamed "the British Cuvier" because his talent and discoveries led people to compare him to the great Baron Georges Cuvier, the founder of comparative anatomy and vertebrate paleontology.

By 1836, he was appointed the Hunterian Professor of the Royal College of Surgeons, and by 1849 he was the curator of that collection. In 1856, he was appointed curator of the natural history collections of the British Museum, and he spent decades working hard for a new building to house the natural history collections separate from the original British Museum in the Bloomsbury section of London (which focuses on art and archaeology). Finally, in 1881 his great "cathedral to science" was dedicated in the South Kensington section of London and became the British Museum (Natural History), now called the Natural History Museum. It is still one of the world's greatest museums with collections unrivaled by almost any other museum in the world.

Owen's work on zoology and paleontology was amazing for his time, and he is responsible for many important discoveries, from describing the chambered nautilus, the African lungfish, the dodo, the great auk, the giant elephant birds (*Aepyornis*) of Madagascar, and huge moas (*Dinornis*) of New Zealand, to the first description of the skeleton of *Archaeopteryx*. Owen described the South American fossil mammals that Darwin brought back from the *Beagle* voyage in 1836, the extinct fossil marsupials of Australia including the rhino-sized wombats known as diprotodonts, gigantic kangaroos, and the marsupial "lion" *Thylacoleo*, the anatomy of the platypus, and formally named the major groups of even-toed (Artiodactyla) and odd-toed (Perissodactyla) hoofed mammals. Even though very few of his somewhat bizarre theoretical ideas have stood the test of time, his basic talent for descriptive anatomy and comparison with other animals was unparalleled. He published more than a dozen books and scientific monographs, and over 100 scientific papers. For this he received many honors in his lifetime.

Despite these accomplishments, his standing in the scientific community was marred by his personal behavior and his peculiar ideas. By all accounts, he was a very ambitious, driven man who was determined to succeed and reach the top (which he did more than once at the Hunterian Museum and then at the British Museum). He was especially jealous of people who disagreed with him, and Owen used all sorts of malicious and underhanded methods to battle his rivals or to undermine their scientific position. According to one biographer, he was described as a malicious, dishonest, and hateful individual. Another biographer called him a "social experimenter with a penchant for sadism. Addicted to controversy and driven by arrogance and jealousy." The biographical passages written by Deborah Cadbury describe Owen as possessed by an "almost fanatical egoism with a callous delight in savaging his critics." One of Owen's Oxford colleagues called him "a damned liar. He lied for God and for malice." Gideon Mantell claimed it was "a pity a man so talented should be so dastardly and envious." Richard Brooke Freeman said it this way: "Owen: the most distinguished vertebrate zoologist and palaeontologist . . . but a most deceitful and odious man." In an 1860 letter to Asa Gray, Charles Darwin described him thus: "No one fact tells so strongly against Owen . . . as that he has never reared one pupil or follower."

Owen did everything he could to steal credit from Gideon Mantell (see chapter 2). Owen claimed that he and Cuvier discovered *Iguanodon*, completely ignoring the man who first found and described the actual fossils. He used his influence in the Royal Society to prevent some of Mantell's papers from being published. Eventually the Royal Society threw him out of their Zoological Council for plagiarizing Mantell's work. When Mantell was crippled and near death and unable to write or publish much, Owen had the audacity to rename some of Mantell's dinosaurs and claim that he had found them. After Mantell's death, Owen wrote a scathing and demeaning obituary that dismissed Mantell as a mediocre amateur naturalist and doctor with discoveries of no consequence. Even though it was published anonymously, everyone knew it came from Owen. Eventually Owen was denied the presidency of the Geological Society of London because of his petty and nasty treatment of Mantell.

When Darwin's ideas about evolution came out in 1859, Owen was insanely jealous because his own weird ideas about nature had never gotten any traction. He did everything he could to battle against Darwin's ideas in scientific circles, and especially with his powerful friends in high places. Darwin wrote of him: "Spiteful, extremely malignant, clever; the Londoners say he is mad with envy because my book is so talked about." "It is painful to be hated in the intense degree with which Owen hates me." Owen's nastiest attack was an anonymous book review that distorted and savaged Darwin's book while praising his own work in the third person.

Owen's enmity soon extended to all of Darwin's friends and supporters, including zoologist Thomas Henry Huxley, botanist Joseph Hooker, and many others. They soon became the powerful elite of British science, and Owen and his ideas became increasingly marginalized and irrelevant. When Huxley debated Archbishop "Soapy Sam" Wilberforce at Oxford in 1860, Owen coached Wilberforce on what to say because the archbishop knew no science—and Huxley could see the fingerprints of Owen's coaching on Wilberforce right away.

When Huxley confidently showed the connection between humans and other great apes, Owen claimed that only human brains had a region called the hippocampus, which set humans apart from all other animals. Huxley did his own dissections and showed that both apes and humans had a hippocampus, and Owen never forgave him. When Owen, as the official paleontologist of the British Museum, first described *Archaeopteryx* in 1861, he did everything possible to minimize how good a transitional form it was between birds and dinosaurs. Huxley corrected that in print as well. Owen also did everything he could to end government funding of Kew Gardens, Hooker's botanical marvel, and have it absorbed into the British Museum.

Despite Owen's constant attacks, the importance of the revolutionary ideas of Darwin, Huxley, Hooker, and their allies made them the center of British natural history in the second half of the 1800s. Huxley, in particular, soon was in the top position of power in British science. He used his clout and worked hard not only to support evolution but to modernize English science education and promote a new generation of scientific labs and permanent research programs. Meanwhile, Owen and his peculiar ideas were seen as more and more of a quaint, outdated notion that no one took seriously. Owen's reputation and his ability to hurt other scientists also declined. Owen died in 1892 at age 88, having outlived his wife and children, an embittered man whose scientific descriptions were first-rate but whose theoretical ideas have not held up well.

THE WHALE LIZARD

But what about the giant mystery bones? After more than 15 years, quite a few of the huge bones of the "whale" or "crocodile" had accumulated. Naturally they drew the attention of Richard Owen, who traveled to Oxford and Yorkshire to see them all. In 1841, he gave a presentation to the Geological Society of London in which he called all these bones the remains of some huge marine crocodile or other type of saurian. His anatomical skills demonstrated that although it was whale-sized, it was definitely reptilian, hence the name he gave it: *Cetiosaurus*, or "whale lizard" in Greek. All he had were some trunk and tail vertebrae, fragments of limb bones, and a few other odd pieces, so there was not really much he could say about what type of creature *Cetiosaurus* was. The uncertainty was so great that he made no attempt to have it reconstructed for the Crystal Palace Exhibit. All he could say was that it was a huge marine reptile, possibly an enormous marine crocodile, which is how he classified it.

A year later Owen formally coined a name for the huge land creatures that had come to light, "Dinosauria," which means "fearfully great lizards" in Greek (not "terrible lizards" as so many books report). He meant the grouping to indicate a new class of gigantic, majestic land reptiles, totally unlike any reptiles alive today. Three genera were the original basis for Owen's Dinosauria: Buckland's *Megalosaurus* and Mantell's *Iguanodon* and *Hylaeosaurus*. Surprisingly, Owen did not include his own *Cetiosaurus* because he believed it was some kind of crocodile or marine reptile, and all his Dinosauria were giant land reptiles.

In later papers Owen named new species of *Cetiosaurus* willy-nilly, without any adequate diagnosis or description. First, in 1842, there was *Cetiosaurus hypoolithicus* and *C. epioolithicus*, depending on whether the fossils came from above (epi) the Oolitic layer (Kingdon's specimens) or below (hypo) the Oolitic layer (the Yorkshire specimens). In a later publication that same year, Owen named four species, ignoring his previous names and also failing to diagnose the new names. These names were *Cetiosaurus brevis* (short), *C. brachyurus* (short tailed), *C. medius* (medium), and *C. longus* (long). These new species were unusual in that Owen named new species based on the same bones without justifying the change. In fact, he didn't have enough material to reconstruct one animal, let alone four distinct species that could be reliably identified. Other scientists soon jumped into the fray, and every new find of a large saurian bone was given a new species of *Cetiosaurus*, making it a taxonomic wastebasket for every sauropod found in Europe. This was the norm for science at the time, and nearly every new fossil received a new scientific name whether it could be distinguished from previously named fossils or not.

OWEN VERSUS MANTELL REDUX: *PELOROSAURUS*

One of Owen's inadequate species names was *Cetiosaurus brevis*, which was based on specimens not from the Jurassic Oolitic layers but from the Lower Cretaceous beds of the Weald—Gideon Mantell's favorite hunting ground. They were found near Cuckfield (although not necessarily in Tilgate Quarry) in 1825 by the same John Kingdon who found the first *Cetiosaurus* specimens near Oxford. However, they came from a different layer, the Tunbridge Wells Sands of the Hastings Group (figure 2.1). Some of the bones in the collection, however, belonged to *Iguanodon*. Recognizing the mistake, Alexander Melville renamed the sauropod bones *Cetiosaurus conybeari*.

Meanwhile, Gideon Mantell was looking at more bones of the Wealden sauropod. He decided that the bones were so different they did not belong in *Cetiosaurus* but in their own genus. In 1850, he coined the name *Pelorosaurus* for these fossils. He was originally going to call it "Colossosaurus" until he learned that *kolossos* means "statue" not "giant" in Greek. Instead, he chose the Greek word *pelor*, which means "monster," in coining the name *Pelorosaurus*. In addition to the original "*Cetiosaurus brevis*" bones, for £8 Mantell bought another upper arm bone (humerus) found at the same site by a local miller, Peter Fuller. The humerus showed a central medullary cavity and other structures typical of large land animals that must support their great weight against the pull of gravity. It was not the dense solid bone structure of marine animals that can take advantage of the buoyancy of water. From this Mantell correctly realized that *Pelorosaurus* and also *Cetiosaurus* were giant land animals, not marine reptiles as Owen had asserted.

Owen was angry with Melville and Mantell for correcting his mistakes, and rather than admit them he went even further. In an 1853 publication (after Mantell had died), Owen rationalized his choice of the name *Cetiosaurus brevis* and dismissed Melville's attempt to rename it *Cetiosaurus conybeari*. In answer to the late Gideon Mantell, he restricted the name *Pelorosaurus* to Mantell's new humerus fossil and put all the rest of his Wealden fossils in *Cetiosaurus brevis*. However, in 1859 he repeated the mistake that Melville had corrected by assigning more *Iguanodon* bones to *Cetiosaurus*. He also assigned those bones to *Pelorosaurus*, further confusing the picture.

The mess was not straightened out for over a century. In 1970, John Ostrom did the detective work to unscramble the complicated and confused picture of these bones. He concluded that *Cetiosaurus brevis* was still valid, although Owen had done it wrong and his reasons were inadequate. *Cetiosaurus conybeari* was based on exactly the same fossils, so that name is not valid. Over the years, most of the Jurassic bones were routinely assigned to *Cetiosaurus*, and Cretaceous bones were called *Pelorosaurus*. In 2007, Mike Taylor and Darren Naish tried to clear up the status of the proper species for *Pelorosaurus*, recommending the use of the name *P. conybeari* for it, although the International Commission on Zoological Nomenclature has not yet ruled on it. The status of *Pelorosaurus* itself is still open for debate. It is based on the original humerus that Mantell described, plus many other bones that may or may not belong to it. The name still exists in the literature, although like many sauropods it is so incomplete that little can be said about its validity.

MORE PIECES OF THE PUZZLE

About 25 years later, more specimens of *Cetiosaurus* were found that helped clarify what kind of animal it was. A worker digging near Bletchingdon in

Figure 3.2 ▲
The limb bones of *Cetiosaurus*, on display in the Oxford Natural History Museum. There were so few bones of the skeleton that neither Owen nor most later scientists could imagine a huge long-necked, long-tailed sauropod. Owen thought it might be the remains of a giant marine crocodile. (Photograph by M. Wedel)

1868 found a huge right thighbone. This inspired the geologist John Phillips to dig further in this locality between March and June of 1870, and they found parts of five different skeletons. They were still mostly vertebrae, but also included most of the forelimbs and hind limbs (figure 3.2). No skull or neck was known, and the parts they had were so incomplete that the animal still could not be reconstructed.

This is the way the situation remained for several decades in Europe. No sauropod skeleton was more than 40 percent complete, so paleontologists only knew that it was a gigantic reptile, but they could not imagine the incredibly long neck, the tiny skull with tiny teeth, or the great length of the tail that we all know today. The situation changed dramatically in the 1880s when the first complete sauropod skeletons of *Apatosaurus* were found in Colorado and Wyoming and eventually mounted in a lifelike pose and reconstructed by artists (see chapter 7). Suddenly the huge but incomplete pieces of European sauropods made sense. It was now possible to recon-struct *Cetiosaurus* similar to other large sauropods with a long neck and tail, huge size, and upright limbs—even though there were still no fossils to sup-port this. (No skull of *Cetiosaurus* has ever been found.)

However, work on *Cetiosaurus* was not finished yet. On June 19, 1968, exactly a century after the Bletchingdon discoveries, the driver of an earth-mover hit some huge bones near Rutland. More than 200 bones were found, including most of the neck vertebrae, most of the back vertebrate, the front half of the tail, and part of the hips and the right thighbone. These specimens were excavated in the next few months, then stored at the Leicester City Museum. Unfortunately, no one had the time or resources to work on them further until 1980. At that time, they finally got the specimen prepared and studied, and it has been on display at the New Walk Museum in Leicester since 1985 (figure 3.3). As mounted, the display uses most of the real bones, although some of the fragile ones are replaced by replicas, as are all the bones that still have never been found. The Rutland *Cetiosaurus* is

Figure 3.3 ▲

The reconstruction of the most complete skeleton known of *Cetiosaurus* in the New Walk Museum in Leicester. (Courtesy of M. Evans, New Walk Museum)

the most complete sauropod, and one of the most complete dinosaurs ever found in Europe. As mounted, it is over 15 meters (50 feet) long, finally giving a true impression of Britain's best-known sauropod dinosaur.

What kind of animal was *Cetiosaurus*? As sauropods go, it was in the medium-size range for a Jurassic form, reaching about 15–16 meters (50–55 feet) in length. Its neck was relatively short for a sauropod, but it had a relatively long tail. Its bones were robust and heavy, so it does not have all the hollowing and lightening of the bones and numerous air chambers seen in the gigantic sauropods such as *Brachiosaurus*. As a relatively generalized and unspecialized sauropod without a really long neck (by sauropod standards), it was probably a generalized feeder. It does not have the extreme neck and forelimb length of brachiosaurs, the extreme neck length of *Mamenchisaurus*, or the long necks and tails of diplodocines. It probably browsed on leaves and vegetation from low to medium heights in the tree canopy.

Most analyses of *Cetiosaurus* tend to suggest that it is related to the more advanced Neosauropoda, along with the Chinese taxon *Shunosaurus, Omeisaurus,* and *Mamenchisaurus*; the Argentinian *Patagosaurus; Barapasaurus* from India; and *Chebsaurus* from Africa—although it is slightly more advanced and closer to Neosauropoda than most of the rest of these primitive sauropods. *Cetiosaurus* was found in the same beds that yielded *Megalosaurus*, which may have preyed upon it when the opportunity arose. It lived in a floodplain or open woodland environment on the fringes of the shallow seas that covered most of Europe in the Middle Jurassic, and not in the marine beds where it had apparently washed out as a carcass (the same goes for *Megalosaurus*). Contrary to Owen's ideas that it was an enormous marine reptile and therefore not a dinosaur, it was found in marine beds simply because those huge carcasses could float some distance.

So *Cetiosaurus* holds the distinction of the first sauropod discovered. But a true understanding of sauropods would not come until more complete skeletons were found in North America in the 1870s, and then in Africa and eventually China and South America in the twentieth century.

FOR FURTHER READING

Cadbury, Deborah. *The Dinosaur Hunters: A True Story of Scientific Rivalry and the Discovery of the Prehistoric World.* New York: Harper Collins, 2000.

——. *Terrible Lizard: The First Dinosaur Hunters and the Birth of a New Science*. New York: Henry Holt, 2001.

Colbert, Edwin. *Men and Dinosaurs: The Search in the Field and in the Laboratory*. New York: Dutton, 1968.

Desmond, Adrian. *Archetypes and Ancestors: Palaeontology in Victorian London 1850–1875*. Chicago: University of Chicago Press, 1982.

Maddox, Brenda. *Reading the Rocks: How Victorian Geologists Discovered the Secret of Life*. New York: Bloomsbury, 2017.

McGowan, Christopher. *The Dragon Seekers: How an Extraordinary Circle of Fossilists Discovered the Dinosaurs and Paved the Way for Darwin*. New York: Perseus, 2001.

Naish, Darren. *The Great Dinosaur Discoveries*. Berkeley: University of California Press, 2009.

Rudwick, Martin. *Bursting the Limits of Time: Reconstructing Geohistory in the Age of Revolution*. Chicago: University of Chicago Press, 2007.

——. *The Meaning of Fossils: Episodes in the History of Paleontology*. New York: Science History, 1976.

——. *Worlds Before Adam: Reconstructing Geohistory in the Age of Reform*. Chicago: University of Chicago Press, 2010.

Rupke, Martin. *Richard Owen: Biology Without Darwin*. Chicago: University of Chicago Press, 2009.

——. *Richard Owen: Victorian Naturalist*. New Haven, Conn.: Yale University Press, 2004.

Spaulding, David A. E. *Dinosaur Hunters: Eccentric Amateurs and Obsessed Professionals*. Rocklin, Calif.: Prima, 1993.

HADROSAURUS

Tired! Not so long as there is an undescribed intestinal worm, or the riddle of a fossil bone, or a rhizopod new to me.

—JOSEPH LEIDY, ANSWERING A QUESTION ABOUT WHETHER HE WAS TIRED OF LIFE

THE LAST MAN WHO KNEW EVERYTHING

The study of dinosaurs emerged in England in the 1830s through 1850s, but comparable discoveries of dinosaurs were not yet found across The Pond. American engineering and science was still in its infancy, focused mainly on practical topics such as agriculture and canal building. Fossils of the first mastodonts had been found in the mid-1700s, and they puzzled scholars from America to France. So had the giant claws studied by Thomas Jefferson, who thought they belonged to a giant lion and instructed Lewis and Clark to look for such a feline on their trip across the continent. (The claws came not from a cat but from a giant ground sloth, now named *Megalonyx jeffersoni* in his honor.) Aside from these rare Ice Age mammals, few fossil bones had yet been found in North America. There were no American vertebrate paleontologists yet, so doctors and other scholars who had some biological training occasionally had a fossil bone sent to them for identification. John James Audubon and many other naturalists were focused on discovering, collecting, and describing the living species of North American animals for the first time—not looking for fossils.

Into this void stepped America's first great naturalist, Joseph Leidy (figure 4.1). His work spanned many different fields of natural science, from paleontology to geology to parasitology to embryology to forensic medicine. In an obituary for Leidy, Henry Fairfield Osborn wrote:

Among zoologists he was the last to treat of the whole animal world from the protozoa to man, rendering in every branch contributions of permanent value. From his researches among the minerals, plants, infusorians, entozoa, and mollusks, he ranged into comparative anatomy as well as into his greatest field of research, vertebrate palaeontology. In the year 1852 we find him writing upon fossils from the West, the geology of the Badlands, the life history of bees, their anatomy and the physiology of their reproductive organs, as well as entering the discussion of some new fungi with an English microscopist, the specific determination of various parasites, as well as numerous plants, the investigation of some new points in comparative anatomy, the observation of the movements of some new Rhizopods. His encyclopaedic knowledge,

Figure 4.1 ▲
Joseph Leidy in mid-career at the time he began his work on dinosaurs, about 1865. (Courtesy of Wikimedia Commons)

broad grasp of the whole field of natural history, precision and originality of observation in every field, present a combination of endowments which will never reappear in a single individual.

No wonder Leonard Warren's excellent 1998 biography of Leidy is subtitled "The Last Man Who Knew Everything." By the time Leidy died in 1891, the amount of science information (to which Leidy was a key contributor) had exploded, and no later natural historian could claim equal expertise in so many different fields. Since then, scientists have become specialists with limited knowledge of other fields of science due to the huge amount of information that science has accumulated.

Born September 9, 1823, in Philadelphia, Joseph Mellick Leidy's father, Philip Leidy, was a hatmaker of German ancestry. His mother, Catharine Mellick Leidy, died giving birth to his younger brother Thomas. Joseph's father soon remarried, choosing his former wife's first cousin, Christiana Mellick, to help him raise the young boys. His father wanted young Joseph to become a sign painter, but his stepmother supported him in more ambitious career goals, so he went to the University of Pennsylvania to study medicine. He was trained by the noted anatomist Dr. James McClintock and the pioneering microscopist Paul B. Goddard. Once he earned his medical degree in 1844 at age 21, his father expected him to set up a lucrative medical practice and earn his keep. Instead, he stayed around Penn medical school, helping the others teach medicine and microscopy and beginning a research career. In 1845, he was commissioned to do dissections and illustrations for Amos Binney's book on American snails. The publication of this book brought Leidy many accolades, and he was soon elected to the scholarly societies of Philadelphia and Boston. By the end of his life, Leidy was a member of more than 50 scientific societies. He finally received a permanent post in 1846 as a curator at Philadelphia's Academy of Natural Sciences.

Leidy rose quickly through the scholarly ranks as he began his research career, publishing more than 553 scientific papers in his lifetime. Leidy's skill in dissection and microscopy and brilliance in anatomy soon elevated him to a post in the University of Pennsylvania medical school where he taught anatomy in 1853. Later he also occupied posts at Swarthmore College and Wagner Free Institute, where he taught many subjects. His textbook on medical anatomy became the standard that instructed almost all

American medical students for more than 40 years. During the Civil War, Leidy served as an Army surgeon, pioneering many innovative practices in surgery that improved the outcomes for wounded soldiers.

Long before Sherlock Holmes or real forensic detective work existed, Leidy's skills as a microscopist made him one of the first scientists involved in detective work and forensic medicine. In 1846, the police brought him a case of a man accused of murdering a Philadelphia farmer. The suspect claimed that the blood on his clothes and hatchet was chicken blood that had spattered on him when he had killed some birds earlier. Leidy used his microscope to determine that the red blood cells had no nuclei, a characteristic of human blood and not animal blood. He further ruled out chickens by leaving samples of their blood out to dry, and they never lost their nuclei. This evidence forced the suspect to confess.

In addition to being the founder of American vertebrate paleontology, Leidy was one of the founders of parasitology and protozoology. His 1879 monograph on protozoans was the beginning of that field of research. Using his microscope, he was able to identify many of the human parasites for the first time. His most famous find was discovering the worm that infects pigs and causes trichinosis in people who eat undercooked pork.

As one of the few people at that time with any knowledge of comparative vertebrate anatomy, fossils also came to Leidy. His many responsibilities in Philadelphia left him little time or resources to mount his own expeditions. Instead the specimens were sent to him; they were mostly fragmentary fossils picked up by the early explorers of the American West. Beginning in 1847, he received a steady stream of fossils from what is now Badlands National Park in South Dakota, including the first American camel *Poebrotherium*, the first American rhinoceros *Subhyracodon*, and the first horse found in North America, *Mesohippus*, plus many other extinct archaic mammals such as oreodonts, entelodonts, and brontotheres that have no living descendants. He also was sent Ice Age horse teeth, which confirmed that horses had been present in North America and then vanished long before Columbus reintroduced them in 1493. Leidy received numerous other Ice Age fossils, so he was the first to name and describe the dire wolf, the giant American lion, the short-faced bear, the Ice Age camel, bison, and horse, and many others. For many American paleontologists, a trip to the Academy of Natural Sciences is essential because it is necessary to see Leidy's original fossils to understand whether a particular genus or species is valid or not.

In the 1860s and early 1870s, Leidy embarked on a few collecting trips to the Rocky Mountains to obtain his own fossils. However, by 1871, vertebrate paleontology became a crowded and contentious field as the warfare between Edward Drinker Cope and Othniel Charles Marsh (see chapter 7) soon escalated to outright stealing of fossils, bribery, and fraud. Leidy was too much of a gentleman to get into this nasty feud, so he gave up fieldwork and most of his research in fossil vertebrates rather than get between Cope and Marsh and their self-destructive war on each other. He did publish on some fossils from Florida (including the hippo-like rhino *Teleoceras*) later in life, but he did not try to compete with Cope or Marsh, who had much more free time and more resources to obtain all the good fossils from the emerging fossil beds out west.

THE NEW JERSEY MARL PIT

Long before the Cope-Marsh wars ended opportunities for collecting fossils in the field, Leidy got some specimens from just down the road in a marl pit near Haddonfield, New Jersey. Marl is a mixture of lime and clay or muddy limestone that is commercially mined to add lime to soil and make it less acid. John Estaugh Hopkins was digging in a marl pit on a tributary of the Cooper River in 1838 and hit some large fossilized bones, which he brought home and put on display. Many locals picked up bones as well.

Twenty years later, a prominent lawyer and abolitionist, William Parker Foulke, was visiting Hopkins's home and saw the huge fossil bones. Since Foulke was also an avid amateur naturalist and geologist, he got the directions to the marl pit where Hopkins had gotten the bones decades earlier. He began digging and found many more bones there. Soon he contacted the eminent Dr. Leidy and convinced him to join the excavation. Together they recovered many more bones until they had an almost complete set of limbs, a partial pelvis, parts of the feet, and 28 vertebrae (including 18 from the tail), but there was almost no skull—only eight teeth and part of the jaw. (The complete skull is still unknown.)

By the end of 1858, Leidy had finished his study and a short description of the bones and formally named it *Hadrosaurus foulkii*, in honor of the man who brought it to him. *Hadros* means "heavy" in Greek (as in the heavy nuclear particle known as "hadrons"), so its name means "heavy lizard." The giant size and the robust shafts of the limb bones unquestionably indicated a very heavy reptile of some sort.

Leidy kept up with the latest scientific papers imported from England, and he had been following the discovery and description of *Iguanodon* for some time. He knew that *Hadrosaurus* must be another member of Owen's Dinosauria, the first to be found in North America. The parts he had, however, were more complete than anything Mantell had found for *Iguanodon*. His specimen definitely had strong robust hind limbs and relatively delicate forelimbs, so it was primarily bipedal. Finally, he wrote up all the detailed descriptions and had detailed illustrations of the bones prepared for his monograph on *Cretaceous Reptiles of the United States*. It was due to be published in 1860, but the Civil War delayed its publication until 1865.

In 1868, the British sculptor Benjamin Waterhouse Hawkins, who had built the famous Crystal Palace models of British prehistoric animals, came to the United States with the intent of restoring *Hadrosaurus*. At that time, most museums kept their specimens in the storage area for research and rarely displayed them to the public. *Hadrosaurus* was the most complete dinosaur yet found, and Hawkins was prepared to produce the first-ever mounted skeleton that could be used for display purposes so the general public could see the skeletons of dinosaurs for the first time. Hawkins had to cast additional bones to replace the missing limbs, ribs, vertebrae, tail bones, and especially the head and neck. He had to guess at what the plaster reconstruction of the skull would look like. Naturally, he made it similar to what was known of *Iguanodon* (figure 4.2A) rather than giving it the more duck-like bill that was later found on other duck-billed dinosaurs related to *Hadrosaurus*. There are many pictures of the *Hadrosaurus* mount in a scaffold of wood and rope, holding the bones in place. Once the mount was open to the public, it was a sensation, and long lines of people came to the academy to marvel at the gigantic extinct reptile. America had its first taste of dinomania, although the modest mount did not match the elaborate giant skeletons and models that would become popular at the beginning of the twentieth century. With the exception of the skull model, all of Hawkins's reconstructions of *Hadrosaurus* have disappeared.

Hadrosaurus became iconic as America's first dinosaur. The marl pit is now a state monument, and there is a statue of the dinosaur in the center of Haddonfield. In 1994, it was formally named as New Jersey's state fossil. In other respects, however, *Hadrosaurus* fell into the shadows. Within 50 years after its discovery, dozens of additional duck-billed dinosaurs were discovered in North America. Leidy himself named some duck-billed

dinosaur teeth *Trachodon*, a name applied to many duck-billed dinosaurs until scientists determined that the specimens were not complete enough to assign to any known dinosaur; the name is now invalid. Cretaceous beds in Wyoming, Montana, and especially the Red Deer River badlands of Alberta soon yielded many complete articulated skeletons of duck-billed dinosaurs, and most of them came with skulls attached. The skull turned out to be the most diagnostic feature because some duckbills (the Lambeo-saurinae) such as *Lambeosaurus, Corythosaurus*, and *Parasaurolophus*, have spectacular crests on the top of their head. Others, such as *Gryposaurus, Brachylophosaurus*, and *Edmontosaurus* (the Hadrosaurinae), apparently had no crest. The known skeletal features of *Hadrosaurus* usually place it among these noncrested duckbills, so it is assumed that its unknown skull had no crest.

The lack of a skull presents problems as well. Because it is missing the most important features that paleontologists use to analyze the relation-ships and family tree of duckbills, *Hadrosaurus* was often ignored or left out of the analysis. Some even suggested that the name was invalid because it is not diagnostic enough to determine where it belongs among duckbills. But a 2011 study focusing just on the limbs and other parts of Leidy's and Foul-ke's skeleton has shown that *Hadrosaurus* can be placed among the family tree of duckbills, even without a skull, so now most dinosaur paleontolo-gists regard it as a valid genus.

Even today, many people use the term "hadrosaur" or "Hadrosauridae" to refer to the complete gamut of duck-billed dinosaurs, although a more careful analysis suggests that the name only applies to part of the duckbill family tree. Nevertheless, this informal use of the word "hadrosaur" is still widespread among paleontologists.

THE WORLD OF HADROSAURS

Hadrosaurus opened up the world to the existence not just of American dinosaurs but especially to the panoply of different kinds of duck-billed dinosaurs. They first appeared, along with their close relatives the iguano-dontids, in the Early Cretaceous, and then began a huge evolutionary radi-ation of more than two dozen genera. They were among the most common dinosaur fossils found in the Cretaceous, suggesting that they were the most common herbivores during that time and may have even roamed in herds. If you collect in certain Cretaceous beds in the Rocky Mountains, such as the Fox Hills or Lance formations, you invariably find that the most common fossils are little cylinders and rods of bone about as thick as your finger. These are the ossified tendons of hadrosaurs, which once formed an elaborate trusswork of crisscrossing tendons that helped these dinosaurs hold their back and tail out in a rigid straight line. Leidy was clearly correct: like many other dinosaurs, hadrosaurs were primarily bipedal with the back and tail held straight and parallel to the ground, balanced on their hips and hind legs. Yet many hadrosaurs had relatively robust arms, so they could drop down and walk in a quadrupedal posture, and some were primarily quadrupeds that rarely walked on two legs.

Hadrosaurs are unusually well known among dinosaurs because many have been found as complete articulated skeletons, often in death poses. Even more impressive, 10 different hadrosaur species have impressions of their skin preserved as well, and some appear to have dried out and mummified in a death pose before they fossilized. Most of these fossils show a skin with a pebbly texture of rounded scales and other dermal bone, with no indication so far that their preserved skin was feathered. One specimen found in 1999 in the Hell Creek Formation of North Dakota is so completely preserved as a mummy that it has been CT scanned to image all its muscles, ligaments, tendons, cartilage between the vertebrae, and internal organs as well. Using this extraordinarily well-preserved specimen in dynamic computer animation models, it was suggested that they were much more muscular than previously thought, and probably could outrun a predator like *Tyrannosaurus rex.*

Another commonly found duckbill fossil is their teeth. Instead of rows of individual teeth separated from one another (like most animals), most duckbill teeth are shaped like long prisms and were packed together tightly into a giant grinding plate known as a dental battery (figure 4.3). You may have read that some duckbills had almost 1,000 teeth. How is this possible? Their teeth were tiny prisms packed closely together, and each battery might contain 250 teeth all tightly fused. With a 250-tooth battery on the right and left side of both upper and lower jaws, there would literally be 1,000 teeth. The jaws were built so the upper grinding plate of the dental battery scraped down against the grinding surface of the lower battery, giving duckbills an incredible ability to chew up any tough vegetation. A 2009 study showed how these dental batteries worked. The lower jaw is relatively rigid, but still

Figure 4.3 ▲
Jaw of a hadrosaur, showing the hundreds of small prism-shaped teeth packed together into a dental battery. Together they formed a single hard mass of enamel ridges and dentin that would wear down into a flat grinding surface and occlude with the dental battery on the upper part of the mouth. (Courtesy of Wikimedia Commons)

has some movement in the front where the jaws meet. However, both sets of upper jaws are hinged against the skull, so when the lower jaw is brought upward in a chewing motion, the upper jaws flex outward and maximize the amount of grinding between the teeth in each chewing stroke.

In fact, many paleontologists think that the explosive radiation of duck-billed dinosaurs in the Cretaceous (along with some other herbivores) might have co-evolved with the evolution of flowering plants, which also exploded in diversity in the Early Cretaceous. Flowering plants have evolved several mechanisms to regrow quickly after dino damage, including vegetative reproduction (sprouting new plants without forming seeds). If a big radiation of herbivorous dinosaurs occurred in the Early Cretaceous, then fast-growing flowering plants would be favored over more primitive plants. These in turn would provide more food sources for herbivores, and thus the co-evolutionary "arms race" ratchets up the specializations of both the herbivores and their plant food.

Studies of the wear patterns in their teeth and the content of their fossilized feces, or coprolites, confirm that hadrosaurs ate a diet of leaves and low-growing plants, and even some rotting wood and occasionally shelled invertebrates. They probably used their broad bills to graze on low plants close to the ground (although not grass because broad grasslands had not yet evolved). This contrasts with notions from the early twentieth century that the duck-like bill was needed for feeding in the water like a duck and eating soft water plants. Hadrosaurs also could rise up and reach the leaves of tall shrubs and low-growing trees, food that was unavailable to low-slung four-legged herbivores such as ceratopsians and ankylosaurs.

As mentioned previously, one branch of hadrosaurs (the lambeosaurs) bore remarkable crests on top of their skulls, from the tall compressed crest of *Tsintaosaurus* to the helmet-like crest of *Corythosaurus* to the long curved tubes of *Parasaurolophus* (chapter 22). The function of this crest remains controversial. Many of them were hollow and are thought to have served as a resonating chamber for making sounds.

Much more is known about duck-billed dinosaurs than just about any other group because there are so many of them. Many are known from complete articulated skeletons with skin, and they underwent a huge diversification in Asia and North America. By contrast, they arrived in South America only in the latest Cretaceous, were rare in Europe, and we don't have enough Upper Cretaceous rocks in Africa to know whether they lived on that continent or not. And it all started with Foulke and Leidy and *Hadrosaurus*, America's first dinosaur.

FOR FURTHER READING

Colbert, Edwin. *Men and Dinosaurs: The Search in the Field and in the Laboratory.* New York: Dutton, 1968.

Howard, Robert West. *The Dawnseekers: The First History of American Paleontology.* New York: Harcourt Brace Jovanovich, 1975.

Lanham, Url. *The Bone Hunters.* New York: Columbia University Press, 1973.

Naish, Darren. *The Great Dinosaur Discoveries.* Berkeley: University of California Press, 2009.

Spaulding, David A. E. *Dinosaur Hunters: Eccentric Amateurs and Obsessed Professionals.* Rocklin, Calif.: Prima, 1993.

Thomson, Keith. *The Legacy of the Mastodon: The Golden Age of Fossils in America.* New Haven, Conn.: Yale University Press, 2005.

Warren, Leonard. *Joseph Leidy: The Last Man Who Knew Everything.* New Haven, Conn.: Yale University Press, 1998.

EORAPTOR

But it's a desolate place like the Valley of the Moon in northwestern Argentina where Bonaparte and his dinosaurs are likely to be found, anytime from September to April. The conditions are harsh, the equipment primitive. Bonaparte and Sereno drove a rickety Renault with a broken fuel pump during their dig in the Valley of the Moon in 1988. An assistant had to perch on the roof much of the ride, dangling the fuel line. Bonaparte is accustomed to sleeping outdoors or in the sheep-shearing room of remote *estanzias*, the vast Argentine ranches. He explored Argentina's Patagonian mountains on horseback, at least until he was turned back by fierce summer snow squalls.

—DON LESSEM, *OMNI MAGAZINE*, 1993

WHAT IS A DINOSAUR?

Often the first thing a paleontologist has to deal with when talking to the public is their misconceptions about dinosaurs. Many people think any large extinct animal is a dinosaur. Those bags of plastic dinosaur toys mixed with nondinosaurs such as the saber-toothed cat and mammoth reinforce this misconception. Likewise, those same plastic "dinosaur toy" sets often include the fin-backed protomammal *Dimetrodon* because it is also large and prehistoric—even though it is part of the mammal lineage. In movies like *Jurassic World* and TV shows like *Walking with Dinosaurs*, and in many other media, people see images of marine reptiles such as mosasaurs, long-necked plesiosaurs, and dolphin-like ichthyosaurs and assume that these large reptiles of the Age of Dinosaurs must be dinosaurs too. Another common misconception is that all dinosaurs are large, but a great many were small, including a lot that were the size of small birds. A little more excusable is the common misconception that pterosaurs are dinosaurs. The public ignorance is so annoying to paleontologists that Mark Norell of the American Museum of Natural History in New York wrote a cleverly illustrated children's book. Every page shows a nondinosaur that the public thinks is a dinosaur, reinforcing the same point over and over again. On the other hand, if you show an image of a bird, most of the public will *not* call it a dinosaur—but it is.

If none of these creatures are dinosaurs, what do paleontologists mean by the term? When Richard Owen had only *Megalosaurus* and *Iguanodon* and a few others to go on, he defined dinosaurs as a group of huge extinct reptiles with a number of distinctive features (see chapter 2). As the number of new dinosaurs (and tiny dinosaurs) increased rapidly in the late 1800s and early 1900s, that definition was modified. By the time of Edward Drinker Cope's work in the 1880s, and especially when Samuel Wendell Williston published *Osteology of the Reptiles* in 1925, the dinosaurs had been separated from other groups of reptiles, such as the marine reptiles, pterosaurs, and others such as the protomammals (still called "mammal-like reptiles" then). In 1878, Othniel Charles Marsh recognized four groups of dinosaurs that are still valid today: sauropods, theropods, ornithopods, and stegosaurs (figure 5.1). But few of these authors provided an exact anatomical definition of what constitutes a dinosaur.

In 1888, British paleontologist Harry Govier Seeley recognized two groups of dinosaurs: the "lizard-hipped" dinosaurs, or Saurischia (which included the theropods and sauropods), and the "bird-hipped" dinosaurs, or Ornithischia (which include most of the herbivorous dinosaurs except sauropods). These ideas gained widespread acceptance, and for the next 130 years most paleontologists agreed that to be a dinosaur a fossil had to be a member of one of these two groups. In 1974, Robert Bakker and Peter Galton described a number of unique anatomical features that established dinosaurs as a natural group.

In 2017, British paleontologists Matthew Baron, David Norman, and Paul Barrett published a detailed analysis of dinosaur interrelationships. Looking at the most primitive theropods, sauropods, and ornithischians, they found no unique anatomical specializations that could be used to define the Saurischia. In addition, other evidence suggests that the theropods and ornithischians are closely related rather than theropods plus sauropods

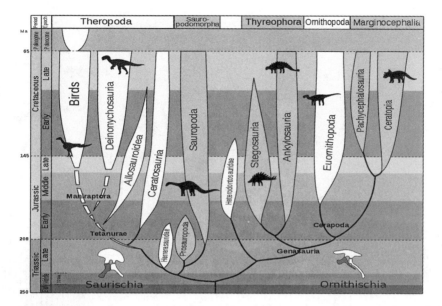

Figure 5.1 ▲

Simplified family tree of the major groups of dinosaurs, showing the hip structure diagnostic of each group. The dark-colored bone on the lower left of each hip diagram is the pubic bone, which points backward in ornithischians (lower right corner diagram). (Courtesy of Wikimedia Commons)

together as Saurischia (figure 5.2). This idea remains controversial, and a recent response by a group of leading researchers led by Max Langer questioned many of the interpretations. These arguments are highly technical and beyond the scope of this book, so we will focus on well-defined groups of genera and families.

As more and more fossils were found, the differences between the two groups (based on hip structure) seemed to be consistent, and the individual groups (Sauropoda, Theropoda, and so on) continued to work well. The Saurischia were the lizard-hipped dinosaurs, with the pubic bone of the hip region pointing forward (figure 5.3A and C). The Ornithischia, or "bird-hipped dinosaurs," had at least part of the pubic bone shifted backward, parallel to the rear bone of the hip region, the ischium (figure 5.3A–B).

However, the major suborders defined by Seeley did not answer the question of how the groups within Saurischia and Ornithischia are interrelated. I remember attending lectures from the early 1970s in which paleontologists were not sure that Saurischia and Ornithischia could be combined

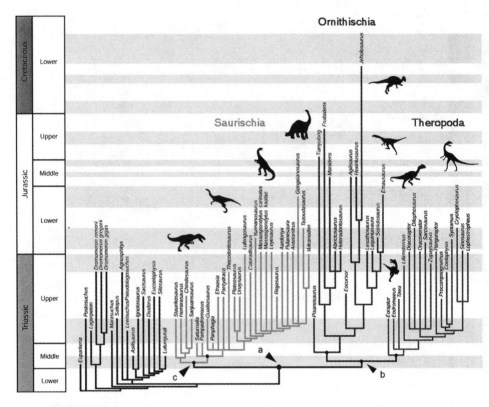

Figure 5.2 ▲

The new family tree of dinosaurs according to the analysis by Matthew Baron, David Norman, and Paul Barrett in 2017. Ornithischia remain a natural group, but "Saurischia" is broken up, placing theropods closer to Ornithischia than to sauropods. (From M. Baron, D. Norman, and P. Barrett, "A New Hypothesis of Dinosaur Relationships and Early Dinosaur Evolution," *Nature* 543 [2017]: 501–506, fig. 1; courtesy of *Nature*)

Figure 5.3 ▶

(*A*) The hip socket or acetabulum of dinosaurs is an open hole between their three hip bones where they join the middle. The "ball" or head of the thighbone (femur) inserts in this hole. On the top left is an ornithischian pelvis, with part of the pubic bone running backward parallel to the ischium. On the top right is a saurischian pelvis, with the pubic bone pointed forward. The head is to the right and the tail is to the left. (*B*) Close-up of the hip structure of a typical ornithischian, showing the pubic bone running beneath and parallel to the ischium (to the left of the thighbone). The head is to the right and the tail is to the left. (*C*) Close-up of the hips of a saurischian, showing the large pubic bone to the right of the thighbone, and the smaller ischium to the left of the thighbone. ([*A*] Redrawn from several sources; [*B–C*] photographs by the author)

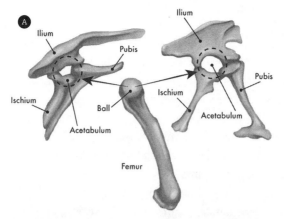

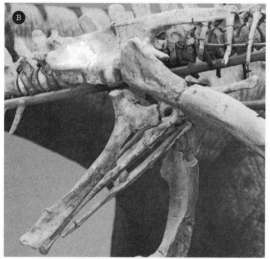

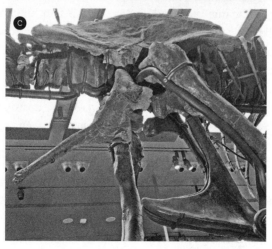

into the Dinosauria. In his 1955 textbook, *Evolution of the Vertebrates*, Edwin Colbert wrote that "the term includes two distinct reptilian orders. Consequently, the word dinosaur is now a convenient vernacular name but not a systematic one." This sad misunderstanding might have been typical of thinking in 1955 and 1969, when the first two editions of Colbert's book were published, but sadly it remained even in the last edition in 2001, by which time this idea had been resoundingly debunked.

However, at that very time these doubts were expressed, a revolutionary new way of thinking about animal classification called "cladistics," or "phylogenetic systematics," came along. First developed by the German entomologist Willi Hennig in 1950, but not translated into English until 1966, it swept through nearly every area of animal and plant systematics in the late 1960s and 1970s and has helped solve many troubling and confusing problems in animal classification. Hennig's ideas were complex, but his principal insight was that we must base classification systems strictly on their evolutionary history, or phylogeny (hence the name "phylogenetic systematics"). For example, if birds are descended from dinosaurs, they must be classified as a subgroup of dinosaurs, not set up in their own "Class Aves" parallel to and distinct from their ancestors among "Class Reptilia."

One of his key insights was that natural groups in classification schemes are defined based on unique evolutionary specializations, or "shared derived characters." For example, our anatomy with four limbs goes all the way back to the earliest tetrapods ("amphibians" in the old sense) that crawled out on land, so we wouldn't use "four limbs" as an anatomical specialization to define mammals or humans because four limbs are primitive features of all nonfish vertebrates (tetrapods). Mammals can be defined by unique evolutionary novelties such as having hair or fur and the presence of milk in mammary glands in females to nurse their young. But we wouldn't use the presence of fur or mammary glands to define a group *within* mammals, such as our own order Primates. Likewise, the "lizard-hipped" condition of the Saurischia is primitive for all reptiles and doesn't serve well to define a natural group.

Once we discarded all the "wastebasket" groups and focused on shared evolutionary specializations as the key to understanding relationships, many previously insoluble problems in classification could be resolved. This also helped redefine what it meant to be a dinosaur in the strict sense. As Jacques Gauthier first documented in 1986, one of the unique evolutionary specializations of all dinosaurs is that there is a hole right through the hip socket ("acetabulum" in anatomical lingo) rather than a pocket in the bone that holds the head of the thighbone as it moves (figure 5.3A). No other animals have this characteristic; it a unique feature that defines Dinosauria.

Additional characteristics include that all dinosaurs held their limbs straight beneath their bodies in an upright posture, just like mammals. Nearly all dinosaurs have just three or fewer fully developed fingers in their hands, with the ring finger and pinky highly reduced or missing. Dinosaurs nearly all walk on the tips of their fingers and toes, with only three toes in their feet (the fourth toe and little toe are reduced or lost). Finally, there are just three vertebrae fused to the upper part of the hip bones, connecting the spine to the hind legs and forming a sacrum. Other animals have fewer or more vertebrae in their hips. Baron, Norman, and Barrett demonstrated that a few of these features show up in the close relatives of dinosaurs, but no other group of animals combine all these features.

RULING REPTILES

Early paleontologists such as Cope and Williston first noticed that a natural cluster of reptiles included crocodiles and dinosaurs, which Cope called the Archosauria, or "ruling reptiles" in Greek. Some of the groups Cope placed in his Archosauria in 1869 have since moved to other parts of the reptilian family tree, but Gauthier's 1986 detailed cladistic analysis of archosaurs found characteristics that define all archosaurs. The most obvious feature is that their teeth (which in most groups are simple conical pegs) sit in sockets in the jaw, whereas the teeth of lizards and snakes emerge from a continuous groove along the inside wall of the jaw. This teeth-in-sockets pattern is known as the "thecodont" condition, and for many years paleontologists used the wastebasket group "Thecodontia" to refer to all archosaurs that were not crocodiles, pterosaurs, dinosaurs, or birds.

If you look at almost any archosaur skull (figure 5.4A), you will find an opening before the eye socket called the antorbital ("ante" means "in front" of the "orbit") fenestra, and in the lower jaw, or mandible, there is an additional hole called the mandibular fenestra. In many archosaurs, these extra holes in the skull serve as points for muscle attachment or allow jaw muscles to bulge or may reduce the bony weight of the skull. Finally, all archosaurs have an additional ridge on the thighbone called the fourth

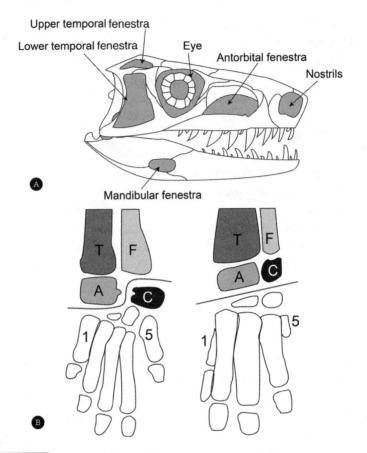

Upper temporal fenestra
Lower temporal fenestra
Eye
Antorbital fenestra
Nostrils
Mandibular fenestra

Figure 5.4 ▲

(A) Archosaur skulls all have distinctive windows, or "fenestrae," in them that lighten the skull and add points for attachment and expansion of muscles, especially jaw muscles. The antorbital ("in front of the orbit") and mandibular fenestrae are unique to archosaurs. (B) The ankle joints of archosaurs come in two main types. The crocodile branch (Pseudo-suchia) mostly has a hinge that runs between the first row of ankle bones, the astragalus (labeled A in the figure) and the calcaneum (labeled C). The dinosaur-pterosaur branch (Ornithodira or Avemetatarsalia) has fused the astragalus and calcaneum to the tip of the shinbone (T and F, or tibia and fibula), and the hinge runs between them and the second row of ankle bones (mesotarsal joint). The number "1" is digit 1 (the big toe) and "5" is the pinky toe. (Redrawn from several sources)

trochanter, which provides an additional anchor point for the muscles that pull the thigh backward. Thanks to these leg muscles, many (but not all) archosaurs have moved away from the completely sprawling posture of liz-ards, who rest on their bellies when not running, to a semi-upright posture seen in crocodilians, or the fully upright posture of dinosaurs. There are other differences as well, but these are the easiest to see on the skeleton.

The archosaurs include not only the living crocodilians and birds and the extinct pterosaurs and dinosaurs but a wide spectrum of other reptilian groups (figure 5.5). These were dominant in the Triassic, pushing the protomammals that had dominated the Permian and Early Triassic aside as they became the ruling reptiles. At one time, all of these extinct creatures were called by the wastebasket name "thecodonts," but now we must discuss each one individually because there is no name for archosaurs that aren't crocodiles or dinosaur and birds. Instead, they come in two main branches (figure 5.5). The Pseudosuchia or Crurotarsi include not only the crocodilians but also their extinct relatives, such as the armored aetosaurs and the crocodile-like phytosaurs. In addition, the largest Triassic land predators were not dinosaurs but a group known as erythrosuchids ("bloody crocodiles" in Greek), rauisuchians or sphenosuchians, which looked a bit

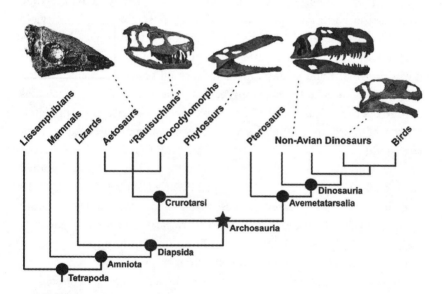

Figure 5.5 ▲

Family tree of the major groups of archosaurs, showing the crocodile branch (Pseudosuchia or Crurotarsi) with the armored aetosaurs, the crocodile-mimic phytosaurs, the huge four-legged predatory rauisuchids, and the true crocodilians. The other half of the archosaurs are Avemetatarsalia or Ornithodira, the dinosaur-pterosaur branch. (From S. Brusatte, M. J. Benton, J. B. Desojo, and M. C. Langer, "The Higher-level Phylogeny of Archosauria [Tetrapoda: Diapsida]," *Journal of Systematic Palaeontology*, no. 8 [2010]: 3–47; courtesy of S. Brusatte)

like Komodo dragons only larger. The other branch, the Avemetatarsalia or Ornithosuchia, includes the pterosaurus, dinosaurs, and birds. These are known from a handful of Triassic fossils that then radiated quickly to produce the dinosaurs and pterosaurs in the Late Triassic.

WHENCE DINOSAURS?

Once the concept of what defines a dinosaur and an archosaur became clearer, we could look for fossils that matched this definition. The search to find actual fossils that most closely resemble the likely ancestors of dinosaurs has been ongoing for more than a century.

All the major groups of dinosaurs—the early relatives of sauropods (*Plateosaurus*; see chapter 6) and the earliest theropods (*Coelophysis*; see chapter 11) and ornithischians—are found in Upper Triassic beds in many places around the world. So the likely place to look for these fossils would be Upper Triassic or even upper Middle Triassic beds. Unfortunately, beds of this age are not exposed in most places, and most of those outcrops do not yield good terrestrial fossils. Nevertheless, a number of fossils have been suggested as close relatives of dinosaurs over the years. Primitive theropods (like *Coelophysis*) and ornithischians were small bipedal animals, and the early sauropods (like the prosauropods) were at least partially bipedal as well, so the likeliest ancestor of the dinosaurs was a small bipedal predator.

One of the early candidates was the small bipedal archosaur *Saltopus* (figure 5.6A), whose name means "hopping foot." It was only about 80-100 centimeters (2.5-3 feet) long (counting the long tail) and weighed about 1 kilogram (2.2 pounds). Described by Friedrich von Huene in 1910 (see chapter 6), it was based on fossils from Upper Triassic Lossiemouth Quarries in the Elgin Formation of northern Scotland. It was frequently featured on the big murals that showed Triassic archosaurs and early dinosaurs by paleoartists such as Rudolph Zallinger, and it appeared in many children's dinosaur books before the 1970s as well. More recent studies, however, have shown it to be a very primitive relative of dinosaurs, and it is not within the Dinosauria.

In 1971, Alfred Sherwood Romer described the little fossil *Lagosuchus* (bunny crocodile) from the Upper Triassic Chañares Formation of Argentina (figure 5.6B). Based on the extremely incomplete remains (just a hind leg, pelvis, shoulder blade, many vertebrae, and part of the skull), it was

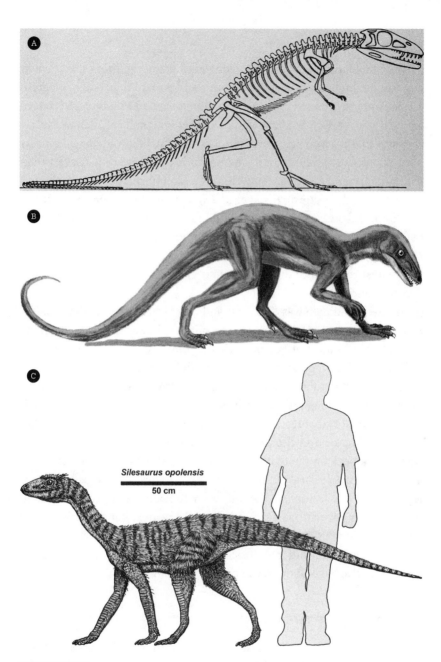

Silesaurus opolensis

50 cm

Figure 5.6 ▲

Early candidates for "dinosaur ancestor": (A) *Saltopus*, a small bipedal predator from the Late Triassic of Germany and Scotland; (B) *Marasuchus* (formerly called *Lagosuchus*), a tiny hopping bipedal predator from the Late Triassic of Argentina; (C) *Silesaurus*, a larger, more quadrupedal form from the Late Triassic of Poland. ([A] and [C] Courtesy of Wikimedia Commons; [B] courtesy of N. Tamura)

a lightly build archosaur with very long hind legs, suggesting that it could hop like a rabbit as well as run on both four legs and two. Later researchers declared the name *Lagosuchus* invalid and stated that the dinosaurian features of the specimen belonged to a different taxon called *Marasuchus*. However, in 1975 José Bonaparte (discussed below) showed that all aspects of its anatomy are dinosaurian, and now it is considered the nearest relative of dinosaurs. A similar fossil from the same formation was described by Romer in the same 1971 paper as *Lagerpeton* (bunny reptile). It too is only known from scraps of the hips, hind limb, and back and tail vertebrae, but many paleontologists pointed at both the "bunny croc" and the "bunny reptile" as likely dinosaurian relatives. Today they are classified as the nearest relatives of dinosaurs (Dinosauromorpha) but are not within the Dinosauria.

Another candidate was *Silesaurus* and its nearest relatives (figure 5.6C), making up the family Silesauridae. The original specimen was found in the early Middle Triassic of Poland. Described by Polish paleontologist Jerzy Dzik in 2003 based on some 20 skeletons (it was one of the best known of the primitive dinosaur relatives), it was the size of a Great Dane, about 2.3 meters (7.5 feet) long because of its long tail, but not as high at the shoulders as a Great Dane. However, it was not primarily bipedal like the other candidates but quadrupedal, and its teeth suggested it was a herbivore, possibly with a beak on the toothless front part of the lower jaw. Most recent analyses place *Silesaurus* and its relatives (*Sacisaurus* from Brazil, *Eucoelophysis* from New Mexico, *Asilisaurus* from Tanzania, and *Lewisuchus* from the Chañares Formation of Argentina) as the nearest kin of the dinosaurs (Dinosauromorpha) but not within Dinosauria itself. So the past 40 years have produced many different fossils that are very close to being true dinosaurs, but none have all the key features that define Dinosauria.

THE VALLEY OF THE MOON

For over a century, paleontologists searched for the earliest, most primitive fossil that could truly be called a dinosaur. They looked almost every place there were good exposures of Upper Triassic beds: North America, Africa, Europe, and Asia. Yet until about 40 years ago almost nothing was known of the dinosaurs of South America. There were Jurassic and Cretaceous badlands exposures in many places in the dry eastern foothills of the Andes in

Patagonia. Thanks to the clouds off the Pacific rising on the Chilean side and raining or snowing out all their moisture, by the time the Pacific winds came down the eastern side, they were stripped of their moisture by the Andean rain shadow, forming miles and miles of desert badlands largely unexplored by paleontologists. The early pioneers of South American paleontology, such as the brothers Florentino and Carlos Ameghino, focused almost exclusively on the bizarre and amazing mammals from the Cenozoic of South America and only rarely found fossils in the Mesozoic beds.

The discovery and explosive growth of research in Argentinian dinosaurs can be attributed mainly to one man, José Bonaparte. He is a distant relative of Napoleon III, but he was certainly the emperor of Argentinian paleontology for many years. Robert Bakker called him "Master of the Mesozoic." Bonaparte is widely considered to be the founder of dinosaurian paleontology in South America. Nearly all the modern dinosaur paleontologists in Argentina were trained and mentored by him.

Bonaparte found and named more than two dozen Mesozoic genera, and 90 percent of these are still considered valid. (This contrasts with people such as Cope, who coined more names but nearly all turned out to be invalid.) His finds include predatory dinosaurs such as the bull-horned *Carnotaurus* and its close relative *Abelisaurus*, plus other predatory dinosaurs including *Ligabueino*, *Noasaurus*, *Piatnitzkysaurus*, and *Velocisaurus*. Most of the South American sauropods were found by Bonaparte (see chapter 10), including *Argentinosaurus* (possibly the biggest of all land animals), *Andesaurus* (another giant close to *Argentinosaurus* in size), *Amargasaurus* (the sauropod with spikes on its back), *Agustinia* (a brachiosaur), *Coloradisaurus* and *Dinheirosaurus* (from Portugal), *Lapparentosaurus* (from Madagascar), *Mussaurus* (a Triassic prosauropod ancestral to later giants), and *Rayososaurus*, *Riojasaurus*, *Saltasaurus* (an armored titanosaur), *Volkheimeria*, and *Ligabuesaurus*.

While finding an amazing diversity of dinosaurs, Bonaparte also discovered and documented many animals that lived alongside them. These include the peculiar pterosaur with the flamingo-like comb of delicate teeth on its lower jaw known as *Pterodaustro*, and the primitive flightless bird *Patagopteryx*. Despite his reputation for finding dinosaurs, Bonaparte prefers to study Mesozoic mammals. He found some of the only Cretaceous mammals from South America belonging to strange groups known as gondwanatheres (unique to South America and Madagascar), plus many other Mesozoic groups. Some of these mammals were previously known only in the Jurassic of the Northern Hemisphere until Bonaparte found them in Argentina.

Bonaparte was born on June 14, 1928, so he is past his ninetieth birthday as I write this chapter. I was lucky to meet him several times during my career. He was interested in my research on Jurassic mammals called dryolestoids, one of his favorite research topics. In an article in *Omni* from 1993, Don Lessem described him this way: "Distinguished, sixtyish, he is a man of modest proportions. His large glasses and thinning pate, polished manners, neat attire, and scholarly parlance lend him the air of an academic, which he is, by practice if not training, ten months a year."

Bonaparte was born in the provincial town of Rosario, Argentina, and grew up in Mercedes, 100 kilometers west of Buenos Aires. His father was an Italian sailor. Most South American pioneers in paleontology had no formal training because there were no paleontologists in the country at that time, and this is true of Bonaparte. He is mostly self-trained and achieved his goals through incredible hard work. Bonaparte was inspired to collect fossils at a young age when a local retired fossil collector showed him some of his own finds. Bonaparte made collecting fossils his obsession and his passion. His house was soon full of fossils that he found in nearby riverbanks, and he created a museum in his town to house them. (They are now in the collection of the University of Tucuman.) By the late 1970s, he was awarded a job at the National Museum of Natural Science in Buenos Aires, the largest museum in Argentina, where he continues to work long after he officially retired.

Almost all the pioneers in paleontology had to work hard with very limited resources because rich donors and government agencies were not willing to fund research on something as impractical as dinosaurs. Bonaparte was incredibly focused on finding fossils no matter what the suffering and inconvenience for him and his crew. Many of his students complain that he is a hard taskmaster and can be extremely difficult. Bonaparte admits this, saying, "Yes, I can be tough, I suppose, but I work hard." He is famous for being a workaholic, nose to the grindstone 16 hours a day, six days a week. He drives his crews just as relentlessly. It is not surprising that many of his former students have complicated feelings about their hard times with him, yet he helped launch the careers of most of Argentina's current generation of dinosaur paleontologists.

Dinosaur paleontologist Peter Dodson said of Bonaparte, "almost single-handedly he's responsible for Argentina becoming the sixth country in the world in kinds of dinosaurs. The United States is still first, but Bonaparte's shown that Argentina is so rich in dinosaurs from so many time periods that it may yet top us one day." Bob Bakker said, "We wouldn't know anything about South America's dinosaurs without him." Dinosaur encyclopedia author George Olshevsky wrote, "His discoveries are fantastic. On a scale of one to ten of how strange a dinosaur could be, with a ten being the first dinosaur with wings, some of Bonaparte's finds are a nine."

Professional paleontologists appreciate him and his hard work. Not only has received numerous awards and honors in his native Argentina, but in 2008 he was awarded the Romer-Simpson Medal of the Society of Vertebrate Paleontology, the highest award any vertebrate paleontologist can receive (figure 5.7). It is named after Alfred S. Romer and George Gaylord

Figure 5.7 ▲

José Bonaparte receiving the Romer-Simpson Medal for lifetime achievement at the Society of Vertebrate Paleontology meeting in 2008. *From left to right:* Dr. Louis Jacobs of Southern Methodist University; Chuck Schaff of Harvard's Museum of Comparative Zoology; Bonaparte; and Lou Taylor of the Denver Museum of Natural and Science. (Courtesy of L. H. Taylor)

Simpson, two of the greatest paleontologists of the twentieth century. In addition, both of them collected many fossils in Argentina during their careers and knew Bonaparte well during the 1930s, 1940s, and 1950s.

THE ORIGIN OF THE DINOSAURS

In 1959, an Andean goatherd by the name of Victorino Herrera was following his flocks when he spotted fossil bones eroding out of the path. He brought in paleontologist Osvaldo Reig, who collected the specimen and described it in 1963 as *Herrerasaurus ischigualastensis*, in honor of the man who found it; its species name comes from the lower Upper Triassic Ischigualasto Formation where it was found. The original specimen had such a weird mixture of sauropod and theropod features that for decades no one was quite sure what kind of creature it was. Reig thought it might be a primitive allosaur or megalosaur, but in 1964 Alick Walker thought it might be a prosauropod. In 1985, Alan Charig noted similarities to both prosauropods and theropods, but Romer's 1966 textbook, *Vertebrate Paleontology*, put *Herrerasaurus* in the prosauropods. Edwin Colbert suggested that it was related to theropods, which was supported by Bonaparte in 1970, and by some later authors. Yet Don Brinkman and Hans-Dieter Sues argued in 1987 that it has features that are primitive for all dinosaurs, including theropods, sauropods, and ornithischians; this was confirmed by Fernando Novas in 1992.

The confusion about *Herrerasaurus* is largely due to the incomplete nature of the original specimens. In 1988, Paul Sereno and his University of Chicago crew were working in the Ischigualasto Formation with José Bonaparte, learning the lay of the land with the master (figure 5.8). Sereno's crew found the first complete skull and skeleton of *Herrerasaurus*, and finally a lot of key features that were missing from the debate were understood. One of the most famous and publicized dinosaur experts alive today, Paul and I were students together at Columbia University and the American Museum of Natural History in the early 1980s. He did one of his first research projects with me working on dwarf rhinoceroses before moving on to dinosaurs. Most authors since then have considered *Herrerasaurus* to be a very primitive saurischian dinosaur, with some favoring closer relationships to the sauropods, and others to the theropods.

For one of the most primitive dinosaurs known, *Herrerasaurus* was clearly bipedal, but not the tiny creature that many paleontologists had expected. Adults were up to 4 meters (16 feet) long, including their very long neck and

Figure 5.8 ▲
The surreal hoodoos and deeply eroded landscape of the Ischigualasto Formation in the Valley of the Moon, Argentina. (Courtesy of Wikimedia Commons)

tail, and weighed about 350 kilograms (770 pounds). However, some adult specimens were only half this size, so there was enormous variability in their adult body sizes, possibly due to the differences between males and females.

Unlike many of the close relatives of dinosaurs we have just discussed, *Herrerasaurus* was completely bipedal with small front limbs and long, powerful running hind limbs. Like many running animals, the thighbones are relatively short and the toes elongated, with some loss of the side toes. There are stiffening features in the tail, showing that it was held out straight behind it to improve balance during running. Its hip bones did not have a large hole through the hips, but instead a bony hip socket with only a small opening (the beginning of the open hip socket seen in all other dinosaurs). Other features of the hip seem to be more like theropods.

The skull of *Herrerasaurus* was long and narrow, with long, recurved teeth with serrated edges, suitable for slashing prey. It had an odd flexible joint in the lower jaw that let the animal slide its jaw back and forth to give it a grasping bite, pulling the prey back after the initial biting motion.

However, the skull was missing nearly all the specializations that are found in nearly all later dinosaurs, another reason that it keeps being considered a primitive relative of all the dinosaurian groups and not a true member of any one of them (figure 5.9A).

Herrerasaurus was a large predator by Triassic standards, and it probably could prey on most of the smaller animals found in the Ischigualasto Formation. A few skulls show marks consistent with the bites of another *Herrerasaurus*, so there were definitely fights among the animals in the group. However, it was not the largest predator in its time. One of the skulls shows bite marks from an animal with teeth very different from *Herrerasaurus*, so it was probably bitten by the huge rauisuchian archosaur *Saurosuchus*.

For all its confusing features, *Herrerasaurus* seems to have a lot of specializations and does not closely resemble the likely ancestor of all the dinosaurs. Three years after working with José Bonaparte when they found the first complete specimens of *Herrerasaurus*, Paul Sereno and his crew from the University of Chicago were back in the "Valley of the Moon" in the Ischigualasto beds in the austral summer of 1991. Ricardo Martinez, a University of San Juan paleontologist working with Sereno, found some tiny bones sticking out of the rocks. The specimen was very delicate, so Martinez, Sereno, and crew spent more than 12 hours carefully excavating it from its burial site and sealing it in a plaster jacket. It then spent over a year at the Field Museum in Chicago, where my good friend preparator Bill Simpson expertly removed it from its stony tomb and recovered all the bones that were there. The original specimen was on display in Chicago for a while, then returned to Argentina, and there are casts of it on display in many museums.

Finally, in 1993, Paul and his coauthors Cathy Forster, Ray Rogers, and A. Moneto published the first description of the fossil. They named it *Eoraptor lunensis*, "dawn raptor of the moon," in reference to the Valley of the Moon (figure 5.9B). It made the cover of *National Geographic* and all the news media as the oldest and most primitive known dinosaur. It certainly is very primitive, and it is a tiny bipedal animal as well. It was only about 1 meter (3.3 feet) long, about the size of a turkey, and probably weighed about 10 kilograms (22 pounds). The long bones all have hollow shafts, so *Eoraptor* was very lightly built compared to the much heavier build of other primitive dinosaur relatives. The skull has relatively large eye sockets and a short snout, so it had great vision but was not built with the vicious jaws and teeth of *Herrerasaurus*. It lacked the sliding jaw joint seen in *Herrerasaurus* and many other

Figure 5.9 ▲

(A) The reconstruction of *Herrerasaurus*. (B) Skeleton of the larger *Herrerasaurus* and the smaller *Eoraptor* side by side. ([A] Courtesy of N. Tamura; [B] courtesy of Wikimedia Commons)

theropod dinosaurs, which makes it more primitive in this aspect. Unlike *Herrerasaurus* and later theropods, only its upper teeth curved backward, another primitive feature. Its lower teeth were simple leaf-shaped structures never seen in any theropod, or in most other dinosaurs either.

Like *Herrerasaurus* and many of the other bipedal archosaurs close to dinosaurs, it had short thighbones and long toes, specializations for rapid running. It had large claws on its three main toes, but apparently the fourth and fifth toes were tiny or lost, as in many dinosaurs. The spool-like centra in the spine were hollow, like many close relatives of dinosaurs. However, it was more advanced than *Herrerasaurus* in having three vertebrae in the sacral region attached to the hip, whereas *Herrerasaurus* has only two as in more primitive archosaurs.

The place of *Eoraptor* in the dinosaur family tree is controversial. It is clearly more primitive than *Herrerasaurus* and many other very primitive dinosaurs, but it still has features found in theropods. When Sereno and colleagues first described it in 1993, and again in 1995, they pointed out that it was one of the most primitive dinosaurs known, but they assigned it to the Theropoda, like they did *Herrerasaurus*; in their words, it is closer to "the hypothetical dinosaurian condition than any other dinosaurian subgroup." Phil Currie in 1997 thought it was closer to the common ancestor of all the dinosaurs rather than a primitive theropod. But in 2011, Ricardo Martinez, Paul Sereno, and coauthors described another primitive dinosaur from Argentina, *Eodromeus*, and argued that *Eoraptor* came out closer to prosauropods. This was disputed by Michael Benton, yet confirmed by a study by Alpadetti and coauthors in 2011. Then in 2011 Hans-Dieter Sues, Sterling Nesbitt, David Berman, and Amy Henrici argued that *Eoraptor* is still within the theropods and is not a primitive relative of all the dinosaurs. Two years later Sereno and coauthors reanalyzed the complete skeleton of *Eoraptor* and returned to the idea that it was related to sauropods. Finally, in 2017, the controversial study by Baron, Norman, and Barrett (which broke up the "Saurischia") placed *Eoraptor* at the very base of the theropods.

In short, the confusing mix of features in *Eoraptor* means that it is very close to the ancestral condition of dinosaurs and does not clearly fall within the theropods or sauropods. Paleontologists do not expect to find a fossil that is perfectly ancestral to any group, but *Eoraptor* comes as close as we have to an approximation of how dinosaurs started out.

FOR FURTHER READING

Baron, Matthew G., David B. Norman, and Paul M. Barrett. "A New Hypothesis of Dinosaur Relationships and Early Dinosaur Evolution." *Nature* 543 (2017): 501–506.

Benton, Michael J. "Origin and Early Evolution of Dinosaurs." In *The Complete Dinosaur*, ed. James O. Farlow and M. K. Brett-Surman, 204–215. Bloomington: Indiana University Press, 1999.

Fastovsky, David, and David Weishampel. *Dinosaurs: A Concise Natural History*, 3rd ed. Cambridge: Cambridge University Press, 2016.

Holtz, Thomas R., Jr. *Dinosaurs: The Most Complete, Up-to-Date Encyclopedia for Dinosaur Lovers of All Ages*. New York: Random House, 2011.

Lach, Will, Jonny Lambert, and Mark Norell. *I Am NOT a Dinosaur!* New York: Sterling Books, 2016.

Langer, Max C. "Basal Saurischia." In *The Dinosauria*, 2nd ed., ed. David B. Weishampel, Peter Dodson, and Halszka Osmólska, 25–46. Berkeley: University of California Press, 2004.

Langer, Max C., Martin D. Ezcurra, Oliver W. M. Rauhut, Michael J. Benton, Fabien Knoll, Blair W. McPhee, Fernando E. Novas, Diego Pol, and Stephen L. Brusatte. "Untangling the Dinosaur Family Tree." *Nature* 551 (2017): E1–E3.

Naish, Darren, and Paul M. Barrett. *Dinosaurs: How They Lived and Evolved*. Washington, D.C.: Smithsonian Books, 2016.

Parrish, J. Michael. "Evolution of the Archosaurs." In *The Complete Dinosaur*, ed. James O. Farlow and M. K. Brett-Surman, 191–203. Bloomington: Indiana University Press, 1999.

Sues, Hans-Dieter. "*Staurikosaurus* and Herrerasauridae." In *The Dinosauria*, 2nd ed., ed. David B. Weishampel, Peter Dodson, and Halszka Osmólska, 143–147. Berkeley: University of California Press, 1990.

PART II

THE LONG-NECKED GIANTS
THE SAUROPODS

At midnight in the museum hall
The fossils gathered for a ball
There were no drums or saxophones,
But just the clatter of their bones,
A rolling, rattling, carefree circus
Of mammoth polkas and mazurkas.
Pterodactyls and brontosauruses
Sang ghostly prehistoric choruses.
Amid the mastodontic wassail
I caught the eye of one small fossil.
"Cheer up, sad world," he said, and winked—
"It's kind of fun to be extinct."

—OGDEN NASH, *FOSSILS*, 1949

PLATEOSAURUS

I abide in a goodly museum,

Frequented by sages profound

'Tis a kind of strange mausoleum

Where the beast that have vanished abound

There's a bird of the ages Triassic

With his antediluvian beak

And many a reptile Jurassic,

And many a monster antique.

—MAY KENDALL, *BALLAD OF THE ICHTHYOSAURUS*, 1887

THE TRIASSIC TRIAD

Most of the earliest discoveries of dinosaurs took place in England in the 1820s and 1830s, but other European countries soon began to make their own mark on paleontology. By the mid-1800s, Germany jumped to the forefront in vertebrate paleontology. Germany had a long tradition of scholarship and research, and Germans are famous for their work ethic, respect for scholars and teachers, and dedication to being the best in a subject. Much of the early foundations of geology were laid by Abraham Gottlob Werner in the Freiburg Mining Academy in the late 1700s, and German scientists had been among the pioneers of fossil collecting since the 1600s.

One of the breakthroughs described a series of beds older than the Jurassic rocks first recognized and named by the German philosopher, explorer, and naturalist Alexander von Humboldt. He officially described the first Jurassic rocks and coined the name "Jurassic" in 1795. By 1831, German geologist Friedrich von Alberti identified a threefold succession of rocks lying beneath the typical Jurassic limestones and shales full of ammonites that was widespread across Germany and Switzerland. The lowest unit was a distinctive sandstone called the Bunter, or Buntsandstein. Above it was a limestone full of fossil shells known as the Muschelkalk (clam chalk) for its abundant fossil shells. At the top was a series of typically red sandstones and shales called the Keuper. This threefold pattern was so consistent and widespread over Central Europe that von Alberti called it the "Triassic" ("tri-" for "three").

Soon geologists were studying the German Triassic in greater detail, and they discovered that it had fossils much more primitive than those found in the overlying Jurassic. Naturally this included dinosaurs as well. The Keuper, in particular, consisted of ancient river, lake, and floodplain deposits of Late Triassic age, and in some places it was rich in gypsum and salt from ancient dry lakebeds. River and floodplain deposits are perfect for the preservation of land plants and animals, and indeed a number of different fossil conifers had been found there. Primitive reptile fossils, some of the earliest tiny fossil mammal teeth, as well as the huge flat-bodied crocodile-shaped amphibian *Mastodonsaurus* (once called *Labyrinthodon*) were also found there.

In 1834, about a decade after the naming of *Megalosaurus* and *Iguanodon*, a local doctor, Johann Friedrich Engelhardt, discovered some vertebrae and giant leg bones in the reddish shales of the Keuper in Heroldsburg, near Nuremberg, Germany. Three years later the specimens reached the noted paleontologist Christian Erich Hermann von Meyer. Today von Meyer is famous as the first paleontologist to describe and name *Archaeopteryx* from the Solnhofen Limestone, as well as the Solnhofen pterosaurs including *Rhamphorhynchus* and *Pterodactylus*, but he also worked on many different kinds of fossils, especially the reptiles and amphibians of the Permian and Triassic beds of Germany. He was one of the earliest scientists to recognize in 1832 that the large land reptiles of the Mesozoic were a distinct group, and in 1845 he named the "Pachypodes" or "Pachypoda" ("heavy foot" in Greek) for the previously published *Iguanodon* and *Megalosaurus*. Luckily for history, Owen had coined the name Dinosauria in 1842. If von Meyer's name had been earlier, kids today would be talking about and marveling at pachypods instead of dinosaurs!

Figure 6.1 ▲

Plateosaurus: (*A*) complete articulated skeleton in the Eberhard-Karls-Universität Tübingen; (*B*) reconstruction of *Plateosaurus* in life. ([*A*] Courtesy of Wikimedia Commons; [*B*] courtesy of N. Tamura)

Von Meyer published his description of the specimens in 1837 and gave them the name *Plateosaurus engelhardti* (figure 6.1). The original type material consisted of about 45 bone fragments, of which half are now lost. The species name honors the discoverer, but the translation of the generic name is controversial. The suffix "-saurus" is clearly Greek for "lizard" or "reptile," but von Meyer did not explain what he meant by the prefix "plateo-." Various scientists since then have decided he meant "broad" or "flat" or even "paddle." With that publication, *Plateosaurus* became only the fifth dinosaur to be named that is still considered valid today (the first four were the British dinosaurs discussed in chapters 1–3). Although it was named five years before Owen named the Dinosauria in 1842, Owen did not mention it because it was too incomplete to say much about it other than that it was a large reptile.

DER SCHWÄBISCHER LINDWURM

Unlike most of the early British dinosaurs, which were based on incomplete material and have never been found as a complete skeleton, *Plateosaurus* was remarkably common in the Late Triassic. In fact, *Plateosaurus* has been found in about 50 localities running along the Triassic outcrops in central and southern Germany and in Switzerland and eastern France. Many of these localities produced multiple nearly complete articulated skeletons, so the anatomy of *Plateosaurus* is very well known. Only the huge assemblage of *Iguanodon* from Bernissart, Belgium, comes close to the sample size of *Plateosaurus* specimens among the earliest named dinosaurs.

One of the best localities is near Halberstadt in the Saxony-Anhalt district of northeastern Germany. Found in a clay pit and excavated between 1910 and 1930, this locality produced between 39 and 50 skeletons in a remarkable state of completeness and preservation. Along with them were two skeletons of one of the earliest known turtles, called *Proganochelys*. Most of this material was described in 1914 by Otto Jaekel (just as World War I began to disrupt scientific research) and ended up at the Museum für Naturkunde in Berlin. Sadly, Allied bombers destroyed several specimens during World War II, but the Berlin *Archaeopteryx* and the huge dinosaurs from Tendaguru, including *Giraffatitan*, were not destroyed (see chapter 9). It is impossible to recover additional specimens because the Halberstadt quarry is now covered by a housing development.

The largest *Plateosaurus* locality by far is Trossingen, near the Black Forest of Württemberg in southern Germany, south of all the previous finds. This locality was worked first by Eberhard Fraas in 1911–1912, then by Friedrich von Huene from 1921–1923 after World War I ended, and finally by Reinhold Seemann in 1932. Trossingen has produced 35 partial or complete skeletons and fragments of another 70 individuals, which is one of the largest samples of a single population of dinosaurs ever discovered.

Plateosaurus was so large that paleontologist Friedrich von Quenstedt nicknamed it the "*Schwäbischer Lindwurm*" after the legendary worm-like dragon of Norse and German mythology called the "lindwurm." Sadly, some of the Trossigen specimens were also destroyed during World War II when Allied bombs destroyed the former Naturaliensammlung in Stuttgart (today, the Naturkunde Stuttgart). A recent study of the material by Dr. Rainer Schoch of the Stuttgart Museum has shown that the most important specimens survived the bombing. The Stuttgart Museum has a spectacular modern exhibit with a room of fossils for each Mesozoic time period. The Late Triassic display has the skeleton of *Mastodonsaurus* and other Keuper fossils and also features skeletons and restorations of *Plateosaurus*.

The third large assemblage of *Plateosaurus* came from a clay pit in Frick, Switzerland. First collected in 1976, the skeletons are deformed by the tectonic processes that bent the rocks as the Alps rose up. Nevertheless, numerous skeletons are complete and articulated, just as in Trossigen and Halberstadt.

By far most of the important work on *Plateosaurus* was done by Friedrich von Huene (figure 6.2). Born in 1875, von Huene (baptismal name: Friedrich Richard Baron Hoyningen) was one of Germany's greatest paleontologists

Figure 6.2 ▲

Friedrich von Huene as a young man. (Courtesy of Wikimedia Commons)

of the first half of the twentieth century. He spent his entire career at Tübingen but never held a post higher than curator because he had no interest in administration. He was a tireless worker, and his major accomplishments include describing the tiny primitive dinosaur relative *Saltopus* from the Upper Triassic of Elgin, Scotland, in 1910 (see chapter 5), and the enormous titanosaurian sauropod *Antarctosaurus* from the Upper Cretaceous of Argentina in 1929. Over his prolific career he published on many other fossil reptiles, including pterosaurs, the little aquatic Permian reptile *Mesosaurus*, the Late Triassic predator *Prestosuchus* from Brazil, and many extinct relatives of mammals and primitive amphibians as well. He holds the distinction of naming and describing more dinosaur species still valid today than anybody except for Othniel Charles Marsh.

Von Huene's greatest work, however, was probably his long-term studies of *Plateosaurus*. Before he began, little was known of the dinosaur other than von Meyer's fragmentary type specimens. From the fragments, no one could tell whether it was related to the predatory theropod dinosaurs such as *Megalosaurus* or to the herbivorous dinosaurs or to something else. It was just a large dinosaur by Triassic standards, but otherwise it remained a mystery. Von Huene began studying the original type specimens from Heroldsburg and found that the museum collections of *Plateosaurus* were mixed together with what he believed to be a large predator related to *Prestosuchus*, called *Teratosaurus*. Most of the *Plateosaurus* reconstructions of that time falsely mixed the predatory skull with sharp recurved teeth of *Teratosaurus* with the huge body and broad hips of *Plateosaurus*, making it an incongruous hodgepodge of a predator and a herbivore. When the Trossingen specimens came into his lab, he had excellent material to work with and fully described the entire skeleton of *Plateosaurus*, demonstrating that it was a large-bodied herbivorous dinosaur. In fact, it was then the largest dinosaur of its time (the Late Triassic). He showed that it had a deep, narrow, boxlike skull with eyes facing sideways, which was good for spotting predators around you. *Plateosaurus* had conical fang-like teeth in the front and thick, bluntly serrated leaf-shaped teeth in the jaw suitable for shredding vegetation (mainly ferns, cycads, and conifers) that formed the Late Triassic forests. The sharp teeth in front have led some paleontologists to suggest that *Plateosaurus* was not a strict herbivore and may have been omnivorous. Its low jaw joint gave it the leverage for a powerful bite, so it could crush even the toughest vegetation. With its large size and long neck, it could reach high vegetation that no other animal at the time could access.

It had rather small but robust front limbs for its size, relative to the long hind limbs, so it was clearly bipedal. Jaekel originally thought that *Plateosaurus* sprawled on all fours in a quadrupedal gait, but later he argued that they hopped like kangaroos. However, von Huene showed that the wrist and hand are configured so they cannot rotate and place the palms or the tips of the fingers down (if it had walked on its finger tips as it does on its toes). So he argued that these animals could not walk on all fours. More recently, studies of the hands show that they could not pronate (rotate until they were palms down). This completely rules out their putting their palms flat on the ground, as von Huene had originally argued. In fact, some of the skeletal mounts had switched the two lower arm bones (radius and ulna) in order to bend the wrist and make their palms lie flat on the ground. Instead, *Plateosaurus* hands had large recurved claws on them, which could have been used for tearing down plant branches, digging up roots, or for defense against predators (such as the early *Dilophosaurus* relative *Liliensternus* from the same beds, which von Huene also described).

Von Huene also followed the conventional belief of his time that dinosaurs were just big sprawling lizard-like creatures, so they were shown in a kangaroo-like pose, resting on their tail. They had to put an awkward bend in the tail of the mounted skeletons to allow it to bend where it touched the ground. We now know that *Plateosaurus* was like most advanced bipedal dinosaurs that hold their body parallel to the ground and their tail out straight behind (see figure 6.1). The large rib cage and the broad hip bones suggest that *Plateosaurus* was quite barrel-chested, which is consistent with having a large complex gut for fermenting and digesting tough fibrous leaves.

Most important of all, von Huene realized that *Plateosaurus* had many features that foreshadowed the huge sauropods that were being discovered in North America and elsewhere: a simple narrow skull with peg-like teeth for stripping and crushing leaves, a relatively long neck, a long tail, and a large trunk for digesting plants. In 1932, he created a new group, the Prosauropoda, for the primitive dinosaurs that were ancestral to the great sauropod giants that evolved later in the Jurassic. A number of related species have since been found in other Triassic beds around the world, but *Plateosaurus* is still the largest, most complete, and best known of them. And he named a bigger group, the Sauropodomorpha, to include both Prosauropoda and Sauropoda. This group is still in use today, although Prosauropoda is now considered a taxonomic wastebasket for all primitive sauropods that are not members of the advanced groups of sauropods.

Von Huene went beyond the pure description and anatomical analysis with which most paleontologists contented themselves. He was also fascinated with the way the Trossigen *Plateosaurus* were fossilized and their burial positions. Thus von Huene was a pioneer in the young field of taphonomy: the study of what happens to the bodies of living animals after death. Both Jaekel and Fraas had argued that the skeletons had been mired in a mud hole, trapped, and starved to death. Von Huene noticed that the skeletons were fully preserved in their death poses, suggesting that they had slumped down in their bellies as they died, with their limbs folded beneath them (figure 6.3). Their necks were curved sideways and backward, which is common in many animals when their nuchal ligament (the long elastic

Figure 6.3 ▲

Top view of the articulated skeleton of *Plateosaurus* in its death position, with the legs folded beneath it and the neck curved back and sideways (the skull is missing). This specimen is one of many from Trossingen and is on display in the Staatliches Museum für Naturkunde Stuttgart. (Courtesy of Wikimedia Commons)

band down the back of the neck that holds the head and neck up) contracts after death and pulls the neck back. Von Huene looked at the dry lake deposits of gypsum and salt and the sandstones from the Keuper sandstones and argued that many of them had died crossing a dry desert region, possibly on some migration to better food sources and water. He argued that they were trapped in the mud next to watering holes when they were too weak to move further on their migration. This interpretation is no longer considered plausible; later workers have argued that the skeletons were buried in a rapidly moving mudflow or have come back to the idea that they were mired in thick mud pools. The fact that only the heavy adults are trapped, and the lighter juveniles and the lighter predators are missing, seems to support this. Whatever the final verdict, von Huene can be considered not only one of Germany's greatest vertebrate paleontologists but also one of the pioneers of taphonomy.

PLATEOSAURUS TODAY

Since von Huene's time (he died in 1969 at the ripe old age of 94), much additional work has been done on *Plateosaurus*. Thanks to the large number of skeletons from several quarries, paleontologists can assess variation within a single population, which is impossible for 95 percent of dinosaurs because most are known from only a few specimens (and often partial specimens at that). The first striking thing is the incredible variability of adult body sizes. By looking at adult bones only (bones where the cap, or epiphysis, is fully fused to the shaft, which happens when the bone stops growing), adults ranged between 4.8 and 10 meters (16 and 33 feet) long and weighed between 600 and 4,000 kilograms (1,300 and 8,800 pounds).

Studies show that many bones are hollow or spongy inside, a feature that many later and bigger dinosaurs used to reduce the weight of their bones in their necks, backbone, and tail. The size of the lungs can be estimated because it is possible to move the ribs in and out to their minimum and maximum extent and estimate the change in ribcage volume. From this analysis, it seems that *Plateosaurus* breathed about 20 liters of air each time it inhaled (for a 700 kilogram dinosaur), or about 29 milliliters per kilogram of air. This ratio is typical of birds, but not of mammals.

Many studies have pointed out that *Plateosaurus* was one of the largest dinosaurs to walk fully on the tips of its long toes (digitigrade). However, it does not have the extreme elongation of the ankle and toe bones, nor the bony features of the ankle, that would allow it to be a very fast runner. Instead, paleontologists think *Plateosaurus* was a moderately fast runner, certainly fast enough with its large size and stride length to outrun most predators (most of which were smaller in the Triassic).

Studies of the growth lines in the limb bones of *Plateosaurus* show it had a typical sauropod growth pattern: rapid growth as a juvenile to reach a large enough size to be safe from predators, then slow but steady growth as it reached adulthood. By contrast, birds and mammals grow quickly as juveniles, then stop growing when they reach adulthood. Some paleontologists argue that *Plateosaurus* was endothermic (regulated its internal temperature as mammals and birds do, by burning its food to generate metabolic body heat).

This is interesting because when sauropods became huge their gigantic mass and relatively small surface area relative to their huge volume meant they probably had to abandon endothermy. Instead, large sauropods are thought to have used their huge bulk and limited surface area to resist changes in body temperature, so they heated up slowly in the day and cooled down slowly in the night (but see discussion in chapter 17). For this reason, they did not need the expensive endothermic system of burning their food to generate metabolic body heat (a strategy called inertial homeothermy, or gigantothermy). The contrast between bigger sauropods and their ancestor *Plateosaurus* shows how these physiological strategies can change, even within a lineage.

The growth lines in the bones also allow us to determine the age of a specimen. Some individuals reached adult size at 12 years, others were still growing at 20 years, and one reached the age of 28 years. It is possible that they lived longer because all the specimens apparently died in accidents, and none reached old age. There is a remarkable absence of juveniles in the assemblage, suggesting that the death trap that mired them tended to trap only the heavier adults and did not entrap the lightweight juveniles (and their small predators).

Finally, comparing the size of the eye socket with living animals suggests that *Plateosaurus* was active both during the day and at night, so it was not strictly diurnal or nocturnal. It may have become active in late afternoon as the temperatures cooled, and then took a midday siesta to avoid the noon heat after being up all night.

Plateosaurus was the first of the sauropod relatives to be discovered and is still the best known and best sampled. However, prosauropods were widespread in Upper Triassic and Lower Jurassic beds all over the supercontinent of Pangea, although most of these fossils are incomplete. Other members of the *Plateosaurus* family include *Plateosauravus* and *Euskelosaurus* from Africa, *Jaklapallisaurus* from India, *Unaysaurus* from Brazil, *Yimenosaurus* from China, and *Ruehleia* from Germany. The other branch of prosauropods are related to *Massospondylus*, originally described by Richard Owen in 1854 from fossils found in Lower Jurassic beds of southern Africa. Close relatives of *Massospondylus* include *Prahdania* from India, *Coloradisaurus* from Argentina, *Lufengosaurus* from China, and *Glacialisaurus* from Antarctica. Another branch is the Riojasauridae, including *Riojasaurus* from Argentina and *Eucnemesaurus* from South Africa. The third group is related to *Anchisaurus*, known from the Lower Jurassic beds of Connecticut. Other primitive sauropods include the Chinese *Yunnanosaurus* and *Jingshanosaurus*, and *Melanorosaurus* from South Africa. The more advanced giant sauropod dinosaurs of the later Jurassic and Cretaceous are most closely related to the anchisaurine branch.

So from the original fragments of *Plateosaurus* found in 1837 as the fifth known dinosaur, there was a big evolutionary radiation of similar-looking prosauropods on nearly every part of Pangea, from Europe to eastern North America, to China and India, to South America, Africa, and even Antarctica. These early bipedal long-necks were eventually replaced by the radiation of advanced sauropods with gigantic bodies and extremely long necks and tails, and they were never able to walk on two legs again.

FOR FURTHER READING

Barrett, Paul M. "Prosauropod Dinosaurs and Iguanas: Speculations on the Diets of Extinct Reptiles." In *Evolution of Herbivory in Terrestrial Vertebrates: Perspectives from the Fossil Record*, ed. Hans-Dieter Sues, 42–78. Cambridge: Cambridge University Press, 2000.

Barrett, Paul M., and David J. Batten, eds. *Evolution and Palaeobiology of Early Sauropodomorph Dinosaurs*. Oxford: Blackwell, 2007.

Colbert, Edwin. *Men and Dinosaurs: The Search in the Field and in the Laboratory*. New York: Dutton, 1968.

Curry Rogers, Kristina, and Jeffrey Wilson. *The Sauropods: Evolution and Biology.* Berkeley: University of California Press, 2005.

Farlow, James, and M. K. Brett-Surman, eds. *The Complete Dinosaur.* Bloomington: Indiana University Press, 1999.

Fastovsky, David, and David Weishampel. *Dinosaurs: A Concise Natural History*, 3rd ed. Cambridge: Cambridge University Press, 2016.

Galton, Peter M., and Paul Upchurch. "Prosauropoda." In *The Dinosauria*, 2nd ed., ed. David B. Weishampel, Peter Dodson, and Halszka Osmólska, 232–258. Berkeley: University of California Press, 2004.

Hallett, Mark, and Mathew J. Wedel. *The Sauropod Dinosaurs: Life in the Age of Giants.* Baltimore: Johns Hopkins University Press, 2016.

Holtz, Thomas R., Jr. *Dinosaurs: The Most Complete, Up-to-Date Encyclopedia for Dinosaur Lovers of All Ages.* New York: Random House, 2011.

Mallison, Heinrich. "The Digital *Plateosaurus* I: Body Mass, Mass Distribution, and Posture Assessed Using CAD and CAE on a Digitally Mounted Complete Skeleton." *Palaeontologia Electronica* 13, no. 2 (2010): 8A.

Naish, Darren. *The Great Dinosaur Discoveries.* Berkeley: University of California Press, 2009.

Naish, Darren, and Paul M. Barrett. *Dinosaurs: How They Lived and Evolved.* Washington, D.C.: Smithsonian Books, 2016.

Remes, Kristian, Carole T. Gee, and P. Martin Sander. *Biology of the Sauropod Dinosaurs: Understanding the Life of Giants.* Bloomington: Indiana University Press, 2011.

Spaulding, David A. E. *Dinosaur Hunters: Eccentric Amateurs and Obsessed Professionals.* Rocklin, Calif.: Prima, 1993.

Wedel, Mathew J. "What Pneumaticity Tells Us About 'Prosauropods,' and Vice Versa." In *Evolution and Palaeobiology of Early Sauropodomorph Dinosaurs*, ed. Paul M. Barrett and David J. Batten, 207–222. Oxford: Blackwell, 2007.

APATOSAURUS AND BRONTOSAURUS

The genus *Brontosaurus* was based chiefly on the structure of the scapula and the presence of five vertebrae in the sacrum. After examining the type specimens of these genera, and making a careful study of the unusually well-preserved specimens described in this paper, the writer is convinced that the *Apatosaur* specimen is merely a juvenile of the form represented in the adult by the *Brontosaur* specimen....

—ELMER RIGGS, 1903

BONE WARS

Before 1880 no one could imagine a giant, long-necked, long-tailed sauropod from the limited bones of early finds of *Cetiosaurus* and *Pelorosaurus*—so they reconstructed them as giant reptilian "whales" (see chapter 3). Yet images of *Brontosaurus* and other sauropods are now iconic, found on everything from kids' toys to the Sinclair Oil logo. How did this come about? What discoveries allowed us to imagine *Brontosaurus* and other sauropods correctly for the first time?

The key fossils were found and described thanks to the two most important American vertebrate paleontologists of the late 1800s: Edward Drinker Cope and Othniel Charles Marsh (figure 7.1). They were similar in many ways: strong-willed, brilliant, egotistical, driven, self-centered, quarrelsome, jealous, mistrustful, and always unhappy. Cope was a scientific prodigy with a talent for anatomy, but he was also pugnacious, short-tempered,

Figure 7.1 ▲

(A) Edward Drinker Cope in his study, surrounded by fossils; (B) portrait of Othniel Charles Marsh late in life. ([A] Image #238372, [B] image #328401, courtesy of the American Museum of Natural History Library)

and inclined to irritate and alienate everyone who could help him. Marsh was much slower, more methodical, and more introverted—but even more jealous and paranoid than Cope. In his book *The Bonehunter's Revenge*, David Rains Wallace wrote, "The patrician Edward may have considered Marsh not quite a gentleman. The academic Othniel probably regarded Cope as not quite a professional."

Despite their similarities, they came from very different circumstances. Cope was from a wealthy and influential Philadelphia Quaker family, which helped him land positions at the Academy of Natural Sciences and a teaching post at Haverford College. He showed his genius for anatomy and natural history at age six, drawing incredibly detailed sketches of specimens and learning all their names. Joseph Leidy himself taught Cope comparative anatomy, so Cope is in a real sense the direct successor to the founder of vertebrate paleontology in the United States. Cope used his personal wealth and influence to secure a position on the famous Hayden Survey out west, and he began his career collecting important fossils from the Rocky Mountain region.

Marsh was eight years older than Cope and came from a poor family from Lockport, New York. He had to work the family farm, but he was an avid fossil collector on the side. He never thought of making fossils his career until a local collector and retired Army officer, Col. Ezekiel Jewett, encouraged him in his studies. Eventually he was a star student at Phillips Andover Academy prep school, and then at Yale. Because he started late and was much older than his classmates, they called him "Daddy" or "Captain." His education was only possible because his rich uncle, George Peabody, decided to support him, and then gave Yale the funds to set up the Peabody Museum of Natural History with a position for Marsh as head of the museum and Curator of Paleontology. Marsh was extremely hard-working and industrious, making the best of the incredible opportunity to build the museum and collect fossils from out west that would make it one of the world's best natural history museums.

The two men first met in Europe during the Civil War years when both were urged to go abroad to avoid being drafted into the war. Marsh had bad eyesight and would never have been accepted in the Army, and Cope was a Quaker pacifist and could not bear arms for religious reasons. Both men received the equivalent of a Ph.D. degree by attending lectures of the best European (especially German) scholars and studying the major museum collections. In Berlin, they got along famously and actually spent several days together visiting museums. They even named some of their first new species after each other.

They met again in 1868 when Marsh visited Cope in the Haddonfield, New Jersey, marl pits, where William Parker Foulke had found and given to Leidy the first specimens of American dinosaurs (see chapter 4). After Cope left, Marsh secretly bribed the pit operators to send him all future fossils, a pattern that would continue for the rest of their careers. Their relationship deteriorated further when Cope published a reconstruction of his long-necked plesiosaur from Kansas, *Elasmosaurus*. Marsh pointed out that Cope had put the skull on the end of the tail, not the neck—and Leidy agreed. Cope was humiliated, and their grudges began to grow as each attacked the other in print again and again. These feuds exploded into the all-out "Bone Wars" as both men went to the Wild West to get spectacular fossils and outdo each other. In the process, they and their crews collected an incredible number of specimens and named hundreds of new species—but at the price of nearly destroying each other and paleontology as well.

Marsh was the first to go west. In 1870, he arranged the first of several annual field trips with a number of rich Yale students paying their way to help him in his excavations and prospecting. The photos of these Yale trips, with the rich young dandies dressed up as tough western gunslingers, are unintentionally hilarious (figure 7.2). Through his connections, Marsh had the U.S. Cavalry guarding them all the way, even as the Native Americans were becoming restless. Marsh mounted several expeditions like these, then gave up traveling west altogether, leaving the collecting to his paid assistants. Marsh had large collections and studied the local geology of the bone beds, so he was aware of the age of his fossils. He also cultivated good relations with the tribes who were not hostile, especially Lakota chief Red Cloud, who counseled the tribes not to go on the warpath. In 1880, when Red Cloud complained to Marsh of the poor food and conditions on the reservations, Marsh used his influence to bring Red Cloud to Washington to

Figure 7.2 ▲

Famous image of Marsh (*back row center*, with the bushy beard) and his Yale students on their expedition to collect fossils out west in 1872. Even though they were rich young Yankee gentlemen, they had to dress and look tough for the photo—and there were real dangers from hostile warriors roaming their area at that time. (Courtesy of Wikimedia Commons)

visit the White House and successfully lobbied for better conditions on the reservations (figure 7.3).

As early as 1867, Cope had paid collectors to send him fossils from the chalk beds of western Kansas (such as *Elasmosaurus* on which he mounted the skull incorrectly on the tail). In 1872, Cope took his first expedition out to the Wild West attached to the Hayden Survey (but receiving no money from them). Once out west, he roamed freely without paying much heed to Hayden or his crew, and he collected wherever the fossils were rich. During

Figure 7.3 ▲

Marsh meeting with Lakota chief Red Cloud; they befriended each other out west. Marsh later interceded on Red Cloud's behalf in Washington to get the Lakota better treatment on the Pine Ridge Reservation after corrupt agents gave them bad food and disease-carrying blankets, or robbed them completely. (Courtesy of Wikimedia Commons)

the summer of 1872, all three men—Leidy, Cope, and Marsh—were in the field and often collecting in overlapping areas. Their competition drove Leidy out of field collecting altogether, which ended the supply of fossils that Hayden had originally given to him.

All three men published hasty preliminary descriptions of their fossils, especially the weird six-horned, fanged, rhino-sized mammals known as uintatheres from the middle Eocene beds of Wyoming and Utah. In the rush to beat their competitors, some of the scientific announcements were sent as telegraphs, only to be garbled by telegraph operators. Some of their descriptions were so hasty that the quickly telegraphed names were incomprehensible and could not be used. In their haste and competition, Cope, Marsh, and Leidy ended up creating dozens of names for just a few species because they never bothered to check whether their discovery was similar to what one of the others had already found. Marsh's last Yale trip was in 1873, but Cope continued to collect most summers for several years attached to the Army Corps of Engineers. When Cope was out in the Judith River badlands of central Montana collecting in the summer of 1876, Custer and his men had just been slaughtered at the Battle of Little Bighorn. The Army warned him to get out of the territory before hostile tribes found him. Cope refused because he was having so much good luck collecting. A Quaker pacifist who didn't believe in guns, Cope had often defused the tension when dealing with Native Americans by taking out his set of false teeth and then replacing them, which amazed them.

The critical year for the Bone Wars was 1877, which is when the story of the discovery of complete sauropods begins. Marsh got the first lead from a local schoolteacher named Arthur Lakes, an Englishman living in Morrison, Colorado. Lakes had written to Marsh, and Marsh wrote back and urged him to collect more fossils for a generous payment—and to keep his discovery secret from Cope. But before Marsh's reply arrived, Lakes also wrote to Cope and sent him a shipment of bones. Marsh's assistant, Benjamin Mudge, then showed up, paid Lakes, and they began to work in earnest in the Morrison quarry. Meanwhile, Lakes asked Cope to send his bones back because Marsh had just paid for them.

A local naturalist, O. W. Lucas, began finding bones further to the south, near Cañon City, Colorado. He sent these bones to Cope, who immediately secured his services for future collections. Hearing of this, Marsh sent his assistant, Samuel Wendell Williston (who later became a famous paleontologist and entomologist at University of Kansas), to set up their own quarry in Cañon City and try to lure Lucas away. Fortunately for Cope, Lucas was finding much better bones than Williston and refused to be lured over to work for Marsh. Williston then gave up on Cañon City and returned to Morrison, where their small quarry collapsed and nearly killed them. Marsh would have been out of luck in the Bone Wars were it not for another lucky discovery—and a mysterious letter.

BONE BONANZA

In March 1877, Union Pacific railroad worker William Harlow Reed was hunting pronghorns in the flats east of Medicine Bow, Wyoming, to feed railway workers in his group. He shot one and hiked up a ridge to retrieve his kill. On the way down the ridge, he noticed huge bones lying all over the ground and sticking out of the rocks.

Reed had heard the news about some eastern scientist paying good money for such huge bones. He talked to his friend, stationmaster William Edwards Carlin, about the exciting find. They both agreed that such a find might pay very well, especially as they were living on bare subsistence wages in the desolate sagebrush flats of Wyoming. So they wrote a letter to Marsh:

> I write to announce to you the discovery not far from this place, of a large number of fossils, supposed to be those of the Megatherium, although there is no one here sufficient of a geologist to state for certainty. We have excavated one (1) partly, and know where there is several others that we have not, as yet, done any work upon. The formation in which they are found is that of the Tertiary Period. We are desirous of disposing of what fossils we have, and also the secret of the others. We are working men and are not able to present them as a gift, and if we can sell the secret of the fossil bed and procure work in excavating others we would like to do so. We have said nothing to any-one as yet. We measured one shoulder-blade and found it to measure four feet eight inches 4 Ft. 8 in. in length. As proof of our sincerity and trust, we will send you a few fossils, at which they cost us in time and money in unearthing. We would be pleased to hear from you, as you are well-known as an enthusiastic geologist, and a man of means, both of which we are desirous of finding—more especially the latter. Hoping to hear from you very soon, before the snows of winter set in.

They signed the letter "Harlow" and "Edwards" and also hinted darkly that someone else had people in the area looking for bones.

The mysterious letter and specimens reached Marsh at Yale, and he was excited immediately because his other collectors were having little luck in Colorado. He telegraphed Williston, who took the next train to Wyoming. When he arrived at Como Station (which consisted of a couple of shacks on the railroad siding), he was told that the mysterious Harlow and Edwards lived at a ranch out by the bluff. There were crates of bones on the floor of the station, so he knew he was in the right place. Eventually they met, and Williston gave them Marsh's check made out to "Harlow" and "Edwards." But they could not cash it because they had used their middle names to hide their identities.

Williston was elated when Carlin and Reed took him up to Como Bluff (figure 7.4). On November 14, he wrote to Marsh that the bones "extend for seven miles & are by the ton. . . . The bones are very thick, well preserved, and easy to get out." In his next letter on November 16, he told Marsh, "Cañon City & Morrison are simply nowhere in comparison to this locality—both as regards perfection, accessibility, and quantity." Williston urged Marsh to pay Carlin and Reed $75 each (a princely sum in those days) and to get their claim established because they feared Marsh's rival, Cope, might be looking for a chance to poach the locality (as indeed he was).

Marsh had actually passed close to those very bones and did not realize it. In 1868, he was on a field excursion, traveling via the Union Pacific railroad, and stopped at Como Station to collect tiger salamanders. Another station down the line had yielded fossils, so he had paid the stationmaster there to look for more. Up and down the line, the word was out that the rich eastern professor was paying good money for old bones, and Carlin and Reed clearly heard about it.

Marsh sent Carlin and Reed a contract, paying them $90 a month (he was paying poor hard-working Williston only $40 a month), with instructions to keep out all rival collectors. Soon they had opened Quarry 1, started crating up huge bones, and sent them east on the railway. Marsh's collectors kept this up for 10 years, sending east a ton of bones almost every week (figure 7.5).

It was impossible to be so productive every week because the winters are bitterly cold and snowy in Wyoming. On November 30, Williston wrote to March, "The small saurian I have not yet sent and cannot for a few days till

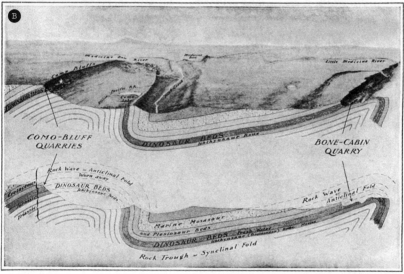

Figure 7.4 ▲

(A) Arthur Lakes's painting of Como Bluff, with William Harlow Reed in the foreground. (B) Lakes's geological cross section showing the Medicine Bow anticline (where the Como Bluff quarries were exposed) and the geologic setting of the region. (Courtesy of Wikimedia Commons)

Figure 7.5 ▲

(A) Arthur Lakes's painting of quarrying at Como Bluff, with William Harlow Reed in the quarry. (B) Lakes painted this scene of working in Como Bluff Quarry 12 in 1879, in a blizzard in the dead of winter, and subtitled it "The Pleasures of Science." (Courtesy of Wikimedia Commons)

the snow blows off so that we can find it." Later Williston wrote, "Inasmuch as mercury in the thermometer during the next two months seldom reached zero—upward I mean—the opening of this famous deposit was made under difficulties." When they could work, they found other surprises. Williston wrote that the quarries "are found containing remains of numerous individuals mingled together in the most inextricable confusion, and in every conceivable position, with connected limb bones standing nearly upright, connected vertebrae describing vertical curves, etc., precisely as though in some ancient mud holes the huge monsters had become mired and died, and succeeding generations had trodden their bones down and then left their own to mingle with them."

By the spring of 1878, the winter snows had receded, and Reed, Carlin, and Williston were back at work. A mysterious visitor who called himself "Haines" was nosing around Como, supposedly to sell groceries, but he showed unusual interest in the bones and where they came from. They suspected it might be Cope, or one of his men. By the fall of 1878, Reed called Carlin lazy. Unhappy with what his boss had said, Carlin moved out and set up his own camp (he called it "Camp Misery") and worked a different quarry. By March 1879, Carlin had opened his own quarry, sent some of the best material to Cope, and smashed up the rest. Marsh then sent his employee, Arthur Lakes, over from the earlier, less productive excavations at Morrison, Colorado, to help out. Lakes and Reed didn't get along well either. Reed was a raw, uncouth, man who just wanted to work fast with the pick and shovel, but Lakes, a former school teacher, insisted on stopping to make sketches of the bones in the quarry before they were removed. His watercolor sketches of the harsh field conditions capture the circumstances better than any camera could (figures 7.4 and 7.5). Lakes even pioneered the first-ever plaster of Paris jackets to protect the bones as they were being jarred by hammers and chisels. In the winter of 1879, Lakes wrote that "collecting at this season is under many difficulties. At the bottom of the narrow pit 30 feet deep into which drift snow keeps blowing, and fingers benumbed by cold from thermo between 20 and 30 below zero and snow often blowing blindingly down and covering up a bone as fast as it is unearthed" (figure 7.5B).

Other workers came and went, but Reed was a constant figure at Como Bluff, even though his pay was unpredictable and so were his supplies. Reed found most of the new quarries and the most important discoveries. Once he almost drowned in an icy creek rafting across it. Some of the time he had no wagons or pack animals, so he had to carry the bones down from the hill, often 8 kilometers (5 miles) to the station. He once dragged a 186-kilogram (400-pound) leg bone 0.8 kilometers (over half a mile) down to the tracks. After six years of hard work with no appreciation from Marsh or his crew, in 1886 he resigned to take up sheep farming and guide hunting parties. But that paid much worse, so he returned to collecting old bones, first as an independent, then for other institutions (see chapter 8). The next group of collectors Marsh hired were so incompatible that they worked in different quarries and argued (even at gunpoint) whenever they crossed paths. The last of Marsh's Como Bluff collectors, Fred Brown, worked from 1884 to 1889; when he quit, the quarries were abandoned.

Despite these hardships, the haul from Como Bluff and its 13 quarries was amazing. Many tons of bones in an excellent state of preservation yielded 26 new species of dinosaurs as well as complete or nearly complete skeletons of others previously named. Many crocodiles, turtles, lizards, frogs and salamanders, and pterosaurs were found as well. In addition, the first-ever Mesozoic mammals from North America were found. These little animals were no bigger than shrews and had jaws with teeth the size of pinheads. I had a personal connection to these fossils when I was a graduate student. Thanks to new excavations at Quarry 9, the main mammal quarry, made by the American Museum and Yale in 1968–1970, I had the great fortune of studying Marsh's Jurassic mammals at the American Museum in the 1970s and 1980s. One of them was a new species I named *Comotherium richi*. The genus was named after Como Bluffs ("Como" plus -*therium*, Greek for "beast"), and the species honors Dr. Thomas Rich, who led the 1968–1970 American Museum-Yale party that found it.

The quarries at Como Bluff yielded the largest and most complete dinosaur skeletons ever found up to that point. By the time all the bones reached Yale and Marsh had a chance to mount the best skeleton of *Brontosaurus*, the mystery of the shape of sauropod dinosaurs such as *Cetiosaurus* was finally solved. The huge bones that had so baffled the early collectors were in fact from creatures they could not have imagined in their wildest dreams. They were the largest land animals that ever lived, with enormous bodies, limbs like tree trunks, long slender necks with small heads and simple teeth, and long whip-like tails. By the early twentieth century, several museums had mounts of large sauropods such as *Brontosaurus* and *Diplodocus* on display (see chapter 8), and the world got to know these huge dinosaurs for the first time.

IS *BRONTOSAURUS* BACK?

Despite its popularity, many paleontologists do not regard *Brontosaurus* as a valid name and haven't since 1903. This is not just a nitpicky issue but goes to the heart of what these names signify, and how we interpret them.

The problem goes back to Marsh, of course. During the Bone Wars, Cope and Marsh hastily gave names to every slightly different specimen that came from the field, without regard to what had already been named—and without looking at the specimens that had been collected and making careful comparisons. Many were named before they were even out of their crates. This was not unusual for its time. Most paleontologists then regarded genus and species names as a kind of "stamp collecting" or "trophy hunting," and the more names you had, the better. They are what we now call taxonomic "splitters"—every slight difference in a new fossil was justification for a new name. Much of the time, the specimens had no bones in common, so there was no way to know whether they were the same or not. This type of hyper-splitting continued into the early twentieth century as well, and most of the early paleontologists created names willy-nilly—without thinking about their specimens as anything but trophies.

Beginning in the 1930s and through the 1950s, paleontologists (led by the brilliant American Museum polymath George Gaylord Simpson) began to think of their fossils as samples of living biological populations. When compared to modern biological populations, an enormous range of variability could be encompassed in a single biological population or a single species—variations that routinely would have been named as new species by earlier scientists. Paleontologists, who had long lagged behind other sciences in using math and statistics, began to incorporate statistics in their thinking wherever possible. They measured and analyzed just how much variation could be expected of a typical population or species sample. From this point onward, paleontology (and biology) began to trend toward "lumping" species together that had once been separated based on trivial distinctions. Even today my colleagues and I, using a much richer and more comprehensive fossil record, spend a lot of time getting rid of old, unjustifiable species from a century ago that were created based on differences that have no biological meaning.

This is especially true of dinosaurs. Nearly every dinosaur name still recognized as valid was probably named more than once by different people (typically Cope and Marsh, if they both got the chance). When such problems arise, biologists are bound by the rule book: the International Code of Zoological Nomenclature. This is the standard set of rules agreed to by all scientists who deal with taxonomic names of species (living or extinct). The details of the code are intricate, but one of the key principles is priority. The first name given to a species or genus in an adequately published form is the valid name, and all later names are called "junior synonyms." The reason for this rule of priority is stability of names and to decrease arguments over whose name is better. Which name came first is usually not in doubt, so the simple rule of "who's on first" makes most decisions simple and noncontroversial. (There are special rules for getting around long-forgotten names that are found to have priority over a well-established name that everyone uses and has used for years.)

How does this apply to *Brontosaurus*? It goes back to the first specimens shipped from the Rockies in 1877. The first large sauropod fossil to arrive at Yale was a specimen with the Yale Peabody Museum catalog number YPM 1860, a partial skeleton (mostly the pelvis and some vertebrae) of a juvenile from Arthur Lakes's Quarry 10 in Morrison, Colorado. Marsh wasted no time; within weeks of receiving it, he had published on it briefly and named it *Apatosaurus ajax*. Two years later he received an even larger, more complete specimen of an adult (YPM 1980) from Como Bluff Quarry 10, one of the richest of all Como Bluff localities. As Marsh (1879) pointed out, it was "the most complete of the Sauropoda ever discovered." To this spectacular fossil, he gave then name *Brontosaurus excelsus*, and this is the specimen that he had mounted as the centerpiece of the Yale Peabody Museum dinosaur exhibits (as it still is today). This spectacular mount influenced Henry Fairfield Osborn of the American Museum to put up a similar mount around 1900 (based on specimens Walter Granger found just northeast of Como Bluff, in Bone Cabin Quarry), which garnered worldwide attention. Both mounts bore Marsh's name, *Brontosaurus*, and so *Brontosaurus* dominated the public consciousness and became the name associated with nearly all sauropod dinosaurs for almost a century.

About 25 years later, Elmer Riggs of the Field Museum in Chicago was studying huge sauropods that had been collected by their crews in the Morrison Formation near Fruita, Colorado, including the first *Brachiosaurus* (see chapter 9). He reexamined Marsh's specimens of *Apatosaurus ajax* and *Brontosaurus excelsus* and thought that the differences Marsh had used could not be justified. In his opinion, they were all the same genus and species, with *Apatosaurus* being based on a partial juvenile specimen. As Riggs wrote (see the epigraph at the beginning of the chapter), "in view of these facts the two genera may be regarded as synonymous. As the term *Apatosaurus* has priority, "Brontosaurus" will be regarded as a synonym."

For most paleontologists, it was "case closed." The name *Brontosaurus* was officially dead after Riggs's 1903 paper, and only rarely did legitimate paleontologists (other than Henry Fairfield Osborn of the American Museum in New York) use the name again (no matter how popular it was in the media and the public consciousness). By the 1980s and 1990s, the copycat outdated children's books were gradually dropping *Brontosaurus* in favor of *Apatosaurus*, and even the U.S. Postal Service got into a hot controversy in 1989 over a stamp with the name *Brontosaurus* on it.

That all changed in 2015 when Emanuel Tschopp, Octavio Mateus, and Roger B. J. Benson published a study of all the diplodocine sauropod dinosaurs. In that paper, they revived the name *Brontosaurus* and claimed that it could be distinguished from *Apatosaurus*. The study generated a huge media frenzy focused on the trivial issue of the resurrection of the name *Brontosaurus*, but it missed the main point of the article about the diversity of sauropods. As many reviewers commented, it was the most thorough study of sauropods ever done and used a wide array of anatomical evidence from more than 81 specimens. No matter what its conclusions, it was a solid piece of work.

But many paleontologists (myself included) are not so sure that this study establishes *Brontosaurus* as a valid name again. As I have commented in print elsewhere, in the Morrison Formation alone (Late Jurassic, mainly Colorado-Utah-Wyoming), Tschopp, Mateus, and Benson record 14 different species clustered in nine genera of diplodocines (*Suuwassea, Amphicoelias, Apatosaurus, Brontosaurus, Supersaurus, Diplodocus, Kaatedocus, Barosaurus,* and *Galeamopus*) from a single formation that covers a limited geographic area and approximately 7–11 million years of time. On top of that, there are nondiplodocid sauropods, including the huge *Brachiosaurus,* plus *Camarasaurus, Haplocanthosaurus,* and possibly several more. If their reasoning is valid, in a single quarry alone (Carnegie Quarry at Dinosaur National Monument, representing a single biological fauna and a short interval of time; see chapter 8), they claim that the distinct species *Apatosaurus louisae, Brontosaurus parvus, Diplodocus carnegii,* and *D. hallorum, Barosaurus* sp. among diplodocines, plus *Camarasaurus* and possibly

Haplocanthosaurus all lived together. That makes at least seven or eight distinct species of huge sauropods from a single interval of time and a single place, all crowding together and sharing common resources.

Paleontologists familiar with the ecology of large land animals have pointed out that this seems like an awful lot of species to all be picking the needles off the same conifers or mowing the same ferns. Huge land animals need lots of room to roam and feed. Ecological theory and empirical data show that larger species require larger home ranges. Sauropods are the largest land animals that ever lived, and they probably required huge geographic ranges to support their needs for food. The principle of competitive exclusion suggests that no two species can compete for the same resources, and that problem is magnified for larger animals, which rarely share territory with their own competing populations let alone closely related genera and species. We know of no examples of giant land vertebrates today that exist in high diversity and compete for the same resources. Nor do prehistoric faunas show more than one or two large-bodied animal species coexisting. Even during the Ice Ages, at most one species of mammoth and one species of mastodon lived in the same area, and they had very different diets, habitats, and ecologies.

As a paleontologist who has worked with the largest land mammals, I have seen this issue close up. For years, paleontologists gave multiple names to the gigantic extinct hornless rhinoceros *Paraceratherium* from Asia (including *Baluchitherium, Indricotherium, Tsungaritherium*, and other genera, plus many different species), but ecological arguments and the high variability of specimens in a single population show that they are all one genus and species. My study of the rhinos *Trigonias, Subhyracodon*, and *Hyracodon* from the Big Badlands showed that dozens of invalid species had been named based on slight differences in the crests of their cheek teeth. When large quarry samples of a single population were analyzed, it became clear that all those subtle crest differences on which so many species were based were in fact just variation within a single population, and none of those names were valid. Matthew Mihlbachler and colleagues studied the huge rhino-like mammals known as brontotheres or titanotheres and found that when all the variation in a population is taken into account dozens of names (*Titanotherium, Brontotherium, Brontops, Allops, Menops*, and *Menodus*) are invalid and only *Megacerops coloradensis* is a valid name for the incredible array of different-looking specimens from the late Eocene.

Based on this reasoning, I'm concerned that all these names for similar-sized sauropods living in the same region are not justified. There is no clear evidence that they had the room or the diversity of food sources or habitats to allow so many species to live close together, all with monstrous appetites as befits their huge size. As a comparable case, Mihlbachler and colleagues found that giraffes (our only living example of a huge, long-necked tree browser) are extremely variable, with lots of differences in the horns on their head, neck vertebrae, and so on, yet they are all one genus and one species (*Giraffa camelopardis*) with many different geographic subspecies. Some scientists have raised these subspecies to species rank because they are genetically distinct as well. Even if there are several different species of giraffes, they live in ranges that do not overlap in Africa, so you would never find two different species in the same place as is claimed for the Morrison dinosaurs. There is another limitation to be aware of: flowering plants had not yet evolved in the Late Jurassic, so these many different huge herbivores had to subsist on slow-growing conifers, which are very low in nutrition and don't recover quickly from heavy browsing, and possibly ferns and cycads on the ground level.

So I (and many other paleontologists) reserve judgment on the Tschopp, Mateus, and Benson study until more evidence shows clearly that such a diversity of sauropods is ecologically plausible. Until then, I use the name *Brontosaurus* with reservations and regard Riggs's arguments as still binding.

WRONG-HEADED DINOSAURS

Not only was the name for this dinosaur a confusing mess, but even worse, most of the museums with mounted specimens had the wrong heads on them! The skulls of sauropods are relatively fragile and apparently are easily lost from the neck vertebrae during burial and fossilization. Nearly all the good sauropod skeletons known in the early days were headless, including Marsh's otherwise nearly complete skeleton of *Brontosaurus* on display at Yale and the nearly complete American Museum specimen that became iconic for *Brontosaurus*. Instead, a model of a skull was constructed based on "the biggest, thickest, strongest skull bones, lower jaws and tooth crowns from three different quarries." We now know that those skull fragments were not associated with the skeleton and, in fact, probably came from a different sauropod, *Camarasaurus* (one of the few dinosaur genera named by Cope that is still valid), or sometimes from *Brachiosaurus*. Adam Hermann, who built the reconstructed skull for the American Museum mount, admitted that it was "largely conjectural and based on that of *Morosaurus*" (Marsh's invalid name for what Cope called *Camarasaurus*—one of the few cases where a Cope name won out over a Marsh name). A similar skull was placed on the Yale skeleton, and this short-faced version of *Apatosaurus* became the one that the public absorbed and believed for most of a century.

Yet there was evidence for which head actually belonged on the skeletons. In 1909, while excavating the Carnegie Museum quarry that became Dinosaur National Monument, Earl Douglass (see chapter 8) found an *Apatosaurus* neck with a skull very close to it. It looked nothing like *Camarasaurus*, but more like the long-faced *Diplodocus*. Douglass and Carnegie Museum director William J. Holland both accepted that it was the proper head for *Apatosaurus*, but the influential Osborn and others refused to accept this evidence. To them it wasn't conclusive because it wasn't directly attached—they were also aware that it would be a lot of work to take their wrong heads off their museum mounts! Holland responded by leaving the Carnegie Museum mounted specimen headless, hoping that a better specimen with a more conclusive head-neck connection would turn up, but it never happened while he lived. When Holland died, his successor hated the headless monster and put what turned out to be a *Camarasaurus* skull on it.

The problem remained unresolved for decades. In the 1970s, legendary sauropod expert Jack McIntosh (who taught theoretical physics at Wesleyan University) and Carnegie curator Dave Berman researched the issue and decided Holland and Douglass were right. Not only was the skull in question near the neck, but all the bones of *Apatosaurus* were very much like those of diplodocines, not camarasaurs. On October 20, 1979, with much fanfare, the Carnegie Museum finally replaced the *Camarasaurus* skull on their mount with the proper skull, and during the 1980s and 1990s Yale and the American Museum and other institutions eventually caught up. Their hunch was confirmed in 2011 when a specimen (now in the Cincinnati Museum) was found with the head attached to the neck: the skull was like that of *Diplodocus*, not *Camarasaurus*. At the same time, another even more complete skeleton was found in the Cactus Park Quarry in western Colorado by "Dinosaur Jim" Jensen. This specimen (nicknamed "Einstein") is now on display at Brigham Young University, and not only does it have the diplodocine skull but a complete cast of the brain inside, which has been CAT-scanned and analyzed (hence the nickname, because it's the brainiest sauropod we know).

Over the years, more and more specimens of apatosaurs have been found, and we have learned more and more about them. Indeed, their story is closely entwined with what we now know is their close relative, *Diplodocus*—the most widely displayed dinosaur in the world. That is the subject of the next chapter.

FOR FURTHER READING

Colbert, Edwin. *Men and Dinosaurs: The Search in the Field and in the Laboratory.* New York: Dutton, 1968.

Curry Rogers, Kristina, and Jeffrey Wilson. *The Sauropods: Evolution and Biology.* Berkeley: University of California Press, 2005.

Davidson, Jane Pierce. *The Bone Sharp: The Life of Edward Drinker Cope.* Philadelphia: Academy of Natural Sciences, 1997.

Farlow, James, and M. K. Brett-Surman, eds. *The Complete Dinosaur.* Bloomington: Indiana University Press, 1999.

Fastovsky, David, and David Weishampel. *Dinosaurs: A Concise Natural History*, 3rd ed. Cambridge: Cambridge University Press, 2016.

Hallett, Mark, and Mathew J. Wedel. *The Sauropod Dinosaurs: Life in the Age of Giants.* Baltimore, Md.: Johns Hopkins University Press, 2016.

Holtz, Thomas R., Jr. *Dinosaurs: The Most Complete, Up-to-Date Encyclopedia for Dinosaur Lovers of All Ages.* New York: Random House, 2011.

Howard, Robert West. *The Dawnseekers: The First History of American Paleontology.* New York: Harcourt Brace Jovanovich, 1975.

Jaffe, Mark. *The Gilded Dinosaur: The Fossil War Between E. D. Cope and O. C. Marsh and the Rise of American Science.* New York: Crown, 2000.

Klein, Nicole. *Biology of the Sauropod Dinosaurs: Understanding the Life of Giants.* Bloomington: Indiana University Press, 2011.

Lanham, Url. *The Bone Hunters.* New York: Columbia University Press, 1973.

Naish, Darren. *The Great Dinosaur Discoveries.* Berkeley: University of California Press, 2009.

Naish, Darren, and Paul M. Barrett. *Dinosaurs: How They Lived and Evolved.* Washington, D.C.: Smithsonian Books, 2016.

Osborn, Henry Fairfield. *Cope, Master Naturalist: Life and Letters of Edward Drinker Cope, with a Bibliography of His Writings.* Manchester, N.H.: Ayer, 1978.

Plate, Robert. *The Dinosaur Hunters: Othniel C. Marsh and Edward D. Cope*. New York: McKay, 1964.

Remes, Kristian, Carole T. Gee, and P. Martin Sander. *Biology of the Sauropod Dinosaurs: Understanding the Life of Giants*. Bloomington: Indiana University Press, 2011.

Schuchert, Charles, and Clara M. LeVene. *O. C. Marsh: Pioneer in Paleontology*. New Haven, Conn.: Yale University Press, 1940.

Shore, Elizabeth Noble. *The Fossil Feud Between E. D. Cope and O. C. Marsh: Spying, Dirty Tricks, Plagiarism—The Exciting Story of the Famous and Bitter Rivalry Between Two of America's Greatest Paleontologists*. Hicksville, N.Y.: Exposition Press, 1974.

Spaulding, David A. E. *Dinosaur Hunters: Eccentric Amateurs and Obsessed Professionals*. Rocklin, Calif.: Prima, 1993.

Thomson, Keith. *The Legacy of the Mastodon: The Golden Age of Fossils in America*. New Haven, Conn.: Yale University Press, 2005.

Tschopp, Emanuel, Octávio Mateus, and Roger B. J. Benson. "A Specimen-Level Phylogenetic Analysis and Taxonomic Revision of Diplodocidae (Dinosauria, Sauropoda)." *PeerJ* 3 (2015): e857.

Upchurch, Paul, Paul M. Barrett, and Peter Dodson. "Sauropoda." In *The Dinosauria*, 2nd ed., ed. David B. Weishampel, Peter Dodson, and Halszka Osmólska, 259–322. Berkeley: University of California Press, 2004.

Wallace, David Rains. *The Bonehunters' Revenge: Dinosaurs, Greed, and the Greatest Scientific Feud of the Gilded Age*. New York: Houghton Mifflin, 1999.

Wilford, John Noble. *The Riddle of the Dinosaur*. New York: Knopf, 1985.

DIPLODOCUS

When it walked the earth trembled under the weight of 120,000 pounds, when it ate it filled a stomach enough to hold three elephants, when it was angry its terrible roar could be heard ten miles, and when it stood up its height was equal to eleven stories of a skyscraper.

—*NEW YORK HERALD*, 1898

REVENGE!

Apatosaurus was not the only dinosaur found in 1877 in Colorado (see chapter 7). Even though Marsh's assistants Samuel Wendell Williston and Benjamin Mudge had little success with their Cañon City quarries when compared to what Lucas was finding for Cope, they did find important fossils and immediately shipped them to Marsh. These included some relatively large long-tail vertebrae from Felch Quarry near Cañon City, which Marsh received in 1878 and immediately published (with no illustrations) as a new genus and species of dinosaur, *Diplodocus longus*. The name *Diplodocus* means "double beam" in Greek, a reference to the V-shaped bones called chevron bones located beneath the tail vertebrae. The specimens were very poor and hardly the basis for a new genus because V-shaped chevron bones occur in most diplodocines (including *Apatosaurus*) and in a number of other sauropods as well. At the time, however, this was the first dinosaur known to have this feature, and Marsh thought it was sufficient to diagnose a new genus. Today we would say that Marsh's original specimen of *Diplodocus longus* (called the "type specimen") is insufficient to know what animal it belonged to. Some paleontologists argue that the name *Diplodocus* is no longer valid. In the interest of stability, they have proposed that a different specimen of a different species of *Diplodocus* be designated as the type specimen of the genus. But that discussion is beyond the scope of this book.

Meanwhile, the Bone Wars, which had peaked in the late 1870s, began to enter a new phase in the 1880s. Originally, Cope and Marsh had resorted to spying on each other and bribing people to switch sides or to reveal the location their rival's fossil sites. Both men and their crews occasionally destroyed fossils they could not remove from the ground or take with them, and when they finished a quarry they took great pains to fill it in with rocks and dirt to hide it and make it hard to reopen. In one instance, the rival crews even threw rocks at each other. Luckily, they didn't resort to gunplay as was common in the Wild West back then. By the 1880s, Marsh was flush with tons of bones arriving at Yale every week. Cope had only limited field crews working for him, and he already had so many bones in his house that there was no place for more.

Cope's fortunes took a turn for the worse when he spent most of his inheritance to buy the journal *American Naturalist* (still a famous journal today). He also had a hard time landing a paying position in a college or a museum, possibly because of Marsh's influence and also because of his legendary temper and grouchy attitude. Cope undertook several field seasons alone by himself, which damaged his health, especially after a bout of malaria. In desperation, Cope risked the remainder of his inheritance in various gold and silver mining investments, only to lose it all when they turned out to be worthless.

Meanwhile, Marsh had established powerful contacts in Washington and eventually became the official paleontologist of the newly created U.S. Geological Survey, headed by his friend John Wesley Powell. A one-armed Civil War veteran, Powell was now famous as the leader of the first expedition to float down the Colorado River and the Grand Canyon. Yet most of Marsh's assistants hated working for him because Marsh never gave them any credit in the publications, paid them very little, or forgot to pay them entirely.

In 1884, the bad situation between Cope and Marsh reached a critical point. Powell told Cope that all his specimens obtained during the Hayden Survey and other government surveys belonged to the U.S. government, and he had to give them up. Cope disagreed, of course, because he had not taken government money while in the field but financed his work with the Hayden Survey out of his own pocket. Cope's hatred of Marsh boiled over, and he decided it was time to get revenge. As Congress was looking into the U.S. Geological Survey, Cope sent them a long list of mistakes made by Marsh, accusations of plagiarism and theft of government specimens, and charges that Marsh had misappropriated government funds. The congressional inquiry didn't get very far, so Cope then gave the damaging charges and evidence to a reporter, William Hosea Ballou. He wrote a series of sensational and shocking articles in the *New York Herald* that raised the issue to the level of a national scandal. The scientific community had known about the battle between the two men for years, but now their dirty laundry was causing public disdain for science. Elizabeth N. Shor wrote about it:

> Most scientists of the day recoiled to find that Cope's feud with Marsh had become front-page news. Those closest to the scientific fields under discussion, geology and vertebrate paleontology, certainly winced, particularly as they found themselves quoted, mentioned, or misspelled. The feud was not news to them, for it had lurked at their scientific meetings for two decades. Most of them had already taken sides.

The papers filled with more of Cope's charges, then rebuttals by Marsh and Powell accusing Cope of similar misdeeds. Cope called Marsh and Powell "partners in incompetence, ignorance, and plagiarism," and called the survey a "gigantic politico-scientific monopoly next in importance to Tammany Hall." The controversy played out for a few more months, then died as public interest moved to other topics. Both sides were severely wounded, but their status remained unchanged.

In 1892, drought conditions and water problems out west caused Congress to launch another investigation of the U.S. Geological Survey, which led to close scrutiny of how Marsh had spent government funds. Many were upset about his support of Darwinian evolution and his description of creatures not in the Bible. Typical of these was a giant scientific monograph Marsh wrote on the "Odontornithes," the early toothed birds *Hesperornis* and *Ichthyornis* from the chalk beds of Kansas. The inquiry led to such incredible scenes as fundamentalist Alabama Congressman Hilary Herbert saying on the floor of Congress, "Birds with teeth! That's where your hard-earned money goes, folks—on some professor's silly birds with teeth."

As a result, the appropriation for the U.S. Geological Survey was severely slashed, and Powell was forced to fire Marsh.

Meanwhile, Cope obtained a professorship at the University of Pennsylvania, then support from the Texas Geological Survey, and succeeded Marsh as head of the American Association for the Advancement of Science when Marsh stepped down. Marsh, however, received the Cuvier Medal, one of the highest awards in paleontology.

The rivalry did not die until Cope himself died in 1897 at the young age of 56, all alone in his house crowded with unstudied fossils. As his health began to decline, Cope was befriended by a young paleontologist who started at Princeton and was now teaching there, Henry Fairfield Osborn. Soon to become the head of the burgeoning American Museum of Natural History, Osborn championed Cope's work after his death. Desperate for money, Cope sold most of his fossil collection to Osborn, which became the core of the future American Museum collection, eventually the largest in the world. Many of Cope's specimens were still in their crates when he died, so (for example) the famous *Allosaurus* skeleton now on the display in the American Museum was untouched in field wrappings until the American Museum preparators rediscovered it and worked on it.

Marsh lived two years longer than his nemesis, but he died in 1899 at age 67. Most of the people who chafed under his supervision and mistreatment did not mourn him much. In the end, he too was broke and asked Yale for a salary. He didn't live long enough to see much of his collection removed from crates, let alone published. The Secretary of the Smithsonian, famous paleontologist Charles Doolittle Walcott, laid legal claim to much of the collection made by Marsh while in the employ of the U.S. Geological Survey, and the fossils were sent to the Smithsonian, where they still reside.

With the death of both men as the century ended, the field was soon taken over by the next generation: Osborn and many others at the American Museum, William Berryman Scott of Princeton, William J. Holland and others of the Carnegie Museum, plus Riggs at the Field Museum in Chicago, Knight and Reed at the University of Wyoming, and other rising museums. They soon developed their own competition for dinosaur bones to fill their halls, but it was a civil, friendly rivalry, not the bitter death-struggle of Cope and Marsh.

Even though the Bone Wars were unseemly and unprofessional, the consequences were mostly beneficial. Paleontology was just a tiny hobby science before the Bone Wars, with only Joseph Leidy publishing obscure articles for specialists. By the time Cope and Marsh died, paleontology was a major science, with a huge public profile, especially as giant dinosaur mounts came to the museums and captivated the public. Dinosaurs entered the public consciousness for good, and many of the poorly understood and incomplete specimens from Europe (such as *Cetiosaurus*; see chapter 3) finally could be reconstructed accurately. This would not have happened so quickly without Cope and Marsh.

Both men wrote and published papers at an astounding rate. Cope published over 1,400 scientific articles in his lifetime, describing more than 1,300 new species. These included living fishes, amphibians, and reptiles as well as fossils. As a result, he became one of the pioneers of American herpetology and ichthyology, and the primary journal, *Copeia*, is named after him. Marsh wasn't quite as prolific, but he published dozens of short articles without illustrations each year, naming and staking claim to every fossil that had just emerged from a crate. Before the Bone Wars, only nine dinosaurs had been named from North America (based on isolated teeth named by Leidy, most of which are invalid now). By the time the Bone Wars were over, more than 150 dinosaurs had been named, including most of the famous ones that every kid recognizes. We will look at some of these in later chapters.

DINOSAUR MEN: THE NEXT GENERATION

"Most Colossal Animal Ever on Earth Just Found Out West" read the headline of the *New York Herald* in 1898, 14 years after it had run the nasty Cope-Marsh feud in its columns. With typical journalistic exaggeration, it described the monster in purple prose. "When it walked the earth trembled under the weight of 120,000 pounds, when it ate it filled a stomach enough to hold three elephants, when it was angry its terrible roar could be heard ten miles, and when it stood up its height was equal to eleven stories of a skyscraper."

The specimen described in the article was a new sauropod they called "Brontosaurus giganteus," which had just been collected from the Morrison Formation just south of Laramie. The photograph of the thighbone identified the collector as "Bill Reeder," who had recovered the specimens for the new museum at the University of Wyoming in Laramie. In actuality, it was the same William Harlow Reed who had started out as the Como railroad stationmaster and then worked at Como Bluff for Marsh for several years after 1877 (figure 8.1). He had been working for Wilbur Clinton

Figure 8.1 ▲

(A) Carnegie's crew at "Camp Carnegie" (*from left to right*): Paul Miller, Jacob Wortman, William Harlow Reed, and Reed's son Willie. (B) Legendary American Museum dinosaur hunters in 1895. *From left to right*: Olaf A. Peterson (later with the Carnegie Museum), Jacob Wortman, Walter Granger, and Albert Thomson. ([A and B] Courtesy of Wikimedia Commons)

Knight of the University of Wyoming since 1894 and had amassed a collection there weighing more than 72 metric tonnes (79 tons), almost as large as the collection he sent to Yale. That collection still fills their remarkable exhibit hall.

The newspaper headlines caught the attention of a very important man: Andrew Carnegie. Carnegie was the one of richest men in the world at that time, rising from an impoverished background in Scotland. Starting as a lowly telegraph operator, he used his hard work, determination, and business savvy to invest in railroads, bridge building, bond trading, and then got in on the ground floor of the growing steel industry. By age 33, he was a millionaire, and when he sold Carnegie Steel to J. P. Morgan in 1901 for $480 million, he was unquestionably the richest man alive.

Carnegie had already decided that he would give away or invest most of his wealth in the public good, spending 90 percent of his fortune to endow hospitals, universities, and especially to build public libraries in most cities and towns. As Pittsburgh's leading citizen, he felt that his adopted home should be civilized as well. He had already spent $24 million to found the Carnegie Institute, with its concert hall, art gallery, and small natural history museum. But the newspaper headlines made him determined that his Carnegie Museum of Natural History should also have gigantic dinosaurs to match those at Yale, the American Museum, the Field Museum, and the University of Wyoming.

For the director of the Carnegie Museum, he hired William Jacob Holland, a versatile and ambitious clergyman, entomologist, and paleontologist (figure 8.2). When Carnegie saw the newspaper article about the "Most Colossal Animal," he clipped it and sent it to Holland with this instruction: "Buy this for Pittsburgh."

Holland quickly figured out that "Bill Reeder" was actually William Harlow Reed and offered him a job at the Carnegie Museum. Reed agreed to find more fossils for the Carnegie Museum and gave the University of Wyoming his notice. Holland then hired former American Museum collector Arthur Coggeshall and another young paleontologist, Jacob Lawson Wortman (see figures 8.1 and 8.2), who would become famous for other discoveries. The three men met in Medicine Bow to see if they could find more of the "Most Colossal Animal." The original "Brontosaurus giganteus" quarry was mostly exhausted, but they poked around the area to the north of Como Bluff in Sheep Creek, Wyoming. On July 2, 1899, they found

The giants of paleontology in 1899—a meeting of the American Museum paleontologists as they take a break from their own excavations to visit the Carnegie Museum Sheep Creek *Diplodocus* quarry. *From left to right:* William Harlow Reed, Albert "Bill" Thomson, William J. Holland (Carnegie Museum Director), Henry Fairfield Osborn (American Museum DVP Head), William Diller Matthew, Walter Granger, Jacob Wortman, and squatting in the foreground, Richard Swann Lull. Osborn, Thomson, Matthew, Lull, and Granger were with the American Museum at the time, and Holland, Reed, and Wortman were with the Carnegie Museum. (Courtesy of Wikimedia Commons)

toe bones and eventually nearly complete skeletons of several individuals of *Diplodocus*. Holland then sent Olof Peterson, a Swedish paleontologist working in their museum, and Charles Whitney Gilmore (who had studied at University of Wyoming and worked with Reed before) to join the field crew for the massive excavation. By patching together parts of four skeletons, they were able to mount a nearly complete skeleton (again, missing the skull, as happens so often with sauropods). It was over 26 meters (84 feet) long, the longest dinosaur ever found even today, with an extremely long whip-like tail.

By 1901, the nearly complete skeleton (nicknamed "Dippy") was back in Pittsburgh and being described and published by John Bell Hatcher (another legendary paleontologist discussed in later chapters). Hatcher officially named it *Diplodocus carnegii*, in honor of their benefactor, who was immensely pleased. (Paleontologists know that rich donors love having fossils named after them and often become more generous as a result.) It was Carnegie's turn to have a dinosaur that made worldwide headlines, and he was not going to miss the opportunity to publicize his new museum and the dinosaur named after him. He hired a team of Italian plasterers to make numerous replicas of the original fossils in Pittsburgh, and then gave the extraordinary replicas to museums all over the world. Holland went with each gift to be feted and celebrated and then to supervise the mounting of the plaster replicas. In this way, museums in England, France, Germany, Austria, Italy, Russia, Spain, Argentina, and Mexico all had copies of Carnegie's dinosaur on display. Until recently, for example, the central hall in the Natural History Museum in London was decorated not by any British dinosaur (which were all too incomplete to feature in the main halls) but by Carnegie's gift (figure 8.3). Coggeshall wrote that "to *Diplodocus carnegii* goes the credit of making 'dinosaur' a household word . . . presidents, kings, emperors, and czars besieged Andrew Carnegie for replicas to be installed in their national museums."

THE REAL "JURASSIC PARK"

The Carnegie Museum's luck with sauropods did not end with the Sheep Creek find. Another one of their paleontologists, Earl Douglass (figure 8.4), spent most of every summer in the early 1900s scouring the rocks of Montana, Utah, and Wyoming, mostly looking for fossil mammals. Holland was visiting Douglass in the field in the Uinta Range of northeastern Utah when they decided to follow up on a hunch. As Holland described it:

> We decided that we would set forth early the next day with our teams of mules and visit the foot-hills, where Hayden had indicated the presence of Jurassic exposures. We started shortly after dawn and spent a long day on the cactus-covered ridge of Dean Man's Bench, in making our way through the gullies and ravines to the north. . . . The next day we went forward through the broken foot-hills which lie east and south of the great gorge through which the Green River emerges from the Uinta Mountains on its course to the Grand Canyon of Arizona. As we slowly made our way through the stunted groves of pine we realized that we were upon Jurassic beds. We tethered our mules in the forest. Douglass went to the right and I to the left, scrambling up and

Figure 8.3 ▲

Mounted replica of *Diplodocus carnegii*, donated to the British Museum by Andrew Carnegie and featured in the main entrance hall for over a century. It has since been moved elsewhere and replaced by a blue whale. (Photograph by the author)

down through the gullies in search of Jurassic fossils, with the understanding that, if he found anything he was to discharge the shotgun which he carried, and if I found anything, I would fire the rifle, which I carried. His shotgun was presently heard and after a somewhat toilsome walk in the direction of the sound I heard him shout. I came up to him standing beside the weathered-out femur of a Diplodocus lying in the bottom of a very narrow ravine in which it was difficult to descend. Whence this perfectly preserved bone had fallen,

Figure 8.4 ▲
Earl Douglass standing next to the bones of a dinosaur at the Carnegie Quarry, which later became Dinosaur National Monument. (Courtesy of Wikimedia Commons)

from what stratum of the many above us it had been washed, we failed to ascertain. But there it was, as clean and perfect as if it had been worked out from the matrix in the laboratory. It was too heavy for us to shoulder and carry away, and possibly even too heavy for the light-wheeled vehicle in which we were traveling. So we left it there, proof positive that in that general region search for dinosaurian remains would probably be successful.

Holland's prediction came true a year later on August 17, 1909. Douglass was working in the same area with a local Mormon farmer, George "Dad" Goodrich. Douglass climbed up the ridge above where the femur had been found the year before and looked down. "At last, in the top of the ledge where the softer overlying beds form a divide—a kind of saddle—I saw eight of the tail bones of *Brontosaurus* in exact position. It was a beautiful site [*sic*]." He and Goodrich went back to town and recruited more helpers, then they began to quarry out the bones. "I have discovered a huge dinosaur *Brontosaurus* and if the skeleton is as perfect as the portions we have exposed, the task of excavating will be enormous and will cost a lot of money, but the rock is that kind to get perfect bones from." This message brought Holland to Utah, and when he saw what Douglass had, he immediately telegraphed Carnegie to get him to agree to fund the excavation.

The nearest town was tiny Vernal, Utah, over 32 kilometers (20 miles) away, so for the next 13 years (1909–1922), Douglass took up permanent residence in the area near Carnegie Quarry. There he and his crews lived and worked year round except when the weather was unbearable. He even brought his young wife and one-year-old baby out to live with him, first in a small tent heated by an iron stove, but eventually in a homesteader's log cabin with a garden and cow and chickens and everything the family needed. First they exposed the *Brontosaurus* he originally discovered, only to find the neck twisted back into the rock—and the skull missing. Nevertheless, it was nearly complete and about 30 meters (98 feet) long, with a tail over 9 meters (30 feet) in length. When it was shipped to Pittsburgh, cleaned and mounted, Holland described it and named it *Apatosaurus louisae* in honor Carnegie's wife Louise. Holland and Douglass both thought that a nearby skull like that of *Diplodocus* was probably the correct skull for *Apatosaurus*, but they were overruled by people who followed Marsh's reconstruction and by Osborn, so this skeleton remained headless in Pittsburgh until after Holland died (see chapter 7).

Once the first skeleton had been removed, Douglass and his men found three other sauropod skeletons nearby and realized that the bone bed was a thick bed of sandstone that was tilted almost vertically. They blasted away the overburden of soft Morrison shales and trenched down to expose the top surface of the tilted sandstone layer. Eventually the trench was 180 meters (600 feet) long and 24 meters (80 feet) deep. Between 1909 and 1922, Douglass and his men removed the top half of the sandstone wall, which was over 90 meters (300 feet) long and 23 meters (75 feet) high. They also excavated the east and west side of the huge wall of sandstone.

They recovered 315 tonnes (347 tons) of fossils and took them by buckboard wagon to the nearest rail stop in Dragon, Utah, over 80 kilometers (50 miles) away. Altogether they found more than 20 skeletons and additional fossils representing about 300 additional individual dinosaurs. The sandstone layer apparently had once been a river channel in a Jurassic river, and portions of carcasses had floated down to that spot and then become buried.

In 1922, William Holland retired and Andrew Carnegie died, and the funding dried up. The Carnegie Museum was crammed to the limit with over 270 metric tonnes (297 tons) of bones that had not yet been prepared or cleaned, so the museum decided to end the excavation and close the quarry. In 1923, Charles W. Gilmore (now at the Smithsonian) reopened the quarry and hauled out about 50 tons of material, including the *Diplodocus* long on display at the Smithsonian, along with the complete, articulated skeleton of a baby *Camarasaurus*. In 1924, Douglass then began working for the University of Utah and collected another 33 crates of specimens, including a complete *Allosaurus*, the last large-scale excavation of the quarry. However, they never offered him a position, and he died in poverty in 1931 without seeing his vision realized.

Douglass's dream was to see the Carnegie Quarry made into a national monument. He knew that the locality had amazing potential because they had removed only half of the original wall of sandstone. He wanted the rest to be left in place as a permanent monument for people to see dinosaur bones as they are found in the field. As he wrote, "I hope that the Government, for the benefit of science and the people, will uncover a large area, leaves the bones and skeletons in relief and house them in. It would make one of the most astounding and instructive sights imaginable." Douglass tried to buy the mineral rights to protect the site, but the courts ruled that dinosaur bones were not minerals. But Holland had a powerful friend, Charles Doolittle Walcott, a paleontologist who was also head of the Smithsonian. Walcott convinced President Woodrow Wilson to designate the quarry area as Dinosaur National Monument in 1915.

The monument was isolated in the middle of the wilderness of Utah. There was almost no way to reach it in the days before cars were common and roads paved, so it remained primitive and undeveloped for years. During the Depression, crews of unemployed men came from the WPA to remove the overburden of Morrison shale from the sandstone layer in anticipation of resuming excavation and display in the future. Nothing much was built on the site during World War II, but in the 1950s the Park Service surveyed the area and determined it was worth developing. A modern glass-sided building was finished in 1958, and its north wall was made of the sandstone layer full of dinosaur bones.

Over the years, the building has become one of the most popular national monuments in the country (figure 8.5). About 400,000 visitors a year come

Figure 8.5 ▲

Dinosaur National Monument Quarry Visitors Center: (A) the glass structure built over the steeply dipping wall of dinosaur bones; (B) a view of the wall of dinosaur bones left where they were found but excavated in relief. ([A] Courtesy of D. J. Chure; [B] photograph by the author)

all the way out to northeastern Utah to gaze at the wall of dinosaur bones that were excavated in relief and left in place. In 2006, the quarry building was closed because it was beginning to rip apart as the swelling clays of the Morrison Formation expanded and contracted after each rain or snow, disrupting the foundation. A much larger Visitor's Center was built down on the flats below the quarry (near where Douglass had his log cabin), and since 1979 a shuttle service has carried the flood of visitors coming to the tiny quarry parking lot. The quarry building was rebuilt with 70-foot steel pilings driven deep into the harder bedrock below, and it reopened in 2011. It houses just the paleontological exhibits; the main Visitor's Center down on the flats has exhibits about the rest of the monument, plus a gift shop, offices, and other support facilities.

Dinosaur National Monument is the enduring legacy of Douglass, Holland, and Carnegie as well. More than 400 individual dinosaurs have been found, more than at any other Jurassic quarry. There are 10 genera of dinosaurs, including at least six kinds of sauropods, three kinds of theropods, plus *Stegosaurus*, an iguanodontid, a dryosaurid, as well as turtles, crocodiles, and even tiny shrew-sized mammals found elsewhere in the monument. Not only is it one of the most productive Jurassic sites ever found, but it is one of the few where visitors can see what the fossils were like as they were found and wonder at the huge number of enormous bones that were discovered here. An excellent website (www.carnegiequarry.com) allows anyone to explore many aspects of science and history at Carnegie Quarry.

FOR FURTHER READING

Brinkman, Paul D. *The Second Jurassic Dinosaur Rush: Museums and Paleontology in America at the Turn of the Twentieth Century*. Chicago: University of Chicago Press, 2010.

Colbert, Edwin. *Men and Dinosaurs: The Search in the Field and in the Laboratory*. New York: Dutton, 1968.

Curry Rogers, Kristina, and Jeffrey Wilson. *The Sauropods: Evolution and Biology*. Berkeley: University of California Press, 2005.

Davidson, Jane Pierce. *The Bone Sharp: The Life of Edward Drinker Cope*. Philadelphia, Penn.: Academy of Natural Sciences, 1997.

Douglass, G. E. *Speak to the Earth and It Shall Teach You: The Life and Times of Earl Douglass, 1862–1931*. 2009, www.booksurge.com.

Farlow, James, and M. K. Brett-Surman. *The Complete Dinosaur*. Bloomington: Indiana University Press, 1999.

Fastovsky, David, and David Weishampel. *Dinosaurs: A Concise Natural History*, 3rd ed. Cambridge: Cambridge University Press, 2016.

Hallett, Mark, and Mathew J. Wedel. *The Sauropod Dinosaurs: Life in the Age of Giants*. Baltimore, Md.: Johns Hopkins University Press, 2016.

Holtz, Thomas R., Jr. *Dinosaurs: The Most Complete, Up-to-Date Encyclopedia for Dinosaur Lovers of All Ages*. New York: Random House, 2011.

Howard, Robert West. *The Dawnseekers: The First History of American Paleontology*. New York: Harcourt Brace Jovanovich, 1975.

Jaffe, Mark. *The Gilded Dinosaur: The Fossil War Between E. D. Cope and O. C. Marsh and the Rise of American Science*. New York: Crown, 2000.

Klein, Nicole. *Biology of the Sauropod Dinosaurs: Understanding the Life of Giants*. Bloomington: Indiana University Press, 2011.

Lanham, Url. *The Bone Hunters*. New York: Columbia University Press, 1973.

Naish, Darren. *The Great Dinosaur Discoveries*. Berkeley: University of California Press, 2009.

Naish, Darren, and Paul M. Barrett. *Dinosaurs: How They Lived and Evolved*. Washington, D.C.: Smithsonian Books, 2016.

Osborn, Henry Fairfield. *Cope, Master Naturalist: Life and Letters of Edward Drinker Cope, with a Bibliography of His Writings*. Manchester, N.H.: Ayer, 1978.

Plate, Robert. *The Dinosaur Hunters: Othniel C. Marsh and Edward D. Cope*. New York: McKay, 1964.

Remes, Kristian, Carole T. Gee, and P. Martin Sander. *Biology of the Sauropod Dinosaurs: Understanding the Life of Giants*. Bloomington: Indiana University Press, 2011.

Schuchert, Charles, and Clara M. LeVene. *O. C. Marsh: Pioneer in Paleontology*. New Haven, Conn.: Yale University Press, 1940.

Shore, Elizabeth Noble. *The Fossil Feud Between E. D. Cope and O. C. Marsh: Spying, Dirty Tricks, Plagiarism—The Exciting Story of the Famous and Bitter Rivalry Between Two of America's Greatest Paleontologists*. Hicksville, N.Y.: Exposition Press, 1974.

Spaulding, David A. E. *Dinosaur Hunters: Eccentric Amateurs and Obsessed Professionals*. Rocklin, Calif.: Prima, 1993.

Thomson, Keith. *The Legacy of the Mastodon: The Golden Age of Fossils in America*. New Haven, Conn.: Yale University Press, 2005.

Tschopp, Emanuel, Octávio Mateus, and Roger B. J. Benson. "A Specimen-Level Phylogenetic Analysis and Taxonomic Revision of Diplodocidae (Dinosauria, Sauropoda)." *PeerJ* 3 (2015): e857.

Upchurch, Paul, Paul M. Barrett, and Peter Dodson. "Sauropoda." In *The Dinosauria*, 2nd ed., ed. David B. Weishampel, Peter Dodson, and Halszka Osmólska, 259–322. Berkeley: University of California Press, 2004.

Wallace, David Rains. *The Bonehunters' Revenge: Dinosaurs, Greed, and the Greatest Scientific Feud of the Gilded Age.* New York: Houghton Mifflin, 1999.

Wilford, John Noble. *The Riddle of the Dinosaur.* New York: Knopf, 1985.

GIRAFFATITAN

Tendaguru! It is a name that carries with it a sort of magic in the story of the search for dinosaurs. It is a name to be linked with two other names that epitomize the excitement and the romance of the hunt of giant dinosaurs—Como Bluff and Dinosaur Monument. For at Tendaguru there were found, in almost overwhelming abundance, the bones of gigantic Jurassic dinosaurs, some of them the same as the dinosaurs excavated at Como Bluff and Dinosaur Monument.... Tendaguru is a record of an extension south of the equator of the same dinosaurian life that had dominated the Northern Hemisphere during the final years of Jurassic history.

—EDWIN COLBERT, *MEN AND DINOSAURS*

Mention Tendaguru to dinosaur enthusiasts and their eyes are likely to sparkle.

—DAVID SPAULDING, *DINOSAUR HUNTERS*

OUT OF AFRICA

In the late 1800s, the major powers of Europe stampeded in a land grab to colonize the "backward" regions of Africa. Some people went to Africa because they felt they must be missionaries to the heathens, others because they thought Africa should be "civilized" like Europe, and others because Africa offered enormous riches and potential for wealth and power. A few countries, such as South Africa, had been colonized much earlier by the British and by the Dutch (whose descendants are known as the Boers) because South Africa was crucial to ships traveling around the continent and also wealthy in gold and diamonds and other resources. But much of

Africa remained unclaimed by the large European powers and was under the rule of local chieftains and sultans.

Much of eastern and southern Africa was taken by the British, western Africa was largely under French or Spanish dominion, and the Congo was ruled by Belgium. Germany came very late to this imperialistic frenzy because it had neither the strength nor the leadership to become a colonial empire until the 1870s when Otto von Bismarck consolidated all the small independent kingdoms into one country. By the 1880s, however, Germany was jostling with the other European empires to get their share. They sent five warships to point their guns on the palace of the Sultan of Zanzibar, who was forced to surrender most of what became Tanzania to them. Southwestern Africa (now Namibia) was also a German colony, as was Rwanda in central Africa, plus Cameroon, Togoland, and a few other parts of western Africa.

Most of the early German colonials were soldiers, hunters, missionaries, government officials, and people exploiting the mineral wealth or ivory market. There were even slave traders as the Germans didn't abolish slave trading in their colonies, although they tried to check its spread. The German colonial empire thrived until they lost World War I in 1918. After the war, they were forced to give up all of their territories as part of the Treaty of Versailles, and other countries (mainly Great Britain) snapped them up.

Germany was one the world's leaders in science and technology in the late 1800s and early 1900s. Their archeologists were at the top of the profession, with Karl Richard Lepsius pioneering modern Egyptology, Heinrich Schliemann finding the ruins of Troy in 1873 and also ancient Mycenae in Greece, and many other major expeditions. The Pergamon Museum in Berlin is filled with some of the best artifacts from Egypt, Babylonia, and Assyria, and many ancient Greek treasures.

Germany's natural historians were equally significant, dating back to the late 1700s when geology and paleontology began with Alexander von Humboldt (1769–1859), Abraham Gottlob Werner (1749–1817), Leopold von Buch (1774–1583), and extending into the 1800s with paleontologists Christian Erich Hermann von Meyer (1801–1869) and Karl Alfred von Zittel (1839–1904), and pioneering embryologists such as Karl Ernst von Baer (1792–1876) and Ernst Haeckel (1834–1919). Both Cope and Marsh spent years learning paleontology in Germany in the 1860s, as did Henry Fairfield Osborn and William Berryman Scott in the 1880s. These Americans listened to lectures by the giants of the field, studied fossil collections, and received the equivalent of a Ph.D.-level education (that degree was not yet offered in the United States).

By the early twentieth century, German dinosaur paleontology was led by Friedrich von Huene (see chapter 6), Erich von Reichenbach Stromer (see chapter 13), and Werner Ernst Martin Janensch (figure 9.1). Born in 1878, Janensch was the fossil reptile specialist at the Museum für Naturkunde in Berlin in 1901. As his fame grew, he received the Leibniz Medal from the Prussian Academy of Sciences in 1911, appointment to the professorship at the Friedrich-Wilhelms-Universität in 1913, and was a founding member of the Paläontologische Gesellschaft in 1913. His most famous accomplishment, however, was to lead a great expedition to Africa.

In 1906, a German pharmacist, chemist, and mining engineer, Bernhard Wilhelm Sattler, was on his way down a path to a mine south of the

Figure 9.1 ▲
Werner Janensch with African workers and a giant dinosaur leg bone. (Courtesy of Wikimedia Commons)

Mbemkure River in German East Africa (now Tanzania) and banged his shin on some enormous bones weathering out of the path near the base of the hill. The hill was called *tendaguru* (steep hill) in the local Wamwera language. Sattler reported the news to German paleontologist Eberhard Fraas, who happened to be visiting nearby, and together they excavated two gigantic skeletons. They shipped the skeletons to the Royal Natural History Collection in Stuttgart where Fraas was a curator, and they were described as *Gigantosaurus robustus* and *Gigantosaurus africanus* in 1908. Unfortunately for Fraas, the name "Gigantosaurus" had already been used by British paleontologist Harry Gover Seeley in 1869 to name some fragmentary dinosaur bones from England (whose correct genus cannot be determined today). So the name "Gigantosaurus" was invalid, not diagnostic, and could not be used. Today Fraas's dinosaurs have been renamed *Janenschia robusta* and *Tornieria africana*. Fraas contracted dysentery while he was in Africa, which made him sickly and killed him in 1915.

Other German scientists were impressed with Fraas's finds and were told that there was much more to be obtained there. The director of the Berlin Museum, Dr. Wilhelm von Branca, was desperate to send another field expedition to obtain bones for his museum. He formed a committee with the Duke Mecklenberg, Regent of Brunswick, and together they raised money from the German Imperial Government, the City of Berlin, the Academy of Learning, and hundreds of private citizens. Altogether they raised about 230,000 marks, or about US$50,000, a considerable sum for 1912.

Once the funds were obtained, Janensch was appointed to lead the expedition as the curator of fossil reptiles for the Berlin Museum. His second in command was Edward Henning of the museum. In 1909, the expedition reached Tendaguru and transformed the quiet little area of forest into a busy village. Hundreds of Africans (170 in the first year, and 500 by the third year) were doing all the heavy manual labor, digging the holes, and carrying the gigantic bones out. All these hard-working Africans came with their families, so a small village of huts sprang up filled with women and children. Because domestic animals would require constant protection against predators such as lions, most of the food had to be brought in from the seaport of Lindi, a four-day march. Photographs of the expedition show long caravans of people carrying enormous loads on their heads to bring supplies from Lindi to Tendaguru (figure 9.2). On the return trip, the men carried the

Figure 9.2 ▲

Images of African workers: (A) excavating huge dinosaurs; (B–C) carrying loads between Lindi and Tendaguru. (Courtesy of Wikimedia Commons)

immense bones covered in plaster jackets and suspended between poles. Often as many as four of them were required to carry just one fossil. In just the first three years, thousands of loads of bones were carried out in 5,400 trips, packed in 800 boxes in Lindi, that weighed over 180,000 kilograms (400,000 pounds) or about 180 metric tonnes (200 tons).

The excavations continued throughout 1910 and 1911, forcing the crews to dig larger and larger pits and to shore up the sides with wooden walls. Most of the specimens were not only huge and heavy but also broken, and most were scattered around rather than being found as partial skeletons such as the finds in the American Rockies. One shoulder blade, for example, was 1.8 meters (6 feet) long but broken into 80 pieces and required 160 hours to clean, piece together, and harden. A single vertebra might consume 450 hours of labor to prepare. Henning wrote:

> The remains of our giants, that is the bones of the legs, vertebrae, and the teeth, were almost without exception strewn around confusedly in the embedding rock. Whenever we thought we were gradually assembling parts of an animal, some saurian goblin would play us a trick: three like thighbones, two pelvises, or something of the sort would almost always suggest the presence of the same species on one spot—but for ever so long the skull, which we particularly wanted, would not turn up. Gradually however, the legendary creature beginning to arise from the graves became stranger and stranger; they kept us constantly in suspense. Thus several elements of legs appeared, and before we could properly free the joints they looked like strong femurs. After days of widening the shaft, a considerably larger piece would be added, and then it would turn out that what we had been dealing with was only the metacarpal bone. Quite a glove those hands would have worn! These bones would now be joined by the tibia. Then after a while the humerus belonging to the same animal would turn up.

After three years (1909–1911), most of the German scientists—Janensch, Henning, and the others—returned to Berlin to begin work on the fossils. They put 26-year-old Hans Reck in charge of the excavation, and he spent 1912 recovering another 45 metric tonnes (50 tons) of bones. Reck would later become famous as the first to find human fossils at Olduvai Gorge. By this time, the excavations spanned a strip over 5.6 kilometers (3.5 miles) long in the north-south direction, and 2.4 kilometers (1.5 miles) wide in the east-west axis. There were 50 separate excavations with sauropod bones and at least 13 with stegosaur fossils. The expeditions formally ended after 1912; war broke out only two years later, making any further work impossible—especially after German East Africa became British Tanganyika in 1919.

In Berlin, however, Janensch and the others were busy writing up the immense reports on the fossils, which were published in a variety of journals over the next decade. They noted that the fossils had been found interbedded with marine beds full of shells. One skeleton was found with marine fossils that had washed inside it. In other places, the huge sauropods were found with their limbs buried upright, as if trapped in water-logged sediment, while the rest of their bones were scattered nearby, probably after having been torn apart by scavengers or washed away as they rotted. Most of the dinosaurs appeared to have been trapped on sand bars and river mouths near the ocean, where they sank into river sediments after they died, then were covered by the next flood of sediments from the river, or by occasional rises in sea level drowning the coastline.

The Tendaguru fauna that lived along this African shoreline during the Jurassic was remarkable. In addition to the gigantic brachiosaur *Giraffatitan* (figure 9.3A) discussed in the next section, there was also a diplodocoid called *Dicraeosaurus* (figure. 9.3B), just about half the size of *Diplodocus*. It had an extremely long tail and a relatively short neck, reaching 13 meters (43 feet) in length and weighing about 5.9 metric tonnes (6.5 tons). Another specimen originally called "Gigantosaurus," but renamed *Tornieria* in 1911, was also a diplodocoid.

The Tendaguru stegosaur known as *Kentrosaurus* had two rows of spikes along its back and only a few smaller plates, but otherwise it was closely related to the American *Stegosaurus* (see chapter 20). There were also spikes sticking out from the hips (or possibly the shoulder) and from the end of the tail. *Kentrosaurus* reached about 4.9 meters (16 feet) in length and was about 25 percent smaller than *Stegosaurus*. One locality produced the disarticulated remains of about 50 individuals in a pit that was 15–18 meters (50–60 feet) deep and over 915 meters (3,000 feet) across. Another find, described by Rudolf Virchow in 1919, was *Dysalotosaurus*, a small ornithischian dinosaur closely related to American *Dryosaurus*.

These herbivores were preyed upon by numerous typical Late Jurassic predators, including animals similar to *Allosaurus* and possibly *Ceratosaurus* (both known from the Morrison Formation in the Rockies). There was

Figure 9.3 ▲

Sauropods from Tendaguru in the Museum für Naturkunde, Berlin: (*A*) current mount of the *Giraffatitan*; (*B*) the small diplodocine *Dicraeosaurus*, dwarfed by *Giraffatitan* next to it. (Courtesy of M. Wedel)

also a small relative of *Ceratosaurus* called *Elaphrosaurus*. A number of other theropod dinosaur groups are represented by fragmentary remains that cannot be assigned to a genus.

Finally, there were several fragments of pterosaurs that resembled *Pterodactylus* and *Rhamphorhynchus* from the Upper Jurassic Solnhofen Formation of Bavaria, and crocodilians, turtles, and even a toothless jaw of a very primitive mammal (one of the earliest known) called *Brancatherulum*.

Many scientists noted how similar most of the Tendaguru fossils were to dinosaur beds of the same age from America's Morrison Formation: allosaurs, stegosaurs, and dryosaurs, but especially the huge sauropods including diplodocoids and brachiosaurs. For a long time, they were puzzled by how such similar dinosaurs could have traversed the Atlantic Ocean to travel between East Africa and Wyoming. They even postulated land bridges across Antarctica. But the answer to the puzzle became clear when plate tectonics came along in the late 1960s. Although these localities were thousands of miles away from one another, they were still on the supercontinent of Pangea in the Late Jurassic, which was just beginning to break up. Most of the Jurassic faunas elsewhere in the world, whether in the Northern or Southern Hemisphere, look very similar, with only some small regional differences.

Like many other famous German fossils (see chapter 20), some of the huge Tendaguru bones were destroyed by Allied bombers in World War II. Fortunately, the spectacular mounted brachiosaur skeleton has survived, as have many other specimens scattered in other museums (although some in other German museums were also destroyed by bombing). There are also replicas of the German specimens in many other museums around the world. In addition, the British recovered specimens when they reopened the Tendaguru quarries when Tanganyika became a British colony after World War I—although most of those bones have still not been formally described or published.

THE "ARM LIZARD"

The most spectacular and abundantly preserved of Tendaguru dinosaurs were the enormous brachiosaurs, which were the largest dinosaurs known in the world for most of the twentieth century (recently surpassed by Argentinian titanosaurs; see chapter 10). The African fossils were immediately compared to an American genus, *Brachiosaurus*, "arm lizard," so-called because they had unusually long arms and short hind legs that raised their shoulders high and elevated their heads (already at the end of long necks). Brachiosaurs are now famous thanks to their appearance as "veggie-saurs" in the first *Jurassic Park* movie, being featured in the TV show *Walking with Dinosaurs*, and through numerous mounted replica skeletons found in many museums.

However, the original *Brachiosaurus* specimen was much less impressive. It was named and described by Elmer Riggs of the Field Columbian Museum (now the Field Museum of Natural History) in Chicago when he was working in the Morrison Formation in what is now Fruita, Colorado. In 1899, S. M. Bradbury, a local dentist, wrote to tell Riggs of dinosaur bones (figure 9.4A) near Grand Junction, Colorado, which brought out the Field Museum scientists. The original *Brachiosaurus* specimens were found by Riggs's field assistant, H. W. Menke, on the Fourth of July, 1900, when prospecting around a hill that became Riggs Quarry 13 (there is now a monument at the spot). The specimen was only 20 percent complete, consisting of a right humerus (upper arm bone), right femur (thighbone), ilium (upper part of the pelvis), coracoid (shoulder bone), sacrum (fused vertebrae of the hip), seven trunk vertebrae, two tail vertebrae, and some ribs, but the enormous vertebrae gave an indication that it was larger than any sauropod then known. Even though the rest of the dinosaur was missing, Riggs could tell that it had long forelimbs, high shoulders, and short hind limbs. He published the original description in 1903, naming it *Brachiosaurus altithorax* (Greek for "tall chest," in recognition of the tall chest cavity suggested by the bones), and published a more detailed description in 1904.

Because the skeleton was so incomplete, in 1908 the specimens were displayed as isolated bones in glass cases by the side of the old dinosaur exhibit hall of the Field Museum. Today only the long humerus (upper arm bone) and two dorsal vertebrae are on display in the "Evolving Planet" exhibit. The center of the hallway is occupied by the complete skeleton of *Apatosaurus* found on a different Field Museum expedition. This was the state of affairs until Janensch described the nearly complete brachiosaur from the Tendaguru. Janensch decided that the African material was also referable to Riggs's genus *Brachiosaurus*, but he gave it a different species name, *Brachiosaurus brancai*, in honor of Dr. Wilhelm von Branca, the head of the Berlin Museum who had made the expedition possible. Now

nearly every reconstruction of *Brachiosaurus* is based on the African fossils on display in Berlin; the original American *Brachiosaurus* material is too incomplete to create a reconstruction or skeletal mount. In 1994, the Field Museum made molds and fiberglass casts of their own bones, then obtained copies of the African specimens, and created a composite reconstruction of *Brachiosaurus*. This impressive replica towered above the visitors at the north end of Stanley Field Hall, the high-ceilinged main room in the Field Museum, from 1994 until it was removed in 1999 to make way for the famous *Tyrannosaurus rex* specimen nicknamed "Sue." It was moved to the B Concourse of United Terminal 1 at O'Hare Airport (figure 9.4B). Ironically, Sue has since been moved to make way for an even bigger sauropod, the *Patagotitan* skeleton from Argentina (see chapter 10). A second copy of the *Brachiosaurus* skeleton was made and placed outside the northwest terrace of the museum.

Janensch himself was not entirely sure that his African specimens were really referable to North American *Brachiosaurus*. In a series of papers from 1929 to 1950 to 1961, he listed what he thought were 13 unique anatomical features the two dinosaurs had in common. Only four are considered significant now, and the rest are shared with other brachiosaurs, other sauropods, or are difficult to assess as the Colorado specimen consists of only a few bones. The huge skeleton on display in Berlin is also incomplete (see figure 9.3A) and made of a composite of at least two or three different individuals, most of them juveniles. The differences between the two dinosaurs bothered Janensch and several other paleontologists for many years. In 1988, dinosaur artist Greg Paul published the name *Giraffatitan* as a subgenus (without detailed justification). The African brachiosaur was called *Brachiosaurus (Giraffatitan) brancai*. In 1994, amateur paleontologist George Olshevsky raised *Giraffatitan* to the level of genus, again without justification. Most scientists ignored these poorly supported name changes made by amateurs as unjustified. However, a computer programmer and amateur sauropod enthusiast, Mike Taylor, did a rigorous study in 2009 and found evidence that the African fossils were indeed a different genus than the Colorado fossils, and he argued that *Giraffatitan* is valid. Since then, the name has come to be accepted by most paleontologists.

Figure 9.4 ◄

Brachiosaurus altithorax: (A) the bones under excavation near Grand Junction, Colorado; (B) the replica now on display in United Airlines Concourse B at O'Hare Airport in Chicago. (Courtesy of Wikimedia Commons)

THE *SUPERSAURUS/ULTRASAUROS* MESS

Since the discovery of *Giraffatitan*, more and more huge sauropod bones have been found, especially by the late "Dinosaur Jim" Jensen of Brigham Young University. One of his best localities is near Grand Junction, Colorado, and the Colorado National Monument. Known as Dry Mesa Quarry, it has yielded some of the largest bones ever found in the United States (figure 9.5). (I knew Jim well, and he once loaned me the tiny arm bone of a shrew-sized Jurassic mammal from Dry Mesa Quarry to describe, which I coauthored with him and published in 1983.) Some of these bones belonged to a huge sauropod found in 1972 but mentioned in the media numerous times before Jensen formally named it *Supersaurus vivianae* in 1985. Even though a lot of bones are known (possibly 30 percent of the skeleton), the most diagnostic fossils suggest that it is a diplodocine slightly larger than the largest *Apatosaurus*.

Figure 9.5 ▲

Jim Jensen lying next to the shoulder blade of *Supersaurus* as it was found in Dry Mesa Quarry. (Courtesy of Wikimedia Commons)

In addition to *Supersaurus* bones found in Dry Mesa Quarry, additional large fossils found there were named *Ultrasaurus* by Jensen in the same 1985 paper. However, this created a huge problem. First, the name *Ultrasaurus* had already been used by Kim Han Mook in 1983 for a completely different Korean dinosaur, so when Jensen finally published the name (after using it informally in the press for years), it was not available. To get around this, Jensen emended the name to *Ultrasauros* (with an "o" instead of a "u" before the final "s," using the Greek rather than Latin spelling of *sauros*, "lizard"). However, there were still problems. The specimen on which the name *Ultrasauros* is based (the "type specimen") turned out to be a backbone vertebra of a *Supersaurus*, so the name *Ultrasauros* is probably a junior synonym of *Supersaurus*.

However, there *are* bones in the Dry Mesa Quarry that belonged to an even larger dinosaur, including a scapulocoracoid (shoulder blade and shoulder girdle) that come not from a diplodocine but from a brachiosaur slightly larger than *Giraffatitan*. This is the specimen that most people mean when they say *Ultrasauros* now, although that name is still not available. If that were not messy enough, Jensen found another backbone vertebra in a nearby quarry that he called *Dystylosaurus edwini*. Most sauropod specialists think that this is just an additional specimen of *Supersaurus vivianae*.

Jensen further compounded the confusion by not only releasing his names to the press without naming them properly in the scientific journals but also by telling the press that this composite (called a "chimera") of a brachiosaur and a diplodocine was an incredibly large sauropod, reaching 30 meters (100 feet) long, 8 meters (25 feet) tall at the shoulder and 15 meters (50 feet) in total height, and weighing 70 metric tonnes (77 tons; estimates has it as large as 150 tonnes/165 tons). Such a huge dinosaur did not exist, and the actual bones show that it is a mix of a large diplodocine plus a very large brachiosaur, not a real animal.

To summarize this complex story, despite all the publicity and multiple names, many of Jensen's Dry Mesa specimens belong to the diplodocine *Supersaurus*. There are *Apatosaurus* and *Camarasaurus* as well, with *Diplodocus* being the most common dinosaur at Dry Mesa. But one shoulder girdle belongs to a very large brachiosaur, larger than any known so far, which could be considered just a large *Brachiosaurus* or maybe *Giraffatitan*. Whatever it is, it cannot be called *Ultrasaurus* OR *Ultrasauros*.

BIOLOGY OF THE BRACHIOSAURS

Riggs realized that his original *Brachiosaurus* has long forelimbs but short hind limbs, suggesting high shoulders and a sloping back. But none of his discoveries gave him any indication of the skull or neck of that animal. Not until Janensch described and mounted a display based on the multiple partial skeletons of *Giraffatitan* from Tendaguru did we get a clear picture of what this animal looked like, how long it was, and how heavy it might have been. When it was finally mounted and displayed in Berlin, it became the largest dinosaur known from a nearly complete skeleton and remained so until fairly recently when the Argentinian titanosaurs surpassed it as the largest land animal that ever lived (see chapter 10).

Some of the early ideas about brachiosaurs were remarkably naïve and even ridiculous. In the early twentieth century, most dinosaur paleontologists viewed sauropods and many other dinosaurs as slow, sluggish lizards living in swamps, needing the buoyancy of water to support their enormous bulk. According to this outdated notion, many dinosaurs dragged their tails and were barely able to walk, let alone keep their bellies from dragging on the ground. Janensch's original mount of *Giraffatitan* had the legs mounted in a sprawling lizard-like posture with its limbs flexed and bowed out sideways instead of with its limbs in the upright vertical posture that we now know they must have had. (The current skeletal mount has corrected this mistake.)

Other paleontologists imagined that the long neck of sauropods was used like a snorkel, allowing them to submerge their bodies and only have their heads above water. Some even put the nostrils on top of their heads to allow them to be almost completely immersed. We now know that the nostrils faced forward, not upward. A famous old painting by the Czech paleoartist Zdenek Burian shows a brachiosaur walking in a deep fjord completely under water except for its head (figure 9.6). This notion is absurd. Any animal this deep below the surface would be subjected to so much water pressure on its body that it could not expand its lungs. Brachiosaurs had no special mechanism for pulling air from the surface down their windpipes against so much hydraulic pressure on the lungs and body. Under so much force, their lungs would have collapsed. In contrast, whales can live deep underwater because they have unique anatomical specializations that enable them to control the air in their lungs under huge pressures. In addition, whales only inhale at the surface, then dive.

An early reconstruction of a brachiosaur by artist Zdenek Burian, which imagines the long neck was like a snorkel for living deep in the water. This is biologically impossible because the pressure of water at that depth would have collapsed their lungs. (Courtesy of Wikimedia Commons)

This is not the only absurdity of snorkeling brachiosaurs. We now know that nearly all dinosaurs (and birds) have numerous air sacs in their bodies (especially around the backbone), which lighten their weight considerably. This helps decrease the amount of weight they must carry on their limbs (and eliminates the need for them to float in water much), but it would also make them very buoyant and prohibit them from diving into a deep fjord as the old Burian painting suggested.

During the 1970s and 1980s, our view of dinosaurs underwent a huge change, often called the "Dinosaur Renaissance." The most controversial claim was that all dinosaurs had a physiology like that of birds and mammals: homeothermic (maintained constant body temperature) and endothermic (got their heat from metabolizing food rather than from the outside environment as most reptiles do). This was certainly true of the smaller dinosaurs, but there is still a lot of doubt that huge sauropods were endotherms because their enormous bulk and limited surface area would have made dumping excess body heat very difficult (see chapter 17 for a more detailed discussion). Instead, they are thought to have employed inertial homeothermy or "gigantothermy": their huge body mass gained and lost heat very slowly without an internal heat source. Inertial homeothermy is a great strategy for living in the globally warm greenhouse climates of the Jurassic and Cretaceous when it never got very cold and snow and ice were virtually nonexistent.

The reexamination of dinosaurs during the 1970s pointed out some other major surprising things about them. The old "sluggish lizard" model could be ruled out because dinosaurs had vertical limbs tucked completely beneath their bodies (as birds and mammals do) and did not sprawl. Numerous track sites of sauropods showed almost no tail drag marks, so sauropods rarely dragged their tails; they held their tails almost straight out behind them, supported by a trusswork of tendons along the spine. And the "swamp" notion was ruled out because detailed sedimentary studies showed that the Morrison Formation (which has most of the American sauropods) was not a swampy environment but a savannah-like setting comparable to modern East Africa, with distinct wet and dry seasons, many tall coniferous forests, and very little standing water.

Most of the outdated notions of sauropods were ruled out by the 1980s, and fortunately the first *Jurassic Park* novel and movie did not illustrate these notions, so modern audiences did not learn about these old, outdated ideas regarding brachiosaurs. But the brachiosaurs did one thing in that movie that we think is impossible—rearing up on their hind legs to reach high branches. Brachiosaurs are so imbalanced and front-heavy that they may well have lost their balance or broken their legs if they reared up on their relatively small haunches; their hips and hind legs were not large enough to support their entire weight. In any case, brachiosaurs would gain very little advantage by rearing up because (like giraffes) they were naturally the tallest animals in their habitat. With their long vertical necks and the high front end of their body, they already enjoyed a height advantage over all the competing sauropods for treetop fodder.

Paleontologists have done numerous studies to estimate the feeding habits of brachiosaurs. They appear to have been specialists on the highest limbs of trees, which were nearly all conifers in the Late Jurassic. (Flowering plants had not yet evolved.) They could bring their necks down to ground level to feed on ferns and cycads (the only other common types of plants as there were no grasses or flowering shrubs and trees yet), but they would have a disadvantage compared to the shorter-necked animals that lived with them. Their simple, small, spoon-shaped teeth were sufficient only to strip branches and pine needles off intact, and they had no ability to chew their food. Their fodder was gulped down whole and must have been ground up in a large gizzard in their chest, then slowly fermented and digested in a huge intestinal tract that took up most of their body cavity. Paleontologists estimate that they needed about 180–240 kilograms (400–550 pounds) of fodder a day to feed their enormous appetites. That kind of intense browsing would have consumed a lot of conifer trees and ferns in relatively short order, so there could not have been large populations of them in any one area, nor was it likely that they coexisted with many other species of sauropods (see chapter 7). *Brachiosaurus* was also a relatively rare animal in the Morrison Formation, with only 12 specimens known, compared to 112 *Apatosaurus*, 179 *Camarasaurus*, and 98 *Diplodocus*. However, *Giraffatitan* specimens greatly outnumber the diplodocoid *Dicraeosaurus* and other sauropods in the Tendaguru beds.

The skulls of *Brachiosaurus* and *Giraffatitan* are quite different; this is one of the main reasons most people place them in different genera now (figure 9.7). Both had relatively short narrow snouts with arches of bone over the top of the head and eyes, but the arches in *Giraffatitan* were much higher. At one time paleontologists suggested that their nostrils were on top, like a snorkel, but now that is thought unlikely; instead, the high arches may have supported a resonating chamber for making sounds (as suggested in *Jurassic Park*). The brain had a volume of only about 300 cubic centimeters, large enough to control their bodies but very small relative to their enormous body mass, so these dinosaurs were not smart. They probably didn't need to be smart because once partially grown they were too large to fear any predators. They must have had enormous hearts to create the blood pressure to pump all that blood uphill to their heads, and they probably had a series of valves in the veins of their neck (as giraffes do) to keep their blood rushing to their head when they bent down to drink, or rushing away from their brain and making them dizzy or passing out (as can happen to humans) when they raised their heads suddenly.

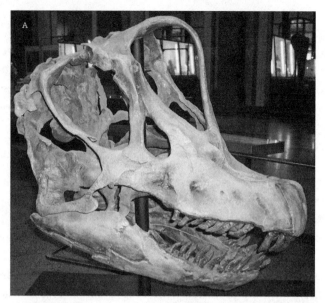

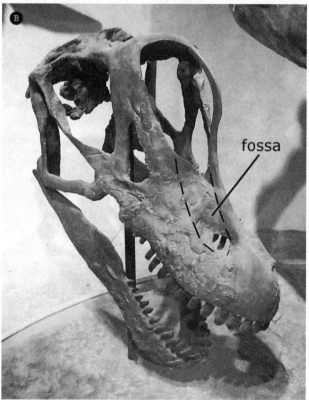

fossa

Figure 9.7 ▲

Comparison of the skulls of brachiosaurs: (*A*) the skull of *Giraffatitan* from Tendaguru; (*B*) the *Brachiosaurus* from Felch Quarry, Colorado. (Courtesy of Wikimedia Commons)

How big were these giants? Estimates vary tremendously because so many of the specimens are incomplete. The nearly complete skeleton of *Giraffatitan* in Berlin was about 22.5 meters (75 feet) long, and 12 meters (39 feet) tall. Its weight has been estimated as low as 15 metric tonnes (16.5 tons) to as high as 78 tonnes (86 tons), but most estimates now place the likely mass in the 20–40 tonne range (22–44 tons) because brachiosaurs were made much lighter by numerous air sacs throughout their bodies. This is not the limit for *Giraffatitan* because the mounted skeleton is not fully grown. There is another limb bone of an adult that is 13 percent larger, suggesting that their maximum adult dimensions were considerably larger.

For most of the twentieth century, brachiosaurs were considered the largest land animals that ever lived. But in the past 30 years, amazing discoveries from Argentina have produced even more enormous animals, the titanosaurs, which are discussed in chapter 10.

FOR FURTHER READING

Colbert, Edwin. *Men and Dinosaurs: The Search in the Field and in the Laboratory.* New York: Dutton, 1968.

Curry Rogers, Kristina, and Jeffrey Wilson. *The Sauropods: Evolution and Biology.* Berkeley: University of California Press, 2005.

Farlow, James, and M. K. Brett-Surman. *The Complete Dinosaur.* Bloomington: Indiana University Press, 1999.

Fastovsky, David, and David Weishampel. *Dinosaurs: A Concise Natural History*, 3rd ed. Cambridge: Cambridge University Press, 2016.

Hallett, Mark, and Mathew J. Wedel. *The Sauropod Dinosaurs: Life in the Age of Giants.* Baltimore, Md.: Johns Hopkins University Press, 2016.

Holtz, Thomas R., Jr. *Dinosaurs: The Most Complete, Up-to-Date Encyclopedia for Dinosaur Lovers of All Ages.* New York: Random House, 2011.

Klein, Nicole. *Biology of the Sauropod Dinosaurs: Understanding the Life of Giants.* Bloomington: Indiana University Press, 2011.

Maier, Gerhard. *African Dinosaurs Unearthed: The Tendaguru Expeditions.* Bloomington: Indiana University Press, 2003.

Naish, Darren. *The Great Dinosaur Discoveries.* Berkeley: University of California Press, 2009.

Naish, Darren, and Paul M. Barrett. *Dinosaurs: How They Lived and Evolved.* Washington, D.C.: Smithsonian Books, 2016.

Remes, Kristian, Carole T. Gee, and P. Martin Sander. *Biology of the Sauropod Dinosaurs: Understanding the Life of Giants.* Bloomington: Indiana University Press, 2011.

Spaulding, David A. E. *Dinosaur Hunters: Eccentric Amateurs and Obsessed Professionals.* Rocklin, Calif.: Prima, 1993.

Upchurch, Paul, Paul M. Barrett, and Peter Dodson. "Sauropoda." In *The Dinosauria*, 2nd ed., ed. David B. Weishampel, Peter Dodson, and Halszka Osmólska, 259–322. Berkeley: University of California Press, 2004.

PATAGOTITAN

The tragedy of the biggest sauropods is that they're so scrappy. *Argentinosaurus*—often cited as being about 100 feet long and in the range of 80 tons—is only known from a relatively paltry collection of vertebrae, ribs, and an incomplete femur. *Bruhathkayosaurus*, a dinosaur that may have been as big or even bigger than *Argentinosaurus*, was only known from limb, hip, and tail elements, and those fossils disappeared (much like the near-mythical dinosaur giant *Amphicoelias*, estimated to be 190 feet long from a long-lost piece of vertebra).

—BRIAN SWITEK, 2014

THE BIGGEST DINOSAUR?

In 2014, pictures began to appear in the newspapers and online of these enormous dinosaur limb bones, with a man lying on the ground next to it for scale (figure 10.1A). The articles hinted that paleontologists in Argentina had found the largest dinosaur ever to live, but most scientists greeted the news with caution. Again and again the media hyped new discoveries as "the biggest" or "the longest" to generate attention for their stories, but when the detailed work was finally done years later, the claim turned out to be false or exaggerated.

However, this discovery was truly exceptional and not just media hype. David Attenborough did a BBC special in 2015 that showed its enormous size and had many shots of the enormous bones, including some with Attenborough in the field at the locality in Argentina. By early 2016, the American Museum in Natural History in New York exhibited a huge replica

Figure 10.1 ▲

The titanosaur *Patagotitan*: (*A*) a limb bone with a human for scale as it was unearthed in Patagonia; (*B*) replica of *Patagotitan* in the American Museum of Natural History as it stands today; (*C*) the replica nicknamed "Maximo" in Stanley Field Hall of the Field Museum in Chicago, first installed in 2018. ([*A*] Courtesy of J. Carballido; [*B*] courtesy of M. Wedel; [*C*] courtesy of B. McGann)

of the Argentinian giant in the Miriam and Ira D. Wallach Orientation Center, the entrance lobby to their fourth-floor halls displaying fossil vertebrates (figure 10.1B). At 122 feet, it was so long that it didn't completely fit inside that enormous room. Its head stuck out through the door frame into the elevator lobby adjacent to the room to the east. It hadn't even been formally named or described yet; it was just called "The Titanosaur." In the spring of 2018, the Field Museum of Natural History in Chicago followed suit, moving their famous *Tyrannosaurus rex* specimen "Sue" to another hall and replacing it with another cast of the Argentinian giant (figure 10.1C). Even though Sue was impressive, the museum staff explained that it looked tiny in the giant central Stanley Field Hall of the Field Museum (which had once housed their replica of *Brachiosaurus/Giraffatitan*; see chapter 9). The giant titanosaur was not dwarfed by this enormous space.

Finally, on August 9, 2017, the long-ballyhooed dinosaur was formally described in the *Proceedings of the Royal Society of London B: Biological Sciences*, a very prestigious journal (and free online). José Carballido, Diego Pol of the American Museum, and nine other Argentinian coauthors named it *Patagotitan mayorum*. The genus name comes from Patagonia, where it was found, plus "titan," and the species honors the Mayo family, on whose ranch it was found. The first bones were found in 2008 by one of the ranch shepherds, Aurelio Hernandez, on the La Flecha ranch, which is in the Patagonian desert about 250 kilometers (150 miles) west of Trelew. The Mayo family put out a call to the paleontologists at the Museum of Paleontology Egidio Feruglio, who sent out seven expeditions from January 2013 to February 2015. Over those two years, the crews found more than 200 bones of sauropods, mostly from six partial sauropod skeletons composed of about 130 of those bones. There were also 57 isolated theropod teeth. Three or more of the partial skeletons make *Patagotitan* one of the most complete sauropods known, with a few neck vertebrae, a nearly complete backbone, many parts of the tail, nearly all the forelimbs and hind limbs, front shoulder girdle and parts of the hips, plus numerous ribs. Like most sauropods, however, there was no skull attached to the neck, so it was another "headless wonder," and the replica mounts have borrowed a skull from a different titanosaur.

The individuals represented by the three most complete partial skeletons did not all die at the same time, and they come from at least three different layers in the same quarry. The way they are disarticulated, with some bones missing, plus the abundance of broken and shed predator teeth suggests that the carcasses came to rest on a river bottom or sand bar and were scavenged by predators and rearranged by currents before being buried by the next flooding event that covered them with a layer of sand and mud.

MEASURING GIANTS

So how big was *Patagotitan*? In the original press releases, Carballido and colleagues gave a length estimate of 40 meters (131 feet) and a weight estimate 77 metric tonnes (85 tons), but by the time the specimen was finally published, those dimensions had shrunk to 37 meters (122 feet) in length and a weight of 69 metric tonnes (76 tons). Other authors estimated that it came in at 33.5 meters (110 feet) and weighed about 45.4 metric tonnes (50 tons).

Another huge titanosaur, *Argentinosaurus* (figure 10.2), had long been considered the largest known land animal, but it was known from much less complete material: a few enormous vertebrae of the back and hip (figure 10.2B), and some of the hind limb bones. The reconstructed skeleton (figure 10.2A) in the Museo Carmen Funes in Plaza Huincul,

Figure 10.2 ▲ ▶

(*A*) Reconstructed skeleton of *Argentinosaurus* in the Museo Carmen Funes in Plaza Huincul, Argentina. For a long time, it was the largest relatively complete dinosaur known. (*B*) A single titanosaur vertebra showing their immense size. (*C*) Compare this to a vertebra of *Giraffatitan*, which was the largest dinosaur known for a long time. ([*A*] Courtesy of F. Novas; [*B*] photograph by the author; [*C*] courtesy of M. Wedel)

Figure 10.2 ▲
(*continued*)

Argentina, is 40 meters (130 feet) long. However, it is so incomplete that estimates of its size range from 26 meters (85 feet) to 30 meters (98 feet) to 30–35 meters (98–115 feet) in length, and weight estimates from different scientists range from 60–88 metric tonnes (66–97 tons), 80–100 metric tonnes (88–110 tons), and 83.2 metric tonnes (91.7 tons). These size estimates might put *Argentinosaurus* back at Number One if we use the smaller numbers for *Patagotitan*.

Earlier in 2014, before *Patagotitan*, the news was filled with the announcement of another giant Argentinian titanosaur. It was named *Dreadnoughtus* after the giant World War I battleships, or "dreadnoughts," that feared no other smaller ship, or "dreaded nothing." The media naturally touted this animal as the largest ever found, claiming that it passed *Argentinosaurus*. It was at that time the most complete of the giant titanosaurs, with 45–70 percent of the bones preserved (depending on which index of completeness is used). Some of the forelimbs are preserved, part of the back, much of the hind legs and pelvis, and part of the tail—but no neck or head. Ken Lacovara and colleagues, who named and described it, estimated that it was about 26 meters (85 feet) in length, and its weight was estimated around 22–38 metric tonnes (24–42 tons).

Why are these numbers so widely divergent? Partial skeletons, especially with material that consists only of a few vertebrae, often have no bones in common that would allow direct comparison. The length estimate is highly dependent on how large and how many vertebrae were in the neck and tail, and none of the specimens have a complete neck or tail yet (most don't have a skull either). As science writer Brian Switek said in 2014 when the first news of *Patagotitan* appeared in the media:

> A great deal of a sauropod's length was in the tail, and how long that tail really was hinges upon how many vertebrae were in that part of the spine. The trouble is that complete dinosaur tails are very rare in the fossil record, and some of those precious fossils even suggest that the number of vertebrae in a dinosaur species' tail could slightly vary from one individual to the next. When you're dealing with dinosaurs that had vertebrae measured in feet, not inches, the number of tail vertebrae paleontologists reconstruct can make a big difference for a size estimate. And given that the new bonebed contains only 150 bones between seven individuals, the length of the sauropod's tail—and other parts—is going to have to rely on what we know about other species.

There is an even bigger problem with estimating weight. All we have are skeletons, so we really don't know how fat or skinny the soft fleshy parts of the living animal were. Even with a complete skeleton, we can only get a rough mass estimate, usually by taking a key limb bone dimension and comparing it to the same bones of animals of known weights. For *Patagotitan*, Carballido and colleagues used the circumference of the humerus (upper arm bone) and femur (thighbone) and got a weight estimate of 69 metric tonnes, with a range from 59 to 86 metric tonnes, whereas the fragmentary femur of *Argentinosaurus* estimates a weight of 66–97 metric tonnes (and that femur is partially reconstructed in plaster, so the measurement is not entirely based on real fossil bone).

The second technique involves constructing a three-dimensional model of the living animal and calculating its volume and thus its mass. In the old days, this meant sculpting an actual physical model, but more recently virtual models have been developed that can quickly estimate volume and mass. Carballido and others got a smaller range of weight estimates for *Patagotitan* this way, ranging from 45–77 metric tonnes. As Brian Switek put it in 2014:

> Weight is another matter altogether. Determining a dinosaur's body mass not only relies on filling in missing bones based upon close relatives, but also a particular researcher's perception of whether the dinosaur in question had a beefy build or was leaner. That's why paleontologists are familiar with shrinking sauropods.

In addition to difficulty estimating weights are problems estimating density. Like birds and apparently many dinosaurs, sauropods had many weight-reducing air sacs in the body (especially along their spine, as the bones suggest). In this case, sauropods might not have been as heavy as would be suggested by taking a simple model of uniform density and calculating its weight. Charlotte Brassey of Manchester Metropolitan University points this out:

> They were so unusual, with long, slender necks and tails, and an air-filled skeleton, that we have no convincing analogues in the modern animal kingdom. And by virtue of being extremely large, we need to extrapolate our understanding of how animals function far beyond the upper limits of living land animals, such as elephants. The more we need to extrapolate, the less confidence we have in our reconstructions.

These are just the most recently discovered and relatively complete species. There are many more just known from a few huge bones. For example, among Argentine titanosaurs, the thighbones of *Antarctosaurus* are the only bones known (figure 10.3), but they are larger than those of *Argentinosaurus* and about the same size as those of *Patagotitan*. Another Argentine

Figure 10.3 ▲

Thighbones of *Antarctosaurus* from the Cretaceous of Argentina, the largest such bones ever found until recently—no other part of this dinosaur is known. Francisco Novas for scale. (Courtesy of F. Novas)

titanosaur, *Argyrosaurus*, is in the same ballpark. As discussed in chapter 7 for *Apatosaurus* and *Brontosaurus*, all these huge titanosaurs may be part of a highly variable population rather than being distinct genera and species. Those from the same time interval in Argentina should not be given new names until this is ruled out.

Another problem with these huge but incomplete specimens is that many of them have been lost. For example, *Bruhathkayosaurus* (South Indian Sanskrit for "huge heavy body" plus the Greek for "lizard") was found in Upper Cretaceous beds in India and described in 1989. When it was first described, its size was estimated at 175–220 metric tonnes (190–240 tons), then later shrunk to 139 metric tonnes (153 tons). It consisted of some hip bones, a partial thighbone and shin bone, a forearm, and a few pieces of vertebrae. However, the shin bone was 2 meters (6.6 feet) long, 30 percent larger than the shin of *Argentinosaurus*, and the thighbone was also larger. So was it really bigger than any of the Argentine fossils? It's hard to know now because all we have are the simple published line drawings of the bones and a few measurements. The fossils themselves were completely destroyed when flash floods inundated the basement storage area and washed them away.

And then there is the mystery of *Amphicoelias fragillimus*. This name was based on a single enormous vertebra found by Cope and named in 1877. His one figure of the specimen shows that it was the incomplete portion of a vertebra (the neural arch) of enormous dimensions, possibly 2.7 meters (8.8 feet) tall if Cope's measurements are accurate! If you compare it to more completely known dinosaurs and estimate its size, you come up with lengths around 40–60 meters (130–200 feet) and weights around 112 metric tonnes (135 tons). That would have dwarfed any of the current record holders.

However, the specimen itself is now lost. According to Ken Carpenter, it was probably stored in Cope's house with no preservatives or hardeners (rarely used in those days) and deteriorated so badly that it crumbled to dust before Osborn bought Cope's collection for the American Museum, or after it reached the museum. Lots of people are skeptical of Cope's measurements because it is difficult to know what landmarks he used when measuring. There may be a typographic error in printing the size or scale (I've found a few of those in older scientific papers, so it's not rare). In 2015, Woodruff and Foster made a strong argument that not only was the name *Amphicoelias fragillimus* invalid but doubt whether we can really say that it was as large as it was once claimed. Cope named another species, *Amphicoelias altus*, in 1878, which is about the same size and shape as *Diplodocus*, and probably just another junior synonym of that dinosaur.

LANDS OF THE TITANS

Throughout this chapter, we have been talking about titanosaurs. This is not just a nickname but an actual group of dinosaurs with their own distinctive characteristics and history. They are distinct from the diplodocines, the camarasaurs, and the brachiosaurs discussed in the three previous chapters. Most of these other groups of dinosaurs flourished in the Jurassic and then vanished, but the titanosaurs were far more diverse and largely flourished throughout the Cretaceous all over Gondwana and Laurasia (figure 10.4).

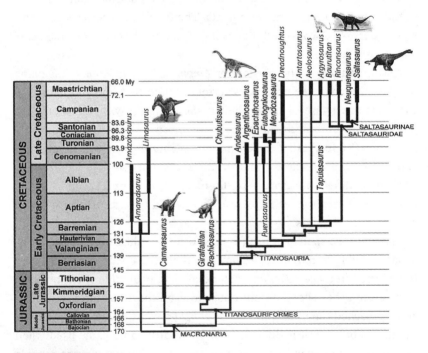

Figure 10.4 ▲

Family tree of sauropods, including titanosaurs, brachiosaurs, diplodocines, camarasaurs, and others mentioned in this book. (Redrawn from several sources)

Brachiosaurs are distinctive because of their long forelimbs and high shoulders and long necks adapted for feeding higher than any other animal. Diplodocoids tended to have slender bodies with long necks and extremely long tails. By contrast, some titanosaurs had relatively shorter necks and tails, and relatively small heads with large nostrils and crests formed by their nasal bones. Most of them had small teeth shaped like little spatulas (broad at the tip, narrow at the root), although a number had teeth shaped like pencils (similar to the teeth of diplodocoids). Although it was dominated by titanosaurs, South America also had diplodocoids, including the odd-looking *Amargasaurus,* which had paired spines down the back of its neck and spine (figure 10.5A).

Titanosaurs also had very broad shoulders and hips with a wider stance than other sauropods, and this can be recognized from their trackways. They tended to have stocky forelimbs, sometimes longer than their hind limbs (although not as disproportionate as in brachiosaurs). Like all other sauropods, they walked on the tips of very long hand and foot bones (metacarpals and metatarsals). Most sauropods have a few stumpy remnants of finger and toe bones as well, and maybe a claw on their thumb, but many titanosaurs lost the bony parts of their fingers and toes completely, replacing it with cartilage. Apparently, they walked on the blunt "stumps" of their metacarpals and metatarsals (covered by pads of cartilage and keratin, as most animals have covering the bones of their fingers and toes). One of them (*Saltasaurus*) had bony plates on its back shaped like large dishes (figure 10.5B), and many of the titanosaurs that are well enough preserved show small dish-like pieces of armor (osteoderms) in their skins that made it harder for a predator to bite into them.

The most diagnostic feature of many titanosaurs, however, was how the centra of the tail vertebrae are convex on the rear surfaces. More advanced titanosaurs have a distinctive peg-and-socket joint in their vertebrae, so even a single vertebra can be identified as titanosaur.

The name *Titanosaurus* was first coined by British paleontologist Richard Lydekker in 1877, who described a gigantic isolated tail vertebra from India. For the next century, the genus *Titanosaurus* became a taxonomic "wastebasket" for large sauropods of unknown relationships. Today the genus *Titanosaurus* is considered invalid because the original specimens are not diagnostic enough to tell how it is related to or distinct from more complete fossils that have been discovered more recently. But

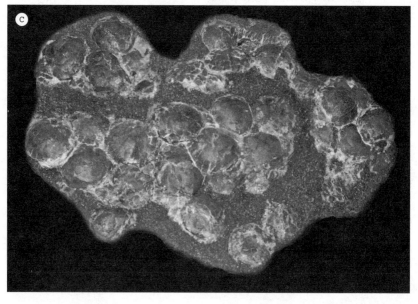

the name of the group Titanosauria is still valid, even if the genus it was based on is no longer in use.

Not all titanosaurs were as huge as *Patagotitan, Argentinosaurus, Antarctosaurus, Argyrosaurus*, and *Dreadnoughtus*. The armored *Saltasaurus* (figure 10.5B) is known from almost complete skeletons and was relatively small for a sauropod: only 8.5–13 meters (28–42 feet) long and about 2.5–7 metric tonnes (00.0–7.6 tons) in weight. It had a barrel-shaped body, shorter limbs, and a relatively short neck compared to most sauropods. *Saltasaurus* is known not only from good juvenile and adult bones but also from nesting grounds full of eggs at Auca Maheuvo in Argentina (figure 10.5C). Each nest had about 25 eggs, roughly 12 centimeters (5 inches) in diameter. These eggs were buried under dirt and vegetation to hide them from predators and to keep them warm and protected. Some of the eggs even had the tiny bones of embryos inside them as well as preserved egg membranes. The embryos had skin impressions, showing that they had a mosaic of tiny armor plates on their bodies even before they hatched.

Titanosaurs were by far the most diverse and widespread group of sauropods, spreading to all the continents when they began their evolutionary radiation in the Early Cretaceous. They clearly replaced diplodocines and brachiosaurs, and they may have competed with and displaced those groups ecologically as well. Once they began evolving, titanosaurs reached a diversity of more than 100 genera known from the Cretaceous (although many of these names are dubious because they are based on fragmentary nondiagnostic fossils).

Titanosaurs were also unusual in that they lived nearly everywhere in the Cretaceous and have been found on every continent that has rocks of the right age and environment. The most primitive titanosaur of all is *Andesaurus*, named in 1991 for specimens found in lower to middle Cretaceous rocks in the Argentinian foothills of the Andes. Although the fossils are incomplete (as always), *Andesaurus* wasn't small just because it was early

Argentine sauropods: (A) the spiny diplodocine *Amargasaurus*, with paired spikes down the back of its neck and spine; (B) the armored *Saltasaurus*, a smaller sauropod from the Cretaceous, with numerous armor plates in its skin; (C) a nest of eggs of *Saltasaurus* from Auca Maheuvo, Argentina. ([A] Courtesy of Wikimedia Commons; [B] courtesy of N. Tamura; [C] courtesy of L. Chiappe)

and primitive. It was about 15–18 meters (49–59 feet) in length. *Andesaurus* was found in beds that also contained the one of the largest predators of all time, *Giganotosaurus*, so they had some fierce rivals to contend with (see chapter 15).

The next most advanced group of titanosaurs is the Lognkosauria, which include most of the giants described in this chapter: *Patagotitan, Argentinosaurus*, plus the enormous *Futalognkosaurus* (only slightly smaller than these other giants and known from more complete skeletons). Most Lognkosauria are known from South America (the previous genera plus *Mendozasaurus, Puertasaurus, Traukutitan, Quetecsaurus*, and *Drusilasaura*), and there are some African fossils. Keep in mind that Africa was attached to South America in the Early Cretaceous, and the South Atlantic didn't really begin to open until the Late Cretaceous, so dinosaurs from Argentina were not that far as the crow flies from dinosaurs in southern Africa.

There are primitive titanosaurs on many continents, such as the numerous genera from Asia, including China (*Jiangshanosaurus, Jiutaisaurus, Qingxiusaurus, Ruyangosaurus, Zuchengsaurus*), Pakistan (*Pakisaurus, Balochisaurus, Sulaimanisaurus, Marisaurus, Khetranisaurus*), India (*Isisaurus* and the original *Titanosaurus*), and *Gobititan* from Mongolia. Other regions are represented by *Savannasaurus* from Australia, *Paralititan* from Egypt, *Rukwatitan* from Malawi, *Lohuecotitan* from Spain, *Macrurosaurus* from England, *Hypselosaurus* from southern France, and *Paludititan* and *Magyarosaurus* from Romania.

Magyarosaurus is particularly interesting because it is a dwarfed titanosaur, only 6 meters (20 feet) long. It is thought that this dwarfing occurred because most of western Europe was flooded with high sea levels, forming a series of small islands. On islands, it is common for larger animals to become dwarfed because they no longer need large body size to cope with big predators, and large body size is a disadvantage when animals need to survive on the much smaller food supply of most islands. This is well documented from pygmy mammoths during the Ice Ages on many islands, from the Channel Islands of California to Crete and Cyprus in the Mediterranean, as well as dwarfed hippos on Madagascar, Cyprus, Crete, and many other places. Apparently this was a phenomenon in larger dinosaurs as well.

The most advanced titanosaurs are called the Eutitanosauria and include the rest of the huge diversity of this group. These include such groups as the small armored saltasaurs mentioned previously, and the aeolosaurs (wind reptiles) of South America, plus the Argyrosauridae of Egypt. Finally, there is the large group of titanosaurs known as the Lithostrotia. Lithostrotians include many South American groups such as *Saltasaurus, Neuquensaurus,* as well as *Rapetosaurus* from Madagascar (a small sauropod but known from one of the most complete skeletons in the group) and its relative *Isisaurus* from India. Other lithostrotian titanosaurs include *Malawisaurus* from Africa, *Nemegtosaurus* from Mongolia, *Diamantinasaurus* from Australia, and *Tapuiasaurus* from Brazil. Clearly the many branches of titanosaurs could get around very easily because they appear over and over again in just about every corner of the world. In December 2011, titanosaur remains were found in Antarctica, and a few years earlier in New Zealand.

But what about North America? You might have noticed its conspicuous absence in our list of cosmopolitan titanosaurs. North America is different from the other regions in that the best-known Late Cretaceous faunas from Montana and Wyoming and Alberta are dominated by duckbills and horned dinosaurs, and lack sauropods. One might get the impression that sauropods vanished from North America in the Cretaceous once the brachiosaurs and camarasaurs and diplodocines died out in the Late Jurassic or Early Cretaceous. But that impression is wrong: they just stuck to the southern parts of North America in the Cretaceous, and for unknown reasons never moved to northern Wyoming and Montana and Alberta.

Only one titanosaur is known: the huge *Alamosaurus,* a close relative of the Argentinian *Saltasaurus.* It got its name after the Uppermost Cretaceous Ojo Alamo Sandstone in the San Juan Basin of northwestern New Mexico. (No truth to the rumor that it is named after the Alamo mission in San Antonio, site of the famous battle for Texan independence.) Although it is known only from fragmentary remains, the available parts suggest that it was almost as large as the South American giants: 28–30 meters (92–98 feet) long, and about 73 metric tonnes (80 tons). It was first discovered in 1921 in New Mexico by Charles W. Gilmore of the Smithsonian. Since then, bones of *Alamosaurus* have been found in the North Horn Formation of Utah, several formations in the Big Bend region of Texas, and even a few scraps in southwestern Wyoming—but none have been found in the Lance Formation of northeastern Wyoming, the Hell Creek Formation of central Montana, or the many Cretaceous formations of Alberta.

Why was this? Paleontologists speculate that there may have been some sort of climatic or geographic boundary between northern Wyoming and southern Wyoming, although nothing is obvious—ankylosaurs, duck-billed dinosaurs like *Kritosaurus*, horned dinosaurs like *Torosaurus*, and large theropods seem to have ranged from New Mexico to Montana to Alberta without a problem.

Others have argued that *Alamosaurus* is a late immigrant to North America and never spread all the way north. After all, it is a member of the most advanced titanosaurs, the Saltasauridae, from the latest Cretaceous of Argentina, and it only shows up in the latest Cretaceous of North America as well. Throughout most of the 80 million years of the Cretaceous, there are no titanosaurs or any other kinds of sauropods anywhere in North America.

Finally, *Alamosaurus* has figured in the debates about the extinction of the dinosaurs as well. Specimens in New Mexico and Texas have been found just a few meters below the boundary that marks the end of the Cretaceous and the mass extinction that occurred then. There are even some specimens in Texas from a unit that spans the Cretaceous-Cenozoic boundary, although there is no clear evidence of *Alamosaurus* living after the Cretaceous extinctions that wiped out all the other nonbird dinosaurs. For a while, however, debates about where to place that boundary in the rock sequence of New Mexico had scientists claiming that there were Cenozoic dinosaurs including *Alamosaurus*. More careful work since then has shown no clear evidence that *Alamosaurus* survived the catastrophe.

The titanosaurs were not only the largest radiation of sauropods and the most cosmopolitan of all sauropod groups but also the largest land animals the earth has ever seen—and among the very last nonbird dinosaurs on earth. With this, we leave the sauropods behind and look at the next group: the predatory dinosaurs, or theropods.

FOR FURTHER READING

Colbert, Edwin. *Men and Dinosaurs: The Search in the Field and in the Laboratory*. New York: Dutton, 1968.

Curry Rogers, Kristina, and Jeffrey Wilson. *The Sauropods: Evolution and Biology*. Berkeley: University of California Press, 2005.

Farlow, James, M. K. Brett-Surman, and Robert Walters. *The Complete Dinosaur*. Bloomington: Indiana University Press, 1999.

Fastovsky, David, and David Weishampel. *Dinosaurs: A Concise Natural History*, 3rd ed. Cambridge: Cambridge University Press, 2016.

Hallett, Mark, and Mathew J. Wedel. *The Sauropod Dinosaurs: Life in the Age of Giants.* Baltimore, Md.: Johns Hopkins University Press, 2016.

Holtz, Thomas R., Jr. *Dinosaurs: The Most Complete, Up-to-Date Encyclopedia for Dinosaur Lovers of All Ages.* New York: Random House, 2011.

Klein, Nicole. *Biology of the Sauropod Dinosaurs: Understanding the Life of Giants.* Bloomington: Indiana University Press, 2011.

Naish, Darren. *The Great Dinosaur Discoveries.* Berkeley: University of California Press, 2009.

Naish, Darren, and Paul M. Barrett. *Dinosaurs: How They Lived and Evolved.* Washington, D.C.: Smithsonian Books, 2016.

Novas, Fernando E. *The Age of Dinosaurs in South America.* Bloomington: Indiana University Press, 2009.

Remes, Kristian, Carole T. Gee, and P. Martin Sander. *Biology of the Sauropod Dinosaurs: Understanding the Life of Giants.* Bloomington: Indiana University Press, 2011.

Upchurch, Paul, Paul M. Barrett, and Peter Dodson. "Sauropoda." In *The Dinosauria,* 2nd ed., ed. David B. Weishampel, Peter Dodson, and Halszka Osmólska, 259–322. Berkeley: University of California Press, 2004.

Wilford, John Noble. *The Riddle of the Dinosaur.* New York: Knopf, 1985.

PART III

RED IN TOOTH AND CLAW
THE THEROPODS

Tho' Nature, red in tooth and claw
With ravine, shriek'd against his creed.
—ALFRED LORD TENNYSON, *IN MEMORIAM A.H.H.*, 1850

COELOPHYSIS

The discovery of *Coelophysis* at Ghost Ranch is a good example of the role of serendipity in paleontological research. According to the Oxford English Dictionary serendipity is a word "coined by Horace Walpole upon the title of a fairy-tale *The Three Princes of Serendip*, the heroes of which were always making discoveries, by accidents and sagacity, of things they were not in quest of." Exactly so, at Ghost Ranch.

—EDWIN H. COLBERT, *THE LITTLE DINOSAURS OF GHOST RANCH*, 1995

SERENDIPITY!

In the summer of 1947, paleontologist Edwin Harris "Ned" Colbert (figure 11.1) was planning a field season at the Petrified Forest of Arizona to look for Triassic fossils. He had spent summers during the war years teaching paleontology at the University of California at Berkeley. While he was there he frequently looked through their collections and found many intriguing fossils that had been collected by Berkeley field trips over the years. Once the war ended in 1945 and wartime restrictions over tires and gasoline had ended, he and other paleontologists were eager to get back in the field and collect new fossils. Colbert had gone through the arduous process of getting a permit from the National Park Service to collect in the Petrified Forest National Monument, which took months of getting signatures in the federal bureaucracy—all the way to the top, with the signature of the Secretary of the Interior himself.

Ned Colbert was one of the more important figures in the history of paleontology in the twentieth century. Born in Clarinda, Iowa, in 1905, he grew up in Maryville, Missouri. As a young student at the University of Nebraska, Lincoln, he worked for Edwin H. Barbour of the Nebraska State Museum. There he got his first taste of working in museums, collecting fossils in the field, and preparing them for display. This led him to get his Ph.D. at the American Museum of Natural History and Columbia University in New York, where he was trained by Osborn's colleagues and protégés, including the great anatomist and paleontologist William King Gregory. His thesis work was not on dinosaurs at all but on Miocene mammals from the Siwalik Hills of Pakistan, which had been collected by the legendary Barnum Brown (see chapter 21) for the American Museum back in the 1920s but never studied or published. As he was finishing his thesis, Barnum Brown retired, and Colbert was named to replace him and study the fossil reptiles in the museum (mostly collected by Brown). Meanwhile, he married Margaret Matthew, the daughter of legendary American Museum paleontologist William Diller Matthew.

A modest, soft-spoken man in a field dominated by outsized personalities like Cope, Marsh, and Osborn, he was a curator at the American

Museum for the next 40 years. There he worked on a wide variety of fossil reptiles and amphibians, wrote many popular books on dinosaurs, and helped design the dinosaur halls that were the highlight of the American Museum until they were updated in the early 1990s. In 1969, he capped his career with the discovery of fossils of the dog-sized protomammal *Lystrosaurus* in Antarctica. This creature was already known from Africa, South America, and India. His discovery confirmed that Antarctica was part of Pangea during the Permian and Triassic (as the new field of plate tectonics had predicted).

Colbert retired from the American Museum in 1969, and then moved to Flagstaff, Arizona, where he was an honorary curator at the Museum of Northern Arizona. He lived to the ripe old age of 96, passing away in 2001. His grandson, Matthew Colbert, is also a paleontologist. He is named after two distinguished paleontologists: his great-grandfather William Diller Matthew and, of course, his grandfather Ned Colbert.

(On a personal note, when I came to the American Museum in 1976 for my own graduate education, Ned Colbert often came back to visit. I met him several times and had nice chats with him about the legendary paleontologists he had known: Osborn, Matthew, Brown, Gregory, Barbour, and so many more. He was once Osborn's assistant as well, and Osborn met and shook hands with Cope, Marsh, Darwin, Huxley, Haeckel, and the great naturalists of the late 1800s, so I'm only two degrees of separation from them.)

As he reached the field in the summer of 1947, Colbert met with fellow American Museum scientist George Gaylord Simpson, then they took the train to Albuquerque and picked up Army-surplus jeeps to begin their field season in the Southwest. Simpson was working on early Cenozoic mammals of the San Juan Basin of northwestern New Mexico, and Colbert intended to drive west and begin his season in the Petrified Forest. However, before he left New Mexico he wanted to check out a locality just north of Abiquiu, which had yielded fossils he saw in the Berkeley collections. It was known as Ghost Ranch (figure 11.2).

Ghost Ranch today has a very popular conference center, with a spectacular view of the scarlet and tan and brick red and brown cliffs of the amphitheater. It was once the home of famous American artist Georgia O'Keefe, who used the spectacular cliffs in the background in several of her paintings. According to the Pack family, who ran the property for many years, Ghost Ranch got its nickname back in the days of the Spanish. One of the

Figure 11.2 ▲

Panoramic view of Ghost Ranch, New Mexico, showing the cliffs of Lower Jurassic dune deposits known as the Navajo Sandstone (first bench midway up the cliff), underlain by slopes of the brick-red floodplain shales of the Upper Triassic Chinle Formation, which yields early dinosaurs and many other fossils. Above the Navajo Sandstone are the Lower Cretaceous Dakota Sandstone and Mancos Shale. (Courtesy of S. Lucas)

sons of the original ranch family, the Archuletas, had killed his brother over a dispute, and his ghost still wanders in the lonely canyons. According to the legend, the "curse of Cain" seemed to have haunted the descendants of that family, and many others who have lived there since. Whatever the truth, the place was certainly not cursed for Ned Colbert. In fact, it gave him extraordinary luck instead.

Colbert and his crew drove up to Ghost Ranch, and the owners let them live in one of the comfortable spare ranch houses. His crew consisted of George Whitaker, a young preparator at the American Museum, and Tom Ierardi, a professor at City College in New York and an enthusiastic fossil collector who had worked with Colbert before. On June 19, 1947, they began prospecting across the badlands exposures of brick-red mudstones of the Upper Triassic Chinle Formation, finding bone scraps and a few better Triassic fossils, such as the crocodile-like archosaur relatives known as phytosaurs. On June 22, they were still prospecting when a very excited George Whitaker came running over the hill to get Colbert and Ierardi to come see what had found. As Colbert wrote:

> As he scrambled up the slope toward us it was evident that he was very excited. He had a right to be; in his hands he had some tiny bone fragments, including a well-preserved claw. They were the fossil remains of *Coelophysis*,

a small Triassic dinosaur—at least that was my identification of these fragments, and I was pretty confident I was correct. . . . We all scrambled down the talus slope where George had joined us, crossed an intermittent rivulet—a tributary of Arroyo del Yeso—and climbed up the opposite talus slope to the level at which George had found his bone fragments. There we found small pieces of fossil bone scattered in moderate abundance.

It seems that paleontologists often make important discoveries just at quitting time, often on the last day of the field season. George found these fragments just before lunch time. So we went back to camp, about a quarter of a mile away, for some nourishment and rest after a long morning, and particularly for some discussion of the discovery that he had made. Needless to say, we returned in short order to the discovery site. We soon found, almost completely covered by loose talus, the stratum of rock that contained the fossils. So we cleared away the soft sand of the slope and began to carefully dig back into the hill to uncover the fossil layer, using small awls and other hand tools. It was clearly not at this stage an operation calling for picks and shovels. The more we dug, the more fossils we found, so that before the afternoon was over we were beginning to realize that here was a most unusual concentration of dinosaur bones—not the huge bones of giant dinosaurs of song and story but rather the tiny bones of ancestral dinosaurs, dinosaurs that lived before the giants.

The excavation continued a few more days until it became clear that it was a huge deposit with many intact skeletons. Colbert realized that Whitaker had made an incredible discovery that would require all their resources, yet he still had the hard-won permit to go work in the Petrified Forest. What to do? He was worried about annoying all the people in the Department of Interior who had approved his permit, only to find out that he did not to use it—but the Ghost Ranch find was the opportunity of a lifetime. On June 30, 1947, Colbert got Simpson to come over from his San Juan Basin camp and inspect the new site, and get his opinion. Simpson was immediately impressed, and according to Colbert:

George most generously declared that "this was the greatest find ever made in the Triassic of North America." Perhaps his statement was hyperbolic—made in the heat of the moment when were all excited by the nature and scope of the discovery. But he did say that—for I made note of it in my fieldbooks at the time.

Simpson and Colbert then put their heads together and made a plan. They sent for another American Museum lab preparator, Carl Sorenson, to help work the big quarry. Colbert, Whitaker, Ierardi, and Sorenson then worked through the rest of the 1947 summer field season, and several of them returned for the next few years until they had all the fossils they could remove. Meanwhile, Colbert had to contact the people at the Park Service and Interior Department and apologize for not using their permit.

(On a personal note, I knew both George Whitaker and Carl Sorensen well when I was a student at the American Museum in 1976-1982. They were both about to retire, but they told me many tales of their exploits when I used to hang out in the lab with them. Whitaker, in particular, was a genuine hero. In 1956, he was with George Gaylord Simpson on an expedition up the Amazon. In camp one day, a crew member cut down a tree, which fell on Simpson and crushed him and broke many bones in his body. Whitaker did his best first aid with splints and bandages, then spent a week canoeing down the Amazon to rush him to the nearest town where he could be flown out to a real hospital. Without George Whitaker, the legendary paleontologist George Gaylord Simpson would have died in the Amazon.)

Sometimes serendipity strikes and you have to snatch the opportunities in front of you rather than doggedly stay with your previous plans. It happens all the time in science. The Bell Lab engineers who were trying to get the background noise out of their microwave antennas accidentally discovered the cosmic background radiation from the Big Bang. Walter Alvarez, doing analyses of rocks from the end of the Cretaceous in Gubbio, Italy, accidentally discovered the evidence for the asteroid impact. Alfred Nobel accidentally found how to put together two chemicals to make a more stable explosive, dynamite. The Raytheon engineers, stuck with excess magnetrons after World War II ended, discovered that they melted a chocolate bar in their lab coats—and the microwave oven was born.

The *Coelophysis* bone bed at Ghost Ranch turned out to be a truly unique discovery. Several thousand bones were found, making up many dozens of articulated skeletons (figure 11.3A). All of these were found before the American Museum had reached its limit of time and money to excavate the bone bed. The quarry floor eventually measured 9-12 meters (30-40 feet) wide and was dug 5 meters (15 feet) into the hillside. The crew had the ranch foreman Herman Hall use a bulldozer to dig a road to the quarry, then scrape away the overburden. They left about 30 centimeters (one foot) of

Figure 11.3 ▲

The *Coelophysis* bone bed at Ghost Ranch: (*A*) an articulated *Coelophysis* skeleton in a death pose, with the neck drawn back by the contraction of the nuchal ligament; (*B*) one of the fossil blocks being jacketed in burlap and plaster by Carl Sorensen; (*C*) the *Coelophysis* quarry as it looks today, just a scooped out bench in the Chinle slope in the center lower left part of the photo. ([*A*] Photograph by the author; [*B*] courtesy of the American Museum of Natural History Library; [*C*] courtesy of S. Lucas)

sediment above the bone layer intact so they could carefully dig down and expose the bones without risking damage from a bulldozer blade. The delicate specimens had to be doused with shellac and other hardeners as soon as they were exposed, then covered with rice paper, and finally placed in plaster bandages (figure 11.3B), making large blocks about 1.5 by 1.5 by 0.6 meters (5 feet by 5 feet by 2 feet) in dimension and covered in plaster jackets that would protect the fossils during transport. As the work proceeded and the heat increased that summer, they even built a temporary wooden roof over the quarry to protect the bones from the rain and the workers from the blistering sun. Once the blocks were completed, they were flipped with a chain hoist, and the workers waited with bated breath, hoping that they would not crumble as they were flipped. The newly exposed surface was covered with additional plaster jackets, then allowed to dry in the sun. Then Hall used his bulldozer to drag the blocks down the road he had dug, where they were loaded onto a flatbed truck for their trip back to New York. The driver of that truck said it was one of the heaviest loads he had ever transported.

"SMALL AND TENDER" DINOSAURS

How did Colbert recognize the scrappy specimens as *Coelophysis*? He had done his homework and studied the Triassic fossils already known to occur in the area. The dinosaur and its name actually go back to Cope and the earliest expeditions in New Mexico. Cope had been collecting in New Mexico for a year or two when Marsh hired local collector David Baldwin to collect from Cope's localities and send them to Marsh. But Cope soon got wind of this, promised to pay Baldwin better than Marsh did—and Baldwin was happy to jump ship because Marsh had shown no appreciation for those fossils. In 1881, Baldwin was working out of Abiquiu near what would become Ghost Ranch. Somewhere in the area, Baldwin found a bunch of small bones, which he sent to Cope with the following information on the label: "Label Sack 2 Box 1 Prof E.D. Cope Contains Triassic or Jurassic bones all small and tender. All in this sack found in same place about four hundred feet below gypsum strata 'Arroyo Seco' Rio Arriba Co New Mexico February 1881. no feet—no head—only one tooth. D. Baldwin—Abiquiu." Another collection was sent later in the same year, labeled thus: "Box 2. Contains sack 3. Part of fossil dug out Gallina Canyon. April 12–May 1. Three reptile teeth. Triassic or Jurassic. 400 feet below gypsum horizon. 180 feet above grey sandstone."

Cope discovered that Baldwin had sent him vertebrae, leg bones, pelvis bones, and rib fragments. He finally published a brief note on the fossils without illustrations in 1887. Based on the fragments, at first he called the specimens *Coelurus*, a small theropod from Marsh's Morrison localities. Instead of honoring the discoverer Baldwin, however, he named it *Coelurus bauri*, honoring one of Marsh's collectors, Georg Baur (who became a noted vertebrate zoologist later in his career). This was his subtle way of flattering Marsh's unhappy minions and possibly luring them away from their miserable employment, while needling Marsh at the same time. Later that same year, Cope redescribed the same fossils, referencing them to the bizarre giraffe-necked reptile *Tanystropheus* from the Triassic of Europe. At this point, he recognized not only the species *bauri* but another he called *longicollis* (long-necked), and a third species named after yet another of Marsh's assistants, Samuel Wendell Williston.

Finally, in 1889 Cope realized that the fossils didn't belong to either of these existing genera but was a new genus, which he dubbed *Coelophysis*. The name means "hollow shape" in Greek, in reference to the hollow ends of the vertebral centra. The fossils were then forgotten and ended up in the American Museum when Osborn purchased Cope's collection. Friedrich von Huene (see chapter 6) redescribed Cope's original fossils in 1915 and illustrated them as well, but still very little was known of the mysterious *Coelophysis*. In 1912, Williston and Ermine Cowles Case (later the founder of the University of Michigan Museum of Paleontology) relocated the site of Baldwin's fossils, but they could not find any new specimens. So the "small and tender" fossils were a mystery until 1947 when Colbert and Whitaker hit the bonanza. Some paleontologists think that the name *Coelophysis* is also invalid because the original Baldwin material is so poor and nondiagnostic. They have recommended that the Ghost Ranch fossils be named *Rioarribasaurus* because it is not clear to them that the Ghost Ranch fossils and Cope's original crummy specimens from Baldwin are the same animals. Most paleontologists were horrified of the idea of dumping the familiar name *Coelophysis* for the Ghost Ranch fossils. In the end, they petitioned the International Commission on Zoological Nomenclature to suppress the name *Rioarribasaurus* and officially designate the Ghost Ranch fossils as the type specimens of the genus *Coelophysis*, and the commission agreed.

Setting the dispute on names aside, the *Coelophysis* quarry samples were truly amazing in their quantity and state of preservation. Most of the specimens were complete and articulated individuals in death poses (figure 11.3A). Nearly all had their necks drawn backward in an arc over their backs. This is a common effect caused when the nuchal ligament (which pulls the neck and head up) starts to shrink and pull backward after the animal dies. Full-grown adults of *Coelophysis* were about 3 meters (9.8 feet) in length and varied in weights from 15–20 kilograms (33–44 pounds). They were lightly built with long running legs and a long tail with interlocking vertebrae that kept the tail rigid and sticking straight out behind the animal (figure 11.4).

Even though it looks superficially similar to the small bipedal fossils of *Eoraptor* and *Herrerasaurus* from the Late Triassic of Argentina (see chapter 5), *Coelophysis* is a much more advanced dinosaur. The shoulder girdle is fully theropod in its anatomy, and it is the earliest dinosaur known to have its collarbones fused into a furcula, or Y-shaped "wishbone." (At one time, some scientists argued that birds were not dinosaurs because dinosaurs seemed to have lost their collarbones, and therefore birds would not have a wishbone. It turns out that these collarbones are delicate and are rarely preserved. Many dinosaur skeletons with wishbones are now known.) Unlike more advanced theropods, *Coelophysis* still had four fingers on its hand (although it used only three, and the fourth was tiny and embedded in tissue). By contrast, most theropods had only three fingers,

Figure 11.4 ▲

Mounted skeletons of *Coelophysis* in the Denver Museum of Nature and Science. (Courtesy of Wikimedia Commons)

and some just two. The feet have only three toes, with a tiny vestigial fourth toe, and were configured like the classic theropod foot.

Coelophysis had a long narrow head that was lightly built with thin struts of bone. The eyes faced forward, so it clearly had good stereovision for running and catching prey. Combined with the bony ring in the eye to protect it (sclerotic ring), the large eyes suggest that *Coelophysis* was mostly a daytime predator. Further research showed that its vision was much better than that of most lizards, and more like that of an eagle or hawk.

Coelophysis had dozens of small, sharp, recurved teeth with serrated edges on the leading and trailing edge of the tooth, showing that it was a vicious predator that could rip open smaller prey. Together with its well-developed front limbs with a wide range of motion, it clearly could reach out and grab fast prey of many sizes.

One of the surprises of the Ghost Ranch was the presence of small bones of another animal in the gut cavity of *Coelophysis*. Originally it was thought to be a skeleton of a juvenile *Coelophysis*, suggesting the adults were cannibals like some crocodilians, Komodo dragons, and other reptiles. However, more recent study of these remains suggests that these bones were not from *Coelophysis* at all but from a small crocodile relative known from the same beds.

Dinosaurs that resemble *Coelophysis* are known from more locales than Ghost Ranch. A number of similar fossils of a delicate theropod were found in the Upper Triassic beds of Rhodesia (now Zimbabwe). Named *Syntarsus rhodesiensis* by Raath in 1969, that name had to be abandoned when someone found it had already been used for a beetle exactly a century earlier. That's against the rules of priority of the International Code of Zoological Nomenclature, so the name *Syntarsus* cannot be used for a dinosaur. Normally, one would expect the correction to be made by the original author once it is noticed. However, a group of scientists thought that Raath was dead and gave the African fossils the name *Megapnosaurus* instead. In 2012, paleontologist and blogger Mickey Mortimer wrote, "Paleontologists might have reacted more positively if the replacement name (*Megapnosaurus*) hadn't been facetious, translating to 'big dead lizard.'" In 2004, Raath reanalyzed the fossils and concluded that in fact the African fossils were the same as *Coelophysis*, so no new name (neither *Syntarsus* nor *Megapnosaurus*) was needed after all. Since then most scientists have concurred with Raath and regarded the African fossils as a closely related species, *Coelophysis rhodesiensis*.

LATE TRIASSIC LANDSCAPES AND LIFE

To some paleontologists, the large number of individuals from the same bone bed suggests that *Coelophysis* lived in large packs and probably hunted that way as well. Other scientists are not convinced of this and think that the large number of specimens just happened to get trapped in the same spot, possibly around a drying water hole. They clearly died slowly and had time to dry out so their necks were pulled back before they were buried, possibly in a flash flood. However, there is a locality in Zimbabwe where over 30 individuals of *Coelophysis rhodesiensis* are found together in similar circumstances. This suggests that their large numbers are not a coincidence but are indications of true pack behavior.

Whether they lived and hunted together or not, the large number of complete *Coelophysis* specimens at Ghost Ranch provides paleontologists with an unusual window into the growth and development of a population. By analyzing the dimensions and growth lines in the thighbone, paleontologists have discovered that *Coelophysis* babies grew very rapidly during their first year of life. The sample had four distinct size clusters that represented separate growth stages, with one-year-olds, two-year-olds, four-year-olds, and adults more than seven years old. In addition to the differences caused by growth, both juveniles and adults seemed to show two different distinctive forms, one with more robust forelimbs and the other with gracile forelimbs. This is thought to be due to sexual dimorphism, where the males and females differ in some parts of their body. This is also suggested by the fact that the robust and gracile forms are in a roughly 50:50 ratio, typical of many populations of animals with slight differences between males and females.

Coelophysis and other dinosaurs of the Upper Triassic Chinle Formation lived in a subtropical environment about 200 million years ago. Analysis of the sediments, especially in the Petrified Forest, suggests that the environment was a broad floodplain with distinct wet and dry seasons. In addition to the Ghost Ranch bone bed, other deposits suggest that many Chinle animals died and became fossilized when they were trying to survive as the waterholes vanished in the dry season. The vegetation was dominated by primitive conifers, cycads, and ferns, and flowering plants did not exist yet. The giant logs that gave the Petrified Forest its name come from primitive conifers related to the modern genus *Araucaria*, which today includes trees such as the Norfolk Island pine and the monkey-puzzle tree.

The lakes and rivers of that time harbored a wide variety of aquatic animals, including gigantic flat-bodied amphibians and the crocodile-like phytosaurs. Early crocodilians are known from these same beds, but they are small and delicate with long limbs for running on hard ground; only later, when phytosaurs vanished, did they take over the niche of large aquatic predators.

On the land, a variety of reptiles formed a complex community, including large predatory archosaur relatives of crocodiles known as erythrosuchids or rauisuchians (see chapter 5) and herbivorous armored archosaurs known as aetosaurs (see figure. 5.5). The largest herbivores, however, were cow-sized herbivorous animals known as dicynodonts. They are not "mammal-like reptiles" as they say in outdated books but protomammals distantly related to living mammals. In addition to *Coelophysis*, a number of dinosaurs lived in the Chinle forests, including the herrerasaur *Chindesaurus*, the coelophysid *Camposaurus*, the small theropod *Daemonosaurus*, and a dinosaur called *Gojirasaurus* ("Godzilla lizard," using the Japanese spelling of that movie monster).

The American Museum finished its work in the late 1940s, but several groups have been back to reopen the Whitaker Quarry. However, none have found the wealth of perfect skeletons that the American Museum first recovered. Colbert had so many specimens that he gave many away to other museums in exchange for specimens that the American Museum wanted. This was a generous gesture, but it hampers researchers who want to restudy the fossils. Instead of being all in the same museum, as most large collections from a single quarry are stored, they are spread across the globe. Researchers must travel to a lot of different museums at great time and expense rather than completing their research on specimens that are conveniently next to each other for direct comparison. A few were given to institutions that didn't take good care of their gift, and they have been damaged or destroyed.

The study of *Coelophysis* goes on and serendipity can strike at any time. There were so many specimens that many of the blocks remain unprepared and sitting in the American Museum basement. Colbert himself didn't get around to finishing a complete monograph on *Coelophysis* until 1989, almost 40 years after the first specimen was collected. In 2005, American Museum graduate student Sterling Nesbitt was poking around through these unopened field jackets, looking for some more *Coelophysis* to study.

He opened one cast that had sat in storage for almost 60 years and immediately realized that the specimen inside was not another *Coelophysis*. Instead, it was primitive relative of the crocodilians. However, it had a long neck and toothless beak, as well as a bipedal build with long hind legs, so its body form converged on that of the ornithomimid dinosaurs, 80 million years before that group had ever evolved. Nesbitt and his advisor, American Museum dinosaur paleontologist Mark Norell, wrote up and published a study of the fossil in 2006, which generated a lot of publicity in the media, including a mention on the Comedy Central TV show *The Colbert Report*. (There is no relation between Ned Colbert, the paleontologist, and Stephen Colbert, the comedian, so far as I know.) Nesbitt and Norell named the fossil *Effigia okeefae*. The genus name *Effigia* is from the Latin word for "ghost" (as in Ghost Ranch), and the species is named in honor of Georgia O'Keefe, who had spent so many years at Ghost Ranch painting its amazing landscapes.

You never know what you might find in a museum basement. The best fossil hunting isn't always in the field. Almost every museum is full of unstudied or poorly understood fossils that have just been sitting there for decades, waiting for a paleontologist who understands what he or she is looking at to come along. Some of them are still in their field jackets.

FOR FURTHER READING

Colbert, Edwin. *A Fossil Hunter's Notebook: My Life with Dinosaurs and Other Friends.* New York: Dutton, 1980.

——. *The Little Dinosaurs of Ghost Ranch.* New York: Columbia University Press, 1995.

——. *Men and Dinosaurs: The Search in the Field and in the Laboratory.* New York: Dutton, 1968.

Farlow, James, and M. K. Brett-Surman. *The Complete Dinosaur.* Bloomington: Indiana University Press, 1999.

Fastovsky, David, and David Weishampel. *Dinosaurs: A Concise Natural History*, 3rd ed. Cambridge: Cambridge University Press, 2016.

Holtz, Thomas R., Jr. *Dinosaurs: The Most Complete, Up-to-Date Encyclopedia for Dinosaur Lovers of All Ages.* New York: Random House, 2011.

Naish, Darren. *The Great Dinosaur Discoveries.* Berkeley: University of California Press, 2009.

Naish, Darren, and Paul M. Barrett. *Dinosaurs: How They Lived and Evolved.* Washington, D.C.: Smithsonian Books, 2016.

Wilford, John Noble. *The Riddle of the Dinosaur.* New York: Knopf, 1985.

CRYOLOPHOSAURUS

If Antarctica were music it would be Mozart. Art, and it would be Michelangelo. Literature, and it would be Shakespeare. And yet it is something even greater; the only place on earth that is still as it should be. May we never tame it.

—ANDREW DENTON

Antarctica has this mythic weight. It resides in the collective unconscious of so many people, and it makes this huge impact, just like outer space. It's like going to the moon.

—JOHN KRAKAUER

TO THE BOTTOM OF THE WORLD

In December 1990, my good friend paleontologist Bill Hammer was riding in a helicopter headed to a series of mountain peaks sticking above the glaciers in Antarctica. He was already a veteran of polar exploration, having spent time near the South Pole with his graduate advisor John Cosgriff, who found primitive amphibians in Triassic rocks of Antarctica in 1977–1978. Cosgriff, in turn, had collaborated with Ned Colbert (see chapter 11), who had begun the process of hunting vertebrate fossils in Antarctica in 1969 with a discovery that helped confirm that Antarctica was once a part of Pangea in the Triassic. (I know Bill well as he spent his whole career teaching at tiny Augustana College in Rock Island, Illinois, whereas I spent the first three years of my career, 1982–1985, teaching at even smaller Knox College in Galesburg, Illinois, just an hour south of him. I often went up to visit him at the amazing Fryxell Geology Museum.)

Exploring for fossils in Antarctica is not for the faint of heart, nor for those who are afraid of hard work or frightening conditions. Enormous preparation is required, as well as several years of writing grants to the National Science Foundation (NSF) to propose a project that the committees of scientists at NSF will deem worthy of funding. Everything in Antarctica costs a lot more due to the logistics and equipment needed to get people out there safely and back without killing anyone. Once you have been approved for funding in competition with dozens of other scientists, you give up your Christmas holidays to be there in the peak of Austral Summer in December and January, when the temperatures are only a few degrees below freezing and there is less snow on the rocks and lots of daylight. The Austral Winter in June and July is totally dark most of the day, and temperatures are about $-40°C$ ($-40°F$). Howling winds lower the wind chill to $-100°F$, and there is snow on all the outcrops, so few people come in the winter.

The logistical challenges are daunting enough. First you fly to New Zealand with all your winter gear packed and ready to put on when you meet your Antarctic flight. Over your long johns and regular clothes, you wear a bright red hooded polar parka for visibility on the snow, insulated snow pants, thick warm gloves, and "bunny boots" with air pockets in them for insulation. You also need tinted goggles and a knit balaclava to cover almost all of your head except your eyes, nose, and mouth. All the international scientists and staff headed toward Antarctica then board a huge C-17 Globemaster cargo plane (with the back half filled with pallets of gear secured in cargo nets) for the 5–6 hour flight to McMurdo Station, which is on the edge of the Ross Ice Shelf (figure 12.1A). McMurdo Station is a large permanent village and is the hub for all scientific expeditions in the region. It has been going for decades. It is composed of a bunch of permanent huts and buildings with a full-time staff to feed and house all the people living there or passing through, plus large areas to store equipment, supplies, fuel, and a tent city for people who are only visiting for a short time (figure 12.1B).

Before leaving McMurdo to head to your research destination, every scientist must pass a two-day survival course at Snow Camp on the Ross Ice Shelf. In addition to learning how to use all the equipment in the survival packs that you carry with you (including a polar tent and high-powered propane stove and military-grade walkie-talkies), they teach you the techniques of polar survival. You learn how to saw compacted snow into huge blocks to create a windbreak wall for your tent and to weigh down the corners of your tent against howling gales.

(A) A transport plane bringing the scientists to the base camp. (B) The base camp sprawls around the central buildings. (C) Nate Smith poses next to the emergency snow cave shelter that everyone must be able to build before they begin their work in Antarctica. (D) A helicopter removing a block of rock and fossil from the locality. ([A, B, D] Courtesy of P. Makovicky; [C] courtesy of N. Smith)

Figure 12.1 ▲
(*continued*)

Then they teach you how to use a shovel to cut a deep shelter trench in the surface snow and build a roof over it of the ice blocks you removed from the trench. In an emergency, this "snow hut" can help you survive extreme conditions by protecting you from the wind and cold (figure 12.1C). You get to find out how well you built your shelter by spending the night sleeping in it. The warm polar sleeping bags keep you cozy even though you are surrounded by a virtual igloo, but you have to strip down to your underwear to get the full benefit of the sleeping bag. Getting in and out of the bag can be quite a shock! Finally, you have to practice "rescuing" someone in whiteout conditions by walking around with a white bucket on your head. You quickly learn that it is practically impossible to go anywhere when the blizzards are howling enough to produce a whiteout.

After you pass the two-day survival camp experience, you return to McMurdo and join a much smaller group of scientists carrying their own gear who will fly on the smaller C-130 Hercules cargo plane to Beardmore Glacier near the Transantarctic Mountain Range, about 90 minutes of flying time from McMurdo Station, where there is a smaller temporary scientific station. The Hercules is equipped with skis and lands on a snow runway. Sometimes the load in the back of the plane is dropped off as the plane skims along the ground in what the military calls a "combat drop"— the plane acts as if the runway is under attack and drops its load without stopping, then immediately takes off again. Finally, all your essential gear and the scientists are loaded into helicopters and flown to the top of the mountain where the outcrops and fossils are located (figure 12.1D).

After all this work and preparation, you still need to set up your remote camp and make sure the tents are properly pitched, sheltered from the wind, and weighted down. Then you finally get a chance to explore the outcrops on top of the peaks at over 13,000 feet. Unlike many other collecting areas, the fossils are not loose on the ground or easy to collect. They are deeply embedded in the rock, and usually all you can see is a cross section of bone. The air is cold and dry and it never rains, so the rocks are not weathered at all. Instead, they are incredibly hard, and ordinary hammers and picks are insufficient to recover the fossils. The explorers must bring portable rock saws to slice deep grooves around the fossil block with the saw (figure 12.2), then they use a portable jackhammer to chip away the excess rock (figure 12.3). It is important to cut the block to its minimum size to reduce the weight that has to be hauled out. Once the block is trimmed

Figure 12.2 ▲
Cutting the fossil into a block with rock saws. (Courtesy of P. Makovicky)

down as small as possible, ordinary chisels can be used to crack it and pry it loose from the rock. Dynamite is used occasionally to blow off an entire layer of hard rock and get to the bone-rich layer.

Once the scientists have finished collecting all the blocks and preparing them to be removed, they wrap them in a cargo net and carry them to the helicopter landing spot. If the blocks weigh less than 300 pounds, they can be put directly in the chopper. If they are heavier than that, the harness of

Chipping away the excess rock and reducing the weight of the cut block with jackhammers. (Courtesy of P. Makovicky)

the cargo net is hooked under the belly of the chopper and lifts the entire block down to the base camp (see figure 12.1D). The helicopter pilots must be very skilled and very careful. It's dangerous to hover a few feet above a narrow ledge on the side of a mountain because there are lots of unpredictable crosswinds and updrafts that could mean disaster.

Finally, all the blocks are loaded onto choppers for a series of trips to McMurdo where the blocks are loaded into larger wooden crates and shipped directly to the lab at the museum. Many years of hard work are required in the lab for the preparator to slowly chip away the incredibly hard sandstone and remove the bones entombed inside it.

In December 1990–January 1991, Bill Hammer was on one of these expeditions, working with David Elliot of Ohio State University. They were looking at different outcrops when Elliot came across a promising cross section of bone exposed on the top of Mt. Kirkpatrick, about 4,000 meters (13,000 feet) in elevation, only 640 kilometers (400 miles) from the South Pole. He notified Bill, and during that season Hammer excavated numerous blocks weighing 2,300 kilograms (5,100 pounds) and shipped them to Augustana College where they were prepared. By 1994, enough of the blocks had been opened and the fossils cleaned up to enable him to publish a description of the dinosaur. A lot of material had been left behind after the initial brief visit in 1990–1991, and Hammer spent several more winter breaks collecting more material. Finally, in 2011, the last part of his major quarry for dinosaurs was collected. In all, more than 5,000 pounds of rock and bones had been removed from Mt. Kirkpatrick by helicopters, spanning almost 20 years. The last expedition included Nate Smith, who had once been one of Hammer's students at Augustana, then went on to get his PhD at the Field Museum in Chicago with Dr. Peter Makovicky working on the Antarctic dinosaurs. The circle is now complete because Smith is taking over the Antarctic research program as his mentor Hammer has just retired. Today Nate Smith is a Curator of Dinosaurs at the Natural History Museum of Los Angeles County.

ICY CRESTED LIZARD

Even as early as 1991, Hammer could see that his find was a spectacular new dinosaur, which he named *Cryolophosaurus* (Greek *cryo* for "ice," *lopho* for "crest," and *saurus* for "lizard"). Most of the skeleton that had been collected in the first few seasons was finally prepared, described, and published in 1994. The dinosaur's name refers to the icy conditions where it was found and to the odd-shaped crest on its head. It resembled the Spanish comb that señoritas used to wear on the top of their head, or maybe a lock of hair sticking up like a cowlick (figure 12.4). Based on the dimensions of the preserved parts, *Cryolophosaurus* was a relatively large theropod for the Early Jurassic, reaching 6.5 meters (21.3 feet) in length and weighing about 465 kilograms (1,025 pounds). This specimen was a juvenile, so the adult would have been a bit larger. That's not as big as an *Allosaurus* or *Tyrannosaurus rex*, but there were no other predators in the Early Jurassic nearly that big.

So far *Cryolophosaurus* is known mainly from a partial skull, a number of vertebrae and ribs, upper and lower arm bones, a thighbone, part of the hip, and several other bones including parts of the foot and ankle. (The specimens collected from 2011 to 2013 may add to this.) The thighbone is very primitive and resembles other Early Jurassic theropods such as *Dilophosaurus*. However, some features of the skull suggest a more advanced theropod,

Figure 12.4 ▲

Cryolophosaurus: (*A*) mounted skeleton at the Field Museum of Natural History in Chicago; (*B*) reconstruction of the dinosaur with its distinctive crest. (Courtesy of B. McGann)

such as the allosaurs. A recent study found that the cast of the skull cavity revealed a brain that was relatively small and primitive and not nearly as advanced as that of later theropods. The biological relationships of *Cryolophosaurus* has jumped all over the theropod family tree because specialists cannot find a consistent answer when they analyze it.

The most interesting and puzzling feature of this dinosaur is the weird-looking crest on its skull (figure 12.4B). It is composed of extensions of certain skull roof bones above the eye sockets. The crests on most dinosaurs are parallel to the midline of the skull (such as the paired crests in *Dilophosaurus* or the single crest on *Monolophosaurus*), but the crest on *Cryolophosaurus* trends perpendicular to the midline, running across the skull from one eye socket to the other.

How was this weird crest used? We cannot really know for sure how extinct animals behaved because we only have the bones and occasionally a few other clues, such as trackways. However, other animals with headgear (antelopes and deer, for example) use their crests, horns, antlers, and other cranial features for two main functions. The most common is recognition of members of their own species and to attract mates, but some headgear is used for intimidation and in combat against other rivals in the herd. The crest of *Cryolophosaurus* is not very thick or robust, so we can rule out head-to-head butting or wrestling. It is likely that the crest did serve for species recognition, mate attraction, and other social cues within a population.

HOLLYWOOD HOKUM

Most of the world is familiar with the odd-looking dinosaur in the first *Jurassic Park* movie known as *Dilophosaurus*. In the book and movie, it serves as the executioner for the villainous Dennis Nedry (played by the character actor Wayne Knight). Unhappy with his payment for creating the park's computer systems, he shut down the computer system to sneak embryos out and sell them to a competitor—which triggered all the bad events that followed.

Unfortunately, the movies got it all wrong when it came to *Dilophosaurus*. For one thing, the actual fossil is about twice the size of the movie monster; it is not a cute creature smaller than a kangaroo. Ironically, they erred in the other direction when they made *Velociraptor* twice as large as the actual dinosaur, and about the size of the real dinosaur *Deinonychus*. The actual *Velociraptor* is about the size of a turkey.

Both the original Michael Crichton book and the movie show *Dilophosaurus* paralyzing Nedry by spitting venom in his eyes before devouring him. *Dilophosaurus* could not spit venom; such an adaptation appears rarely in the animal kingdom, and the most familiar are different species of spitting cobras of Africa. They fire venom from their mouths through a tiny hole in the front of their tubular fangs, so the venom is sprayed forward under pressure. Fossils of *Dilophosaurus* have normal, thick, blade-shaped serrated theropod teeth, with no internal canals for injecting venom, let alone spraying it forward. Finally, the moviemakers freely confess that they added the frill around its head on a whim, just to make it scarier, but there is no evidence for this feature either. It was inspired by *Chlamydosaurus*, the frilled dragon lizard of Australia, which has a flap of leathery skin around its neck that can create a frightening display when threatened. These frills are held out by giant modified hyoid bones of the throat region. If *Dilophosaurus* had a frill, we would see those bones, but it doesn't have them. Hollywood made *Dilophosaurus* famous—but it created a monster totally unlike the real dinosaur.

The real story of *Dilophosaurus* is interesting enough, and it's a shame that Hollywood felt obliged to change and distort it. During the summer of 1942, University of California Berkeley paleontologist Charles L. Camp was on a field trip in the Navajo country of northern Arizona. A local Navajo man, Jesse Williams, showed Camp where he had seen dinosaur bones weathering out of the rocks in 1940, originally discovered by another Navajo, John Wetherill. Upon further digging, Camp and his crew found three dinosaur skeletons arranged in a sort of triangle in purplish shales of the Lower Jurassic Kayenta Formation about 20 miles north of Tuba City, Arizona. The first skeleton was nearly complete, missing only the front of the skull and a few vertebrae. The second skeleton was highly eroded and consisted of the skull, jaws, some vertebrae, and some limb bones. The third skeleton was almost completely eroded and consisted of some vertebrae. In just 10 days, Camp and his crew removed the first, most complete skeleton in blocks wrapped in plaster and then got the second skeleton. Both were loaded onto a truck and made it safely to Berkeley, where they were cleaned and prepared over the next two years.

The first skeleton was mounted and put on display in the Museum of Paleontology at U.C. Berkeley, where it remained on display for many years. When the paleontology department merged into the Integrative Biology program and relocated inside the Valley Life Science Building, there was only limited space for fossils on display, so it is out of public view now. Camp never got around to formally describing or naming the fossil, and it fell to Berkeley paleontologist Samuel P. Welles to finally publish a description of it in 1954. He originally referred it to the British genus *Megalosaurus*, after the first dinosaur ever named (chapter 1) back when *Megalosaurus* was a wastebasket name for almost all primitive theropods. Welles gave it the species name *wetherilli* after the man who originally discovered the fossils.

In 1964 Welles returned to the Navajo Reservation and found an even more complete skeleton about a quarter of a mile south of the original 1942 discovery. This specimen was much larger, nearly complete, and showed the skull features even better (figure 12.5A). In particular, the paired, blade-like crests (which were badly crushed and deformed in the 1942 specimens) were easier to interpret, so in 1970 Welles renamed his "*Megalosaurus*" *wetherilli* as *Dilophosaurus wetherilli*. *Dilophosaurus* means "two-crested lizard" in Greek, and he kept the original species name. Finally, the new specimen made the shape of the crest clear (figure 12.5B). The function of the paired crests on *Dilophosaurus* have long been controversial, although it probably helped with species recognition, impressing mates, or intimidating rivals (as antlers and horns do in mammals).

Dilophosaurus also had a peculiar "gap" in its upper tooth row where the nasal bones were attached to the maxillary bones of the upper jaw. The function of this particular feature is also controversial, but it was not a hinge in the upper jaws, as once thought. Instead, most scientists have pointed out that *Dilophosaurus* had rather delicate jaws and skull compared to other theropods with heavy robust bony skulls. Some have thought that its rather lightly built skull was better for catching fish and small prey rather than attacking the large prosauropod dinosaurs that lived at the time. Other paleontologists are not so sure. Just as in the pointless argument over *Tyrannosaurus rex* (see chapter 14), some paleontologists have argued that the weak-jawed *Dilophosaurus* was merely a scavenger and was not robust enough to be a predator on large prey. But like that problematic argument with other dinosaurs, the real answer is more complex. Nearly all modern large predatory animals (such as lions and hyenas) are both predators and scavengers and will take any meal they can get when they are hungry. For example, lions are known to dine on rotting carcasses, whereas hyenas are good hunters that prefer fresh kills to carrion. Like every other "either/or" argument in paleontology, nature is far more complex, with multiple causes and reasons for just about any feature.

Figure 12.5 ▲

Dilophosaurus: (*A*) the most complete skeleton shown as it was found in an articulated death pose; (*B*) reconstruction of *Dilophosaurus* showing the distinctive paired crests along the midline of the head. ([*A*] Courtesy of Wikimedia Commons; [*B*] courtesy of N. Tamura)

DINOSAURS OF DARKNESS

Many people wonder how dinosaurs like *Cryolophosaurus* survived the cold and chill of Antarctica. The answer is that during the Mesozoic the Antarctic was not cold at all; it had a mild temperate climate. The glaciers of the South Pole are a relatively recent feature, first appearing in the Oligocene about 32 million years ago, and they did not become permanent ice caps until about 14 million years ago. Before those events, the entire planet had a warm "greenhouse" climate, with high carbon dioxide and little or no ice anywhere. Sea levels were relative high through most of the Mesozoic.

The evidence for the mild Antarctic climates came first from fossil plants, which were collected at the very beginning of Antarctic exploration in the ill-fated dash to the South Pole by Robert Falcon Scott in 1912. Among the objects found in their possession by the crews who discovered their frozen bodies were fossil leaves of the seed fern *Glossopteris*, which was ubiquitous on all the Gondwana continents during the Permian. Many more plants have been found since then. Combined with geochemical evidence from fossil shells, they indicate an average temperature of about 10°C (about 50°F), so they probably experienced a climate much like modern London. The vegetation consisted of many of the trees and plants found in the cold temperate forests of New Zealand and southern South America, including ferns, *Araucaria* trees (such as the modern Norfolk Island pine and monkey-puzzle tree, both relicts still living Down Under in Australia and New Zealand when they were discovered), ginkgoes, cycads ("sego palms" to gardeners, even though they are conifers, not true palm trees), podocarps, and many other typical Mesozoic plants. Although there were some early flowering plants in the Cretaceous of the polar regions, they were far outnumbered by conifers and other evergreens (as they are today in the subarctic forest or taiga and in the high mountains below the tree line). The climate was only cool temperate with rare episodes of freezing, but it was still dark down there—all the localities are from inside the Antarctic Circle, so they experienced total darkness 24/7 during their winter solstice and remained dark for several months and experienced the "midnight sun" during the peak of summer.

Aside from the Early Jurassic *Cryolophosaurus*, and a newly discovered primitive sauropod called *Glacialisaurus hammeri* (discovered by Bill Hammer and crew in 1990–1991 and named and described by Nate Smith in honor of his mentor), only a few other dinosaurs are known from Antarctica. They come from Cretaceous beds and include *Antarctopelta*, a partial skeleton of an ankylosaur from James Ross Island on the Antarctic Peninsula near Argentina, and some fragments of an ornithopod dinosaur that was named *Trinisaura santamartaenesis*. The list is very short because collecting fossils in the Antarctic is dangerous, expensive, and limited to just a few months of the year, and there are not many rocks exposed above the ice cap.

Polar dinosaurs are better known from Australia, which also was inside the Antarctic Circle in the Cretaceous. (Australia has since moved north to its present position.) My good friend Tom Rich and his wife Pat have collected Australian dinosaurs for decades and have found some remarkable specimens. They include the small ornithopods *Leaellynasaura amicagraphica* and *Atlascopcosaurus loadsi*, the small predatory coelurosaur *Timimus hermani*, and other small theropods. The meter-long (39-inch) *Leaellynasaura* (named after Lea Ellyn Rich, the daughter of Tom and Pat Rich) is the most complete fossil of the bunch. It had unusually large eye sockets and large optic lobes in the brain, suggesting that it had huge eyes for seeing in the many months of darkness or semidarkness of the Antarctic winter. Its bones show no growth lines, so it clearly did not hibernate in the winter. Even though polar Australia was dark or dimly lit during the winter, this little herbivore must have found enough food to survive. However, the little coelurosaur *Timimus* (named after Tim Rich, son of Tom and Pat Rich) did show strong growth lines in its bones, suggesting it did hibernate during the polar winters when conditions were dark and prey was scarce. Most peculiar of all is a weird lower jaw of a tiny shrew-sized mammal from an extinct group distantly related to the living platypus. In addition to these dinosaurs, fish, turtles, pterosaurs, birds, and amphibians were also recovered, which represent an entire ecosystem adapted to months of darkness and cool temperatures.

The mild temperatures were not restricted to the Antarctic; the entire planet was warmer. The Arctic was a region with cool but mild climates and abundant forests and dinosaurs during the Cretaceous as well. However, the dinosaurs were very different. As early as 1961, geologists were finding bones of duck-billed dinosaurs in places like the Colville River of Alaska. When major excavations were undertaken by the University of Alaska, thousands of bones were found, representing 12 species of dinosaurs.

Most abundant were the common duckbill from Alberta, similar to *Edmontosaurus*, which has now been renamed *Ugrunaaluk*. There were Albertan duckbills such as *Kritosaurus* and *Lambeosaurus* as well. Horned dinosaurs included *Pachyrhinosaurus*, with the flat bony boss on its snout (star of the recent CG movie *Walking with Dinosaurs*), and the three-horned *Anchiceratops*. Much less common were pachycephalosaurs, dinosaurs with a thick boss of bone over their tiny brains, and the other herbivores including *Thescelosaurus*, an ornithopod dinosaur. Their predators included the Albertan tyrannosaur *Albertosaurus*, plus typical Alberta and Montana theropods such as *Troodon*, *Dromaeosaurus*, and *Saurornitholestes*.

Like the Antarctic Cretaceous dinosaurs, they lived in a world with cool temperatures suitable for conifers such as bald cypress, cycads, ginkgoes, ferns, and horsetails. The temperatures in Cretaceous Alaska (based on plants and also on geochemistry) averaged about 5°C (41°F) with summer means around 10°C (50°F), and winter temperatures around freezing. Like the dinosaurs of the Antarctic, these dinosaurs lived within the Arctic Circle, so they experienced many months of near darkness. Unlike those in the Australia-Antarctic continent, however, it is possible that many of the dinosaurs (especially the duckbills) migrated down to warm Alberta to escape the cold dark winters.

This was the greenhouse world of the Age of Dinosaurs. The planet was warm to the poles with no ice, high sea levels, and lots of different animals living in conifer and fern cool temperate forests within a few miles of the North Pole.

FOR FURTHER READING

Colbert, Edwin. *Digging Into the Past: An Autobiography*. New York: Dembner Books, 1989.
——. *A Fossil Hunter's Notebook: My Life with Dinosaurs and Other Friends*. New York: Dutton, 1980.
Farlow, James, and M. K. Brett-Surman. *The Complete Dinosaur*. Bloomington: Indiana University Press, 1999.
Fastovsky, David, and David Weishampel. *Dinosaurs: A Concise Natural History*, 3rd ed. Cambridge: Cambridge University Press, 2016.
Holtz, Thomas R., Jr. *Dinosaurs: The Most Complete, Up-to-Date Encyclopedia for Dinosaur Lovers of All Ages*. New York: Random House, 2011.

Naish, Darren. *The Great Dinosaur Discoveries.* Berkeley: University of California Press, 2009.

Naish, Darren, and Paul M. Barrett. *Dinosaurs: How They Lived and Evolved.* Washington, D.C.: Smithsonian Books, 2016.

Prothero, Donald R. *Greenhouse of the Dinosaurs: Evolution, Extinction, and the Future of the Planet.* New York: Columbia University Press, 2009.

Rich, Thomas, and Patricia Vickers Rich. *Dinosaurs of Darkness.* Bloomington: Indiana University Press, 2000.

Walker, Sally M. *The Search for Antarctic Dinosaurs.* New York: Millbrook Press, 2007.

SPINOSAURUS

Stromer was a study in irony. Though often considered frail, he survived
the punishing heat (and cold) of Egypt's Western Desert, and eventually
lived to the age of 82. He disliked roughing it, but preferred that to
crowded, noisy cities, so he braved windstorms in flea-infested tents near
tiny towns or in the middle of nowhere. He was landed gentry but always
short of cash. He resented an Egyptian assistant, whom he depended on
completely, for trying to rise above his "station." Years later, however,
Stromer irritated the Third Reich for refusing to join the Nazi party and-
worse-for maintaining ties with Jewish colleagues.

—MICHON SCOTT, HTTPS://WWW.STRANGESCIENCE.NET/STROMER.HTM

STROMER'S RIDDLE

Ernst Freiherr Stromer von Reichenbach experienced both the highest
highs and the lowest of lows in his 82 years of life (figure 13.1). He was born
into an aristocratic family (the "Freiherr" in his name is roughly equivalent
to "Baron"), but his boss, a prominent Nazi, treated him with disdain and
tragically refused to allow Stromer's fossils to be removed to safe storage
during World War II—which meant they were destroyed by Allied bombing.
Meanwhile, the Third Reich sent his three sons off to die in World War II.
(One of them ended up in a Soviet POW camp, survived, and eventually got
home to Germany in 1950.) Stromer made incredible discoveries in Egypt,
then had to wait over a decade to see most of his fossils. He finally got to
describe his fossils, only to see them vanish in a single night. He lived seven
more years after the war ended, but he never had another chance to make
important scientific discoveries.

Figure 13.1 ▲
Ernst Stromer von Reichenbach standing next to one of the Egyptian dinosaur bones.
(Courtesy of Wikimedia Commons)

Stromer's greatest claim to fame was his role in leading one of the first great paleontological expeditions to Egypt. From the 1880s through 1914, Germany was the world leader in science and scholarship, and its archeologists were making amazing discoveries, from unearthing ruins in Mesopotamia to the discovery of ancient Troy and ancient Mycenae. The Berlin Museum and Werner Janensch spent years in German East Africa (now Tanzania) finding the famous Tendaguru dinosaur fauna (see chapter 9), and Friedrich von Huene made important discoveries about *Plateosaurus* (see chapter 6).

Stromer came from the same scholarly tradition and was bitten by the bug of exploration. As early as 1901, when he was only 30 years old, he was already a curator at the natural history museum in Munich, the Paläontologisches Museum München, and had visited Egypt for the first time. He became convinced that important fossils could be found there, and he was particularly interested in finding fossil mammals. Stromer spent several years raising funds and making contacts so that he could undertake a grand expedition to Egypt, primarily to collect mammal fossils as the British had already found some incredible mammals in the Fayûm area just west of Cairo.

Stromer finally arrived in Egypt in November 1910, but his luck failed and he was stuck in port in Alexandria for days while his ship was quarantined for cholera. When he finally arrived in Cairo, he contacted the local German Egyptologist and John Ball, head of the Geological Survey of Egypt, to get the necessary permits and maps. As early as 1910, tensions were rising between Britain and Germany, and it was not easy for Germans to get permits from the British authorities, who ruled Egypt. Many were suspicious of German scientists wandering about in remote areas away from supervision, possibly spying for the Kaiser. After some delays, Stromer finally met his friend and the local fossil collector, Richard Markgraf, who had helped him during his first visit in 1901.

On November 19, 1910, they began to travel across the Gizeh Plateau past the Great Pyramids. Traveling west, they visited Wadi el Natrun, where they found lots of scrappy fossils of turtles and crocodiles but no spectacular mammals. After a month in the field, Stromer returned to Cairo, but Markgraf remained behind. When Markgraf turned up in Cairo a few weeks later, he brought with him an important find: the skull of a small relative of the Colobus monkey that Stromer later named and described as *Libypithecus markgrafi*. The collectors then traveled up the Nile, but they found nothing on this leg of the expedition.

For the final part of the expedition in early January 1911, they took a train all the way to the western edge of Egypt near the Libyan border to the Baharia Oasis. Stromer expected to find Eocene mammals there similar to those the British had found in the Fayûm, but he was off by tens of millions of years. The beds he found there were Lower Cretaceous and were virtually unknown and unexplored by scientists. The explorers were delayed by howling winds, freezing nights, and even major sandstorms, but when the weather finally cleared, they began to explore in earnest. After finding lots of shark and other fish teeth and petrified wood, Stromer finally got his first lucky break. Wandering around the flank of a hill called Jebel El Diest, he found "three large bones which I attempt to excavate and photograph. The upper extremity is heavily weathered and incomplete [but] measures 110 cm long and 15 cm thick. The second and better one underneath is probably a femur [thighbone] and is wholly 95 cm long and, in the middle, also 15 cm thick. The third is too deep in the ground and will require too much time to recover." Then he found part of a dinosaur hip and a huge claw. Stromer had not brought burlap bags and plaster of Paris that most fossil collectors used, so he was forced to cut up mosquito netting and soak it in a mixture of flour and water to form a jacket around the fragile fossils. After exploring some other areas and collecting more fossils, he was on his way home to Germany by February 18, 1911. Over the next few years, Stromer led several more trips to Bahariya Oasis and recovered many more fossils, returning from his last trip in 1914, shortly before World War I broke out.

Unfortunately, Stromer had bad luck again: most of his fossils were still in Egypt when the war started, and he was stuck in his museum without many of his fossils through the four years of war. Even after the war ended in 1918, Stromer had to wait until 1922 before the fossils all finally arrived from Egypt. When he opened the crates, he found that the fossils had been unpacked by the Egyptian authorities and repacked poorly, so they were in fragments. He spent months piecing them all together before finally publishing on them during the 1920s and 1930s.

At the high point of his career, when his bizarre and spectacular finds were finally published, Stromer's luck once again ran out as the Nazis took over. Even though his age and aristocratic background protected him against personal recriminations, all three of his sons were drafted and fought in the war, where two were killed. As the Allied bombing continued to strike deeper into the heart of Germany, museum curators packed their most important objects of art and natural history and hid them in deep caves and salt mines. But the head of the Paläontologisches Museum München, Karl Beurlen, was an ardent Nazi and ignored Stromer's begging and pleading to move the fossils before Munich was bombed. The final blow came on the night of April 24–25, 1944, when the British Royal Air Force bombed Munich; the museum and all its collections were totally destroyed. All of Stromer's dangerous and hard work in Egypt, and years of scientific research and publication, was gone in a single night. Many other important dinosaur fossils in German museums were destroyed in the same way during other nights of bombing, including many of the *Plateosaurus* fossils (see chapter 6) and a lot of the Tendaguru bones in Berlin (see chapter 9). After the war ended, paleontologists have routinely made multiple casts of important fossils and shared them with other museums just in case a disaster, such as a war or a flood, destroyed the museum. (In September 2018, the Brazilian Museum burned to the ground, destroying lots of irreplaceable objects, but some of the fossils have replicas in other museums).

Stromer spent the last eight years of his life a broken man, with only the return of his son Wolfgang from a Soviet POW camp in 1950 as a solace in his old age. But during his career he had described a wide range of important fossils from the Egyptian Cretaceous, including the sauropod *Aegyptosaurus*, the giant crocodile *Stomatosuchus*, and a number of theropods including *Bahariasaurus, Carcharodontosaurus*, and *Spinosaurus*. What struck Stromer as odd was the high diversity of predators, with three large theropods and a huge crocodile but hardly any herbivorous animals except the sauropod. Normally one would expect the prey species to greatly outnumber the predators; in most large animal communities, the biomass of prey species has to be about 10 times the biomass of predators. Later paleontologists called this dilemma regarding the imbalance of Egypt's dinosaur fauna "Stromer's Riddle." The answer would not be forthcoming until the twenty-first century and renewed activity in North Africa.

PIECING TOGETHER THE LOST DINOSAUR

Thanks to the bombing of Munich, all that remained of the original Stromer *Spinosaurus* collection was photographs, illustrations, descriptions, and measurements. In fact, the original dinosaur was very poorly known, with just the tall spines of the backbone vertebrae, part of a lower jaw and a few neck vertebrae, a few ribs, and isolated scraps of other bones (figure 13.2). Stromer also described part of an upper jaw, but it was never illustrated because it was poorly preserved. The size and shape of the dinosaur was highly conjectural because almost no parts of the skull, limbs or limb

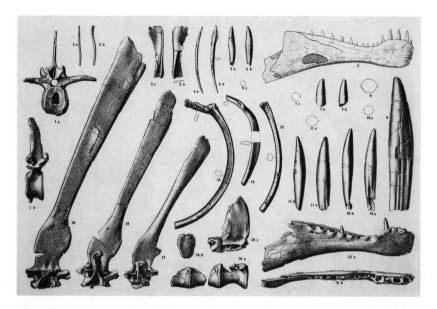

Figure 13.2 ▲

Original illustration of the handful of partial fossils of *Spinosaurus*, from Stromer's monographs. (Courtesy of Wikimedia Commons)

girdles, and tail survived to provide important clues of its real shape and size. So how do we know anything about the dinosaur at all?

One solution would be to go back to Bahariya Oasis and try to find more fossils of *Spinosaurus*. In 2001, an expedition led by Joshua Smith and Ken Lacovara of the University of Pennsylvania did just that, and they found quite a few fossils (figure 13.3). Some of them belonged to a huge titanosaur sauropod they named *Paralititan*, which was about 80 feet long and weighed about 63 metric tonnes (70 tons). They also analyzed the sedimentary environment of the Bahariya Formation and found it represented a nearshore setting of lagoons and rivers and swamps close to the ocean. With this discovery (and the discovery of other herbivorous dinosaurs from elsewhere in North Africa, such as the sail-backed iguanodont *Ouranosaurus*), Stromer's Riddle has been answered. There were plenty of large sauropods, iguanodonts, and other prey species for the large predators to eat. Unfortunately, the 2001 University of Pennsylvania expedition found very little material that would help in our quest to see the entire *Spinosaurus*.

Figure 13.3 ▲

View of Bahariya Oasis locality where *Paraliatitan* was found by the University of Pennsylvania expeditions, and near Stromer's locality for *Spinosaurus*. (Courtesy of Wikimedia Commons)

Modern technology has been used to reconstruct Stromer's lost fossils, carefully digitizing the existing photographs and illustrations and turning them into solid objects again with a 3D printer. This technique does not replace all the information the original fossils yielded, but at least it gives us the size and shape of a replica of the bones.

Fortunately, other *Spinosaurus* fossils have turned up elsewhere in North Africa. In 1996, the Kem Kem beds of southern Morocco yielded vertebrae of an even bigger spinosaur that Dale Russell named *Spinosaurus maroccanus*, although today paleontologists doubt that it is anything more than a larger individual of *Spinosaurus aegypticus*, or possibly the spinosaurid *Sigilmassasaurus*. Another Moroccan specimen ended up in the Museé National d'Histoire Naturelle in Paris and consists of most of the upper jaw and fragments of the lower jaw. Yet another specimen from the Kem Kem beds is now in the Museo di Storia Naturale di Milano (Milan Natural History Museum) and consists of a complete snout originally misidentified as a crocodile. In 2008, an anonymous local fossil collector gave a box of bones to paleontologist Nizar Ibrahim while he was visiting Morocco. This turned out to be parts of a subadult skeleton, with fragments of the skull, many neck and back vertebrae, including the tall spines, some of the shoulder and arm bones, parts of the hip, complete hind limbs, and many tail vertebrae. In 2014, Ibrahim, Paul Sereno, and coauthors published a composite reconstruction of *Spinosaurus* that made use of all the pieces now known and tried to give us a look at the complete animal.

SPINOSAURUS RESURRECTED

What kind of dinosaur was *Spinosaurus*? It wasn't much like the outdated version seen in the third *Jurassic Park* movie, which was a huge biped that terrified even a tyrannosaur. If the reconstruction by Ibrahim, Sereno, and colleagues is valid, it was a quadrupedal swimmer that acted more like a crocodile and spent little or no time walking on its hind legs on land (figure 13.4). Its long, narrow snout was filled with conical teeth for catching fish, not blade-like teeth with serrated edges suitable for ripping flesh, as found in most theropods. And it was nowhere near as big as *Jurassic Park III* had rendered it. At best, it was more slender and only slightly longer than a *Tyrannosaurus*, but it was without the huge body mass or the flesh-ripping teeth or powerful neck and jaws for biting large prey. If it encountered a large theropod, it would almost certainly have backed off or fled for the nearest water.

Almost all of the features of *Spinosaurus* are adaptations for an aquatic lifestyle as well. Midway back on the snout were its nostrils, suitable for breathing while partially submerged. The snout also had channels for nerves that would have helped it sense changes in water pressure caused by the motion of prey in the water. Its stumpy limbs were not suitable for walking on land, but they were good for paddling in the water, and the long thin finger and toe bones suggest that it had webbed feet. Even the geochemistry of the bones and teeth suggest it was an aquatic animal. Finally, the limb bones were very dense and solid, typical of animals such as hippos that need dense limb bones to help as ballast.

The biggest mystery was the enormous sail on the back of *Spinosaurus*, which gave it the name. It was clearly not big enough to be a true "sail" for propelling it through the water under wind power; it was much too small for the huge bulk of the dinosaur. In fact, it's so large and conspicuous that it

Figure 13.4 ▲

(A) Modern reconstructed skeleton of *Spinosaurus* showing it was mostly a quadrupedal swimming animal with subequal front and hind limbs, not a gigantic bipedal predator as in the *Jurassic Park III* movie; (B) modern artistic reconstruction of *Spinosaurus*. ([A] Courtesy of Wikimedia Commons; [B] courtesy of N. Tamura)

would prevent the dinosaur from completely submerging underwater and sneaking up on prey, as crocodilians do. Others have argued that it was a big heat-gathering surface for regulating body temperature, but then why does it not occur in any other theropod (but for some reason, does occur in the African iguanodontid *Ouranosaurus*)? Most paleontologists point to the large conspicuous nature of the sail and consider it some sort of device to advertise its huge size and dominance in competing with other spinosaurs in its territory, as the horns and antlers of many deer and antelopes do today.

So how big was *Spinosaurus*? Estimating length on many dinosaurs is problematic when we have only a few bones, and weight estimates are even more speculative. Remember, even the fairly complete 2014 reconstruction by Ibrahim, Sereno, and others did not include much of the tail among the bones recovered, so the length cannot be measured directly but only estimated based on the proportions of other bones.

In 1926, Friedrich von Huene (who was able to see the original fossils before they were destroyed) put its length at 15 meters (almost 50 feet) and weight at 6 metric tonnes (6.6 tons). But this was based on the fragments that Stromer had at that time, which was mostly the sail and lower jaw and not much else. Dinosaur artist Greg Paul in 1988 pegged its length at 15 meters (49 feet) and a mass of 4 metric tonnes (4.4 tons), but again this was when almost nothing was known of the dinosaur other than Stromer's original fragmentary fossils. In 2005, Del Sasso and colleagues estimated its length at 16-18 meters (52-59 feet) and weight at 7-9 metric tonnes (7.7-9.9 tons) using comparisons with other spinosaurs such as *Suchomimus*, but this has been criticized because it is based on the incomplete skull of *Spinosaurus*. Based on comparisons with a wide range of theropods, Francois Therrien and Donald Henderson's 2007 estimate considered the previous numbers to be too large. They put the body length at 12.6-14.3 meters (41-47 feet) and the weight at 12-21 metric tonnes (13.2-23.0 tons). The lower end of this estimated range would make *Spinosaurus* smaller than not only *Tyrannosaurus rex* but also *Carcharodontosaurus* and *Giganotosaurus*. Finally, the 2014 estimate of Ibrahim, Sereno, and others based on their composite skeleton from many individuals (including some juvenile bones) places its length at 15.2 meters (51 feet). If *Tyrannosaurus rex* was about 13 meters (43 feet) and about 10 metric tonnes (11 tons), it was only slightly smaller than *Spinosaurus*. Clearly, the scene in *Jurassic Park III* in which

Spinosaurus picks up a *T. rex* and tosses it around like a toy is pure fiction, based on what we now know of *Spinosaurus*. It was much more likely to be the other way around: a bulldog predator with the crushing bite of a *T. rex* would easily kill a lightly built fish-eater like *Spinosaurus* in a fight, even if the latter was slightly longer.

THE SPINOSAURS

Spinosaurus is the most famous of the fish-eating group broadly known as spinosaurs, but they are much more diverse than just the famous namesake of the group. Another of the famous spinosaurs is *Baryonyx walkeri*, one of the stars of the recent movie *Jurassic World: Fallen Kingdom*. Its genus name means "heavy claw" in Greek, and its species name honors amateur fossil hunter William Walker, who found an enormous claw in a clay pit near Ockley, in Surrey, England, in early January 1983. These clays were part of the same Lower Cretaceous Wealden beds that yielded *Iguanodon* to Gideon Mantell more than 150 years earlier (see chapter 2). In all the years since then, many people had searched the Wealden for more dinosaurs, so it is remarkable that finally another one turned up after so much time.

British Museum (Natural History) paleontologists Alan Charig and Angela Milner visited his locality a few weeks later in February 1983 and found more bones. By June the museum crew had collected most of the skeleton, with over 2 tonnes of matrix and bones altogether. The owners of the pit, Ockley Brick Company, not only donated the skeleton to the Natural History Museum but even provided heavy equipment to help the crew with the excavation.

Even though the bones were no longer articulated, they were close to their life position, so the skeleton could be reconstructed accurately (figure 13.5). What they discovered was a dinosaur that looked nothing like anything that had been seen at that time. (Remember, *Spinosaurus* was then known from only the few pieces destroyed in the bombing raid, so there were few overlapping parts.) The long-snouted face with fish-catching teeth made some think it was a crocodile, but the rest of the bones clearly showed Charig and Milner that it was a dinosaur trying to live like a crocodile. The limbs were about equal in length, so it was capable of quadrupedal locomotion, as well as bipedality, like *Spinosaurus*. Not only did it have a snout like a crocodile with conical fish-catching teeth, but fish fossils were found

Figure 13.5 ▲
Baryonyx walkeri: (*A*) skeletal reconstruction in the Southwestern Florida Museum of Natural History; (*B*) life-like reconstruction of the animal as a delicately built fish-eater. ([*A*] Courtesy of Wikimedia Commons; [*B*] courtesy of N. Tamura)

among its stomach contents. Its robust front limbs apparently carried the huge claws that Walker had first found, which were suitable for grappling with prey. Charig and Milner were not certain what kind of dinosaur it resembled because it was bizarre and not very complete. However, French paleontologist Eric Buffetaut and dinosaur artist Greg Paul noticed the characteristic shape of the upturned tip of the lower jaw and the distinctive tooth sockets and suggested it might be like the (then poorly known) *Spinosaurus*. Since then, additional fossils have confirmed this. *Baryonyx* has since been reported from Spain, Portugal, and elsewhere in Europe. In contrast to the problem of fragmentary dinosaurs found previously in Britain (see chapters 1–3), *Baryonyx* is the most complete dinosaur ever found in the United Kingdom.

Spinosaurs are found in Europe and North Africa as well as elsewhere in the Early Cretaceous. *Suchomimus* (mentioned earlier) was a much larger spinosaur found in the Elrhaz Formation in the Sahara Desert of Niger by my friend Paul Sereno and crew and described in 1998. At about 9.5 meters (31 feet) and 2.5 tonnes (2.75 tons), it was only slightly smaller than *Spinosaurus*. It also had the characteristic crocodile-like snout and skull, and long spines on its back, not over the shoulders and middle back as in *Spinosaurus* but over the hips. In addition, there is *Cristatusaurus*, also from Niger, and *Sigilmassasaurus* from the Kem Kem beds. *Ostafrikasaurus*, found in the Tendaguru beds of Tanzania by the Germans in 1912, was recognized as another spinosaurid by Eric Buffetaut in 2012. So spinosaurids were widespread over Europe and Africa.

Spinosaurs even occurred in Asia and South America. *Ichthyovenator* comes from Laos and *Siamosaurus* from Thailand. *Irritator* and *Oxalaia* are known from the Cretaceous of Brazil, extending the range of these animals across more of Pangea. *Irritator*, named by David Martill and colleagues in 1996, consists of a partial skull that has a very crocodile-like shape. However, Hans-Dieter Sues and colleagues first realized in 2002 that it was a spinosaur, and it is the most complete skull of a spinosaur found to date. The name *Irritator* comes from the fact that the fossil was found in an old museum collection (with no locality information), apparently bought from dealers who had tried to improve its sales potential. They had added a piece of jaw to "enhance" the specimen, first thought to be a huge pterodactyl, and make the fossil look more marketable. Once the specimen was under study by scientists, it was an irritating task to get rid of the fakery and clean it down to the original bone. Its full scientific name is *Irritator challengeri*, after Professor Challenger of the novel *The Lost World* by Arthur Conan Doyle, also author of Sherlock Holmes.

Although their fossils are not common, the crocodile-like spinosaurs were ubiquitous in the Early Cretaceous on most of the major continents—Africa, Eurasia, Australia, and South America. Only North America has no record of the group yet, although some North American theropod teeth have been suggested to be spinosaurid.

FOR FURTHER READING

Colbert, Edwin. *Men and Dinosaurs: The Search in the Field and in the Laboratory.* New York: Dutton, 1968.

Farlow, James, Mike Brett-Surman, and Robert Walters. *The Complete Dinosaur.* Bloomington: Indiana University Press, 1999.

Fastovsky, David, and David Weishampel. *Dinosaurs: A Concise Natural History,* 3rd ed. Cambridge: Cambridge University Press, 2016.

Holtz, Thomas R., Jr. *Dinosaurs: The Most Complete, Up-to-Date Encyclopedia for Dinosaur Lovers of All Ages.* New York: Random House, 2011.

Holtz, Thomas R., Jr., Ralph E. Molnar, and Philip J. Currie. "Basal Tetanurae." In *The Dinosauria,* 2nd ed., ed. David B. Weishampel, Peter Dodson, and Halszka Osmólska, 71–110. Berkeley: University of California Press, 2004.

Ibrahim, Nizar, Paul C. Sereno, Christiano Dal Sasso, Simone Maganuco, Matteo Fabbri, David M. Martill, Samir Zouhri, Nathan Myhrvold, and Dawid A. Iurino. "Semiaquatic Adaptations in a Giant Predatory Dinosaur." *Science* 345, no. 6204 (2014): 1613–1616.

Naish, Darren. *The Great Dinosaur Discoveries.* Berkeley: University of California Press, 2009.

Naish, Darren, and Paul M. Barrett. *Dinosaurs: How They Lived and Evolved.* Washington, D.C.: Smithsonian Books, 2016.

Nothdurft, William, and Josh Smith. *The Lost Dinosaurs of Egypt.* New York: Random House, 2002.

Sereno, Paul C., Allison L. Beck, Didier B. Dutheuil, Boubacar Gado, Hans C. E. Larsson, Gabrielle H. Lyon, Jonathan D. Marcot, Oliver W. M. Rauhut, Rudyard W. Sadleir, Christian A. Sidor, David D. Varricchio, Gregory P. Wilson, and Jeffrey A. Wilson. "A Long-Snouted Predatory Dinosaur from Africa and the Evolution of Spinosaurids." *Science* 282, no. 5392 (1998): 1298–1302.

Sereno, Paul C., Didier B. Dutheil, M. Iarochene, Hans C. E. Larsson, Gabrielle H. Lyon, Paul M. Magwene, Christian A. Sidor, David J. Varricchio, and Jeffrey A. Wilson. "Predatory Dinosaurs from the Sahara and Late Cretaceous Faunal Differentiation." *Science* 272, no. 5264 (1996): 986–991.

Smith, Joshua B., Matthew C. Lamanna, Kenneth J. Lacovara, Peter Dodson, Jennifer R. Smith, Jason C. Poole, Robert Giegengack, and Yousry Attia. "A Giant Sauropod Dinosaur from an Upper Cretaceous Mangrove Deposit in Egypt." *Science* 292, no. 5522 (2001): 1704–1706.

Weishampel, David B., Paul M. Barrett, Rodolfo A. Coria, Jean Le Loeuff, Xu Xing, Zhao Xijin, Ashok Sahni, Elizabeth M. P. Gomani, and Christopher R. Noto. "Dinosaur Distribution." In *The Dinosauria*, 2nd ed., ed. David B. Weishampel, Peter Dodson, and Halszka Osmólska, 517–606. Berkeley: University of California Press, 2004.

Wilford, John Noble. *The Riddle of the Dinosaur.* New York: Knopf, 1985.

TYRANNOSAURUS

I propose to make this animal the type of the new genus, *Tyrannosaurus*, in reference to its size, which far exceeds that of any carnivorous land animal hitherto described. . . . This animal is in fact the *ne plus ultra* of the evolution of the large carnivorous dinosaurs: in brief it is entitled to the royal and high sounding group name which I have applied to it.

—HENRY FAIRFIELD OSBORN, 1905

OSBORN'S TRIUMPH

Henry Fairfield Osborn was an interesting and pivotal figure in the history of paleontology (figure 14.1). He did more to bring the image of dinosaurs to the public eye than any other person, especially in commissioning huge skeletal mounts and paintings and models of dinosaurs that created the first real wave of public awareness and "dinomania" in the early twentieth century. Modern paleontologists, however, are not so impressed with much of his scientific work, which was tinged with racism and eugenics and odd notions about evolution. They are particularly distressed by his habit of creating too many new species based on meaningless tiny differences between specimens, or his habit of taking credit for other people's work. He was a study in contrasts, but paleontology would not be nearly so popular without him, nor would there be nearly as many wonderful dinosaurs in museums.

Osborn was born in 1857 to prominent railroad tycoon William Henry Osborn, and for his entire life he lived and behaved and thought like an American aristocrat. The late 1800s, also known as the Gilded Age, was

Figure 14.1 ▲

Henry Fairfield Osborn in his later years, showing his confident attitude as an American aristocrat and the director of the American Museum of Natural History. (Courtesy of Wikimedia Commons)

a time when the explosion of industrial might in the United States created a number of tycoons and "robber barons" who acted and believed they were royalty. Osborn carried this haughty, aristocratic attitude throughout his life, which often chafed the people who worked under him.

Stories about Osborn's condescending, pompous, egotistical behavior are legendary. His biographer, Ronald Rainger, describes him as a powerful, well-connected, American aristocrat, running the department and the museum in a strict hierarchy in which he provided the administration and financial means. His subordinates (many of whom were more competent paleontologists than Osborn) did all the grunt jobs (fieldwork, preparation, curation, illustration, writing, and description), for which Osborn took complete or at least partial credit. A retinue of secretaries and assistants attended to his every beck and call, and they even had separate quarters outside his mansion in Garrison, New York, where they did his bidding when he was at home. At social events, only his top scientific assistants and fellow curators could dine with Osborn and his family; the "lesser staff" were consigned to a different room. George Gaylord Simpson, the legendary mammalian paleontologist who overlapped at the American Museum with Osborn's later career, describes an incident when he once tried to apply an ink blotter to Osborn's autograph from a flowing fountain pen. Osborn stopped him and said, "Never blot the signature of a great man." His arrogance and aristocratic ideas may have made his coworkers angry, but it served him well as he mingled effortlessly with the rich tycoons of New York and got them to donate money to the cause of science.

Like the children of many rich eastern families, Osborn grew up in a mansion with servants in Fairfield, Connecticut, and then was sent to Princeton from 1873–1877 to get his college education alongside other sons of rich men. However, at Princeton he heard of the amazing finds of Marsh and Cope during the Bone Wars (see chapters 7–8) and was hooked on the lure of finding amazing fossils on his own. In an act that only a rich, overconfident child of wealth could imagine, he organized his own field trip out west in 1877 with his classmates Francis Speier Jr. and William Berryman Scott (figure 14.2). (Scott later went on to become a famous paleontologist in his own right, founding the Princeton Geology Department where he spent his career.) Osborn and his friends didn't want Yale to have a monopoly on the rich bone beds out west, so they set up their own expedition. The three rich young greenhorns managed to survive the harsh conditions out west, avoid being killed by Indians on the warpath, and did it again with a larger party in 1878. They came back more determined than ever to become important paleontologists. Osborn met with both Leidy and Cope during his college years, and Cope became his mentor—thus making Marsh his enemy. Osborn supported Cope during his battles with Marsh, and as the sickly Cope reached desperate financial straits before he died, Osborn generously bought his entire collection for his American Museum.

Neither Osborn nor Scott could earn a Ph.D. in the United States (such programs did not yet exist here), so they traveled to Europe to experience the cutting edge of scientific research in 1879–1880, taking courses from biologist Thomas Henry Huxley and embryologist Francis Balfour in England (and meeting Charles Darwin) and learning from the leading German paleontologists as well. In 1883, Princeton awarded Osborn its own doctorate and offered him a post teaching biology and comparative anatomy.

By 1891, Osborn's research in both paleontology and embryology—and his connections—landed him a job at the young American Museum

Figure 14.2 ▲

Young undergraduates Henry Fairfield Osborn (*center*), William Berryman Scott (*right*), and Francis Speier (*holding gun*) posing before a photographer as they led their own Princeton expedition out to the Wild West collecting fossils in 1877. (Courtesy of Wikimedia Commons)

of Natural History in New York as curator and head of the newly formed Department of Vertebrate Paleontology (DVP). He was also appointed to a professorship at Columbia University in New York. This established the Columbia/American Museum paleontology doctoral program, which continued for more than a century. I was trained there, as were many important paleontologists across the United States, including Bruce MacFadden (Florida Museum of Natural History), Bob Emry (Smithsonian), Bob Hunt (University of Nebraska State Museum), Rich Cifelli (Stovall Museum of the University of Oklahoma), and John Flynn (first the Field Museum, now at the American Museum). This program was recently transformed into the Richard Van Gelder graduate program, a recent addition to the American Museum, which offered its doctorates independent of Columbia.

Osborn quickly worked to make the American Museum DVP the best in the world. In addition to purchasing all of Cope's fossils as the core of their collections, he used his connections among tycoons to procure funding to send field crews out to the Rockies to find many more dinosaurs (see figure 8.2). Several major expeditions to Wyoming yielded spectacular dinosaurs from Bone Cabin Quarry and other places not far from Como Bluff, and other collectors retraced the steps of Cope and Marsh or found new collecting grounds. By the time Osborn died in 1935, the DVP collections were the largest and most important in the world, especially for dinosaurs and fossil mammals.

All this activity drew many talented people to the DVP. These included not only important collectors Jacob Wortman, Olof Peterson, and Barnum Brown, but also outstanding anatomists and paleontologists William King Gregory and William Diller Matthew. Osborn kept a preparators' lab staffed with numerous talented men who cleaned and prepared the many crates of bones sent from the field, and he invented new techniques for creating large mounts in lifelike poses while hiding most of the supporting structures. Marsh had done this to some extent, but Osborn had a larger vision for paleontology: make these huge skeletons of extinct creatures famous in popular culture and draw big crowds to his American Museum (which he was director of from 1908 to 1933). For this purpose, he hired the first great paleoartist, Charles R. Knight, to create dynamic reconstructions of extinct animals in paintings, murals, and sculptures. Osborn also knew how to create publicity using spectacular accounts of extinct monsters, which were ready-made for reporters to wow the public. The final stroke of luck was promoting a young daredevil, Roy Chapman Andrews, who spearheaded Osborn's legendary Central Asiatic expeditions in the 1920s to find dinosaurs and mammals (and early humans, which they never found). Many people think Andrews was one of the models for the hero/explorer/archeologist characters such as Indiana Jones and Allan Quartermain.

"NE PLUS ULTRA" IN THE EVOLUTION OF LARGE CARNIVORES

In 1900, Barnum Brown (see chapter 21) was collecting in what is now called the Lance Formation of eastern Wyoming, and he found a partial skeleton of a huge theropod dinosaur. By 1905, the skeleton was prepared, and Osborn named it *Dynamosaurus imperiosus* ("powerful imperial lizard" in Greek). In 1902, Brown had begun to collect in the Hell Creek beds of central Montana, which were from the latest Cretaceous as was the Lance Formation of Wyoming, and he found another even more complete skeleton of this huge theropod. In the 1905 paper in which Osborn named *Dynamosaurus*, he called this second specimen *Tyrannosaurus rex*. He soon began to realize that *Dynamosaurus* and *Tyrannosaurus* were the same creature and required only one name. Fortunately for science and dino-enthusiasts everywhere, Osborn published a paper in 1906 in which he used his status as first reviser to suppress the rather confusing name *Dynamosaurus* in favor of the much more vivid *Tyrannosaurus*.

Dynamosaurus was not the only name given to the creature before it acquired the name *Tyrannosaurus*. Numerous fragments of what now appear to be *Tyrannosaurus* were found many times before 1900. In 1874, Arthur Lakes found some giant teeth near Golden, Colorado, that were probably from *Tyrannosaurus*. In the early 1890s, John Bell Hatcher collected parts of a skeleton from the Lance Formation in eastern Wyoming that were thought to be from a large *Ornithomimus*, named *Ornithomimus grandis*. Cope found some broken vertebrae from the Cretaceous of western South Dakota that he named *Manospondylus gigas*. The type locality of this find has been relocated and yields *Tyrannosaurus* fossils, so it's possible to argue that the two names represent the same dinosaur. If so, scholars will almost certainly petition the International Commission on Zoological Nomenclature to suppress *Manospondylus* as a "forgotten name." Luckily for posterity, no one is seriously thinking of replacing *Tyrannosaurus* with *Manospondylus* or *Dynamosaurus*! (*Ornithomimus* is not a problem because it originally was given to a different dinosaur.)

Brown kept finding more of these specimens in the Hell Creek beds, and five partial skeletons were eventually collected. The fourth of these skeletons was by far most complete, and this became the famous mounted skeleton still on display in the American Museum in New York (figure 14.3). The original *Dynamosaurus* skeleton was eventually sold to the

Figure 14.3 ▶

Different exhibits of the same *Tyrannosaurus rex* skeleton in the American Museum: (*A*) the original mount from 1915 in the tripodal "kangaroo" stance, leaning on its tail; (*B*) the current mount, redone in 1992, with the animal balanced on its hind limbs and the tail out straight behind it. ([*A*] Image #327524, courtesy of the American Museum of Natural History Library; [*B*] courtesy of Wikimedia Commons)

National History Museum in London, and the first *T. rex* skeleton (the type specimen) was sold to the Carnegie Museum in Pittsburgh.

In his original 1905 paper, Osborn realized that this discovery was impressive, and he lost no chance to hype it with purple prose. He called it "the *ne plus ultra* of the evolution of the large carnivorous dinosaurs: in brief it is entitled to the royal and high-sounding group name which I have applied to it."Wasting no time in making sure the press publicized his prize dinosaur, Osborn sent them lots of information to write about. On December 3, 1906, an article in the *New York Times* described the creature as "the most formidable fighting animal of which there is any record whatever," the "king of all kings in the domain of animal life," "the absolute warlord of the earth," and a "royal man-eater of the jungle." In another *New York Times* article, *Tyrannosaurus rex* was called a "prize fighter of antiquity" and the "Last of the Great Reptiles and the King of Them All."

But newspaper publicity about the find was not enough for Osborn. He quickly got his preparators working on cleaning the fossil and creating an impressive mount, and he commissioned Charles R. Knight to create both sculptures and paintings of it, which were soon copied in the media around the world. *Tyrannosaurus* became one of the most popular of the dinosaurs and was a cultural icon. It first appeared in the 1918 silent film *Ghost of Slumber Mountain* and later in the 1925 stop-motion version of *The Lost World* (based on the 1912 novel by Sir Arthur Conan Doyle). Tyrannosaurs made an appearance in the original 1933 version of *King Kong* and in the 2005 Peter Jackson remake. They have appeared in everything from movies to TV shows (for example, the "Sharptooths" of the *Land Before Time* series and Barney the Dinosaur on PBS Kids TV). *Tyrannosaurus rex* was the star and the major villain of the first four *Jurassic Park/Jurassic World* movies. It has been featured in parade floats and turned into thousands of items of merchandise. There was even a rock band called "T. rex." It is hard not to be impressed by a huge predator, towering over visitors to the museum. The late paleontologist Stephen Jay Gould said that his first sight of this skeleton at age five both terrified him and inspired him to become a paleontologist.

A *REX* NAMED "SUE"

In the 114 years since it was first discovered and described, a lot has been learned about *Tyrannosaurus rex*. The most obvious change is the posture of the dinosaur, as can be seen in the changing position of the mounted skeletons. The original American Museum mount that Osborn supervised depicted it as very reptilian, dragging its tail and walking in a tripodal posture (figure 14.3A).

During the Dinosaur Renaissance in the 1970s and 1980s, paleontologists began to rethink bipedal dinosaur posture. It made more sense that they were active, fast-moving predators, with their bodies balanced on their hind legs in a bird-like stance. This configuration was confirmed by the fact that large theropod trackways never show tail drag marks, and some theropods (such as *Deinonychus*) have a stiffening network of elongate extensions of their vertebrae preserved in their tails. (This is similar to the ossified trusswork of tendons in duck-billed dinosaurs, and the long extensions of the vertebrae in sauropods.) During the 1970s, paleontologists studying the original, upright *Tyrannosaurus rex* mounts realized that the pose was impossible. It would cause their limbs to become disjointed, the tail to bend to an impossible degree, and weaken the joint between the skull and neck. When the American Museum revamped their dinosaur halls in 1992, they remounted their *Tyrannosaurus rex* (Brown's fourth specimen), which Osborn had put in the kangaroo pose in 1915, with the spine in the horizontal position (only 77 years later). Thanks to the *Jurassic Park* book by Michael Crichton, and the movie by Steven Spielberg in 1993, most people are now familiar with this horizontally balanced version of *Tyrannosaurus rex* (figure 14.3B), and the archaic, tail-dragging version seen on many toys and sculptures and images looks odd to us.

The size estimates of *Tyrannosaurus rex* have also changed over time. Currently, the largest known relatively complete skeleton is in the Field Museum of Natural History (figure 14.4) and is nicknamed "Sue" (after Sue Hendrikson, who discovered it). It measures 3.66 meters (12 feet) at the hips and is 12.3 meters (40 feet) in length, and the number of tail vertebrae is not a guess (as we saw with many incomplete dinosaurs, such as the sauropods, in chapter 10). This is the longest preserved *Tyrannosaurus rex* fossil we have—there's a good chance that some individuals were larger. Given that the maximum length of known specimens is not guesswork, it's surprising that there are a wide range of weight estimates. Over the years the weight numbers have been as low as 8.4 metric tonnes (9.3 tons) to 14 metric tonnes (15.4 tons), but most modern estimates place it between 5.4 metric tonnes (6 tons) and 8 metric tonnes (8.8 tons). A study by Packard and colleagues in 2009 tested the methods used to estimate the dinosaur weights on elephants of known weight and argued that most of

The *Tyrannosaurus rex* specimen known as Sue in the Field Museum of Natural History in Chicago. (Courtesy of Wikimedia Commons)

the estimates are too high—and this does not accounting for the possibility that *Tyrannosaurus rex*, like many dinosaurs, may have had air-filled sacs in much of its body. Most recently the estimates given are around 9 metric tonnes and not much higher, so some paleontologists consider a 10-tonne *Tyrannosaurus* to be very unlikely.

Recent research has made it possible to better understand the skull and what it could do with its powerful jaws. Unlike most other theropods, which had relatively long slender jaws (chapter 13) or short faces with relatively small teeth such as the abelisaurs of South America (chapter 15), *Tyrannosaurus rex* had an extremely robust skull and huge teeth and could generate an enormous bite force. The skull of the largest *Tyrannosaurus rex* was about 1.5 meters (5 feet) long, but it was made lighter due to numerous air pockets and holes in the solid bone, reducing the weight of the head. In cross section, the snout was shaped like an upside-down "U," making it more rigid and stronger than a typical theropod skull, which has a cross section shaped like an upside-down "V." The snout was also narrow enough that the eyes could face fully forward, giving *Tyrannosaurus rex* excellent binocular stereovision and depth perception for hunting. The teeth are enormous, measuring typically about 30 centimeters (12 inches) from tip to root. They curve backward, giving it strength for the teeth to pull back as they rip out hunks of flesh. They were shaped a bit like steak knives, but they were as big as a banana and had serrated ridges on the cutting edges. The front teeth were thicker and deeply rooted, with a D-shaped cross section, giving them strength so they didn't break when *Tyrannosaurus rex* bit down and pulled backward.

Instead of being a slashing predator, *Tyrannosaurus* was a bone-crushing "bulldog dinosaur." Modern techniques for modeling bite forces suggest that *Tyrannosaurus rex* could produce 35,000–57,000 newtons (7,900 to 13,000 pounds) of force. This is 3 times more powerful that the bite of the great white shark, 3.5 times as strong as the Australian saltwater crocodile, 7 times more powerful than *Allosaurus*, and 15 times as powerful as the bite of a lion. Recently that estimate was revised upward to 183,000–235,000 newtons (41,000–53,000 pounds) of force, stronger than the bite of even the giant extinct shark *Carcharocles megalodon*.

So what did *Tyrannosaurus rex* do with those powerful jaws? We know that they fought with each other because several skulls show healed wounds on their faces and other bones that could only have been caused by the bite of another *Tyrannosaurus rex*. Some even have broken teeth embedded in their bones. Clearly they were capable of killing and eating nearly any dinosaur that lived in the Late Cretaceous with them. Unfortunately, the media have given a lot of publicity to the silly argument that *Tyrannosaurus rex* was primarily a scavenger and not a carnivore. In reality, modern large predators are not so picky. Lions, which mostly hunt their food, will gladly eat carrion when they are hungry; and hyenas, which are famous for breaking down carcasses and crushing bones, are very efficient pack hunters who prefer to kill their own meals when they get the chance. There is no reason to think that *Tyrannosaurus rex* was a picky eater; rather, it was opportunistic and ate anything it could, especially when it was hungry!

One of the most famous features of *Tyrannosaurus rex* was its tiny arms. Dozens of cartoons, gags, and even greeting cards have made fun of the fact that its arms seem useless. If you look closely, *Tyrannosaurus rex* also had only two functioning fingers, whereas most other theropods still had three fingers. Many illustrations of dinosaurs show them with five full fingers, which only occurred in the most primitive dinosaurs (and the ring finger and pinky are always highly reduced). This misconception may have been partially due to the fact that Osborn's famous 1915 mount of *Tyranno-saurus rex* in the American Museum was missing its lower arm and fingers, so Osborn made its hands resemble the three-fingered *Allosaurus*. However, in 1914, Lawrence Lambe showed that the closely related *Gorgosaurus* only had two fingers. Final proof of this did not come until 1989, when the "Wankel rex" in the Museum of the Rockies in Bozeman, Montana, was found and had complete forelimbs.

Paleontologists have debated why the forelimbs were so small ever since *Tyrannosaurus rex* was first found. Osborn suggested that they might have been useful to hold a mate during copulation. Others argued that it would help them rise from a prone position, and recent digital models have shown that to be plausible. More recent research has shown that the actual bones of the arms are quite strong and robust and would have had powerful muscles, capable of lifting 200 kilograms (440 pounds), so they were not weak arms. This suggests they could hold a smaller prey animal more easily than once believed, and they were certainly capable of slashing a prey animal or another *Tyrannosaurus rex* in close combat. When compared to *Carnotaurus*, which has even tinier stunted arms with vestigial fingers (chapter 15), the arms of *Tyrannosaurus rex* are not that small. More important, *Tyrannosaurus rex* focused on the powerful head and neck and jaws as its primary weapon. Along with its strong hind feet with long sharp claws, it had a different way of feeding than animals that rely on strong arms to catch prey. The arms of tyrannosaurs are most likely vestigial and less important for a predator that primarily used it head and legs.

The intense public interest in *Tyrannosaurus rex*, plus the ever-increasing number of specimens (over 50 known now), has allowed all sorts of research about them to flourish. For example, several juvenile specimens allow us to look at how they grew up. Unlike many other dinosaurs, which grew rapidly after hatching and then slowed down, tyrannosaur babies grew slowly at first, remaining less than 1.8 metric tonnes (2 tons) until they reached 14 years of age, then growing rapidly to full adult size. During this rapid growth period, they would add about 600 kilograms (1,300 pounds) of weight each year. When they reached an age of 16–18 years, their growth slowed down until they reached full size. The oldest known specimen was about 28 years old, so they had a short life full of slow growth, then fast growth, then a long struggle to survive as a full-sized adult. As Tom Holtz has put it, they "lived fast and died young." In fact, over half of the specimens of *Tyrannosaurus rex* seemed to have died within six years of reaching sexual maturity around age 18.

This is a common pattern among some birds and mammals that have high infant mortality rates, then rapid growth to adult size, and finally high death rates as adults due to battles over mating and the stresses of reproduction.

Many scientists have tried various methods to see if *Tyrannosaurus rex* shows sexual dimorphism (differences in males and females), but no study has been convincing so far. Only one specimen has been conclusively demonstrated to be female as it had medullary tissue in a number of bones. This bone tissue is only found in female birds that are laying eggs; they lay down spongy porous medullary tissue in their bones as they divert calcium to their eggs and embryos.

One of the most significant changes in how we think about *Tyrannosaurus* concerns its body covering. For almost a century, it was rendered as a big scaly reptile, a lizard on steroids. The only known skin impressions of *Tyrannosaurus* preserve a mosaic of small scales. In the 1990s, discoveries in China produced many fossils of dinosaurs, birds, and mammals with soft tissues preserved, especially in lake shales, which are low in oxygen and formed in stagnant water (see chapter 18). These produced a small tyrannosaur called *Dilong paradoxus*, which clearly showed filamentous feathers or fluff on its body. When a larger tyrannosaur, *Yutyrannus halli*, was found in China, it too was covered with a coat of feathers. Given that these animals are very closely related to *Tyrannosaurus rex* and all other tyrannosaurs, it is extremely likely that the iconic dinosaur of the five *Jurassic Park* movies was not the scaly lizard that the movie makers created but a bird-like creature with a coating of down or at least feather tufts over much of its body. Of course, by the second movie (*Jurassic Park: The Lost World*), the filmmakers stopped listening to paleontologists and continued to show all the dinosaurs as scaly lizard-like creatures, without adding feathers to any of them. This was truly sad for many of us in paleontology. The original novel and movie was up to date with the current state of dinosaur research in the early 1990s, but they abdicated their efforts to keep the movies current and just gave the audience the scaly monsters they had come to expect.

TYRANNOSAURS BECOME TSOTCHKES

One of the most spectacular and popular dinosaurs of all, *Tyrannosaurus rex* always makes the front page of major newspapers when a new specimen is found (possibly the only fossil that can make that claim other than human fossils). The pressure is on for collectors (both professional paleontologists and poachers) to find more specimens. The situation reached a crisis level when the famous skeleton of Sue was found on a ranch in western South Dakota. It was collected by a commercial fossil-hunting firm called the Black Hills Institute for Geologic Research. They are a group of people who collect fossils for profit, although they also maintain a well-curated museum. Many scientifically important specimens have been lost forever to science because they were sold to the highest bidder.

The discovery of Sue (figure 14.4) in 1990 completely changed the way tyrannosaurs are viewed in the public eye. They were no longer important scientific discoveries that needed to be accessible to any qualified researcher but were objects to score a big cash payout. The situation became messy when the legal rights to the land where Sue was found were challenged. The police confiscating the fossil from the Black Hills Institute, and its director, Peter Larson, went to jail on charges of tax evasion. Eventually, the ownership issue was resolved, and Sue was on the auction block. Everyone in the paleontological community feared that this scientifically irreplaceable specimen would end up in some millionaire's mansion or in a corporate lobby as an expensive showpiece. The bidding reached $8 million, far above the budget of any public museum. The Field Museum persuaded McDonald's, Disney, the California State Universities, and several other donors to pitch in, and they won the bid at over $8 million. Sue was then prepared and mounted at the Field Museum, where it is still on display (although it was recently moved from its original place on the floor of the Stanley Field Hall).

The $8 million price tag to keep Sue from becoming some millionaire's expensive showpiece caught the attention of poachers, and trade in illegal specimens unfortunately flooded the auction houses. Fossils were suddenly very valuable, and important collecting areas were heavily poached. Not only did the poachers steal the fossils illegally and remove them from further study by science, but in many cases they damaged the fossils by hacking them out carelessly due to their hurry to get the bones out before they were caught in the criminal act. Sometimes poachers "enhanced" the specimens with reconstructed plaster fakery to jack up its price. Worse still, poachers usually don't care about any of the information that comes from the beds that surround the fossil, which can tell geologists the age of the beds, the environment in which they were formed, and provide many other important pieces of information. For years after the Sue auction, hundreds of critical specimens were lost or destroyed in the mad rush by poachers to rape the land and make a buck.

This story highlights a much bigger problem, which most people never hear about: the huge international black market in stolen fossils and antiquities. Bit by bit some governments are becoming better at protecting their national treasures, but the poachers and smugglers are always much better funded and quicker even than Interpol. Not only is there a large black-market trade in stolen artwork and artifacts, but the market in natural objects is equally brazen and profitable. The stories I've heard just make us cry! Famous fossil localities in protected national parks all over the world have been brazenly poached by thieves, not only destroying most of the fossils but ruining the scientific value of the locality as well. Museum research collections, and even specimens on public display with security guards and video cameras protecting them, are stolen or damaged by thieves. One-of-a-kind fossils that are certainly new species and genera and have the potential to revolutionize our understanding of life's history are seen briefly and then end up in some rich person's foyer. Major auction houses and the more reputable fossil dealers have to be careful and identify poached specimens with fraudulent locality data hiding the illegality of their collection. I've heard the horror stories from my colleagues who had the proper permits and found an important bone bed on public land, only to come back a few days later and find that someone had plundered the best material and left the rest in broken piles, hacked out of the ground with no attempt to protect the fossil in a jacket—or record the location and stratigraphic horizon of the specimen, which is a big part of its scientific value. It's common practice now for scientists to bury their excavation and hide it once they leave to prevent this from happen. Museums have ask me not to publish GPS coordinates of my sites, which also might give away fossil localities. My fellow paleontologists in the fossil-rich national parks spend more and more time planning how to prevent poaching and less time doing the science they are better qualified to do. The situation on private land is even messier. It is usually legal to collect on most privately held ranches with the consent of the landowner, but the story of the tyrannosaur named Sue illustrate that handshake agreements, disputes over where the specimen was found, and unclear property rights can make those specimens a legal nightmare.

The Sue story puts museums in a catch-22 situation. They do not have the money to bid on these scientifically important specimens unless they receive funding from rich sponsors (as did Sue), and they don't want to bid on specimens with questionable provenience because these specimens are probably poached and their sale would encourage lawbreakers. However, museums want to make their best effort to keep scientifically important specimens in the public domain where they can be studied and researched and yield their secrets, and eventually be put on display for the public to see.

The issue reached a crisis point in 2012 when some specimens came up for auction at a prominent New York auction house. The buzz was going back and forth among my paleontologist colleagues for weeks: an important tyrannosaur skeleton (*Tarbosaurus bataar*) that had been poached in Mongolia was scheduled to be auctioned on May 20 in New York City. We paleontologists were all outraged and spent days signing online petitions, blogging, and sending letters and emails to the appropriate parties, but the tiny scientific community of vertebrate paleontologists in the United States (no more than 2,000 people) doesn't hold any real power beyond a handful of museum curator positions and top professorships. Our activity did get the attention of the Mongolian government, however, and they sent formal letters of protest to the auction house and to the U.S. government. All the emails and blog posts were full of anger and despair that such a blatant theft could be rewarded with a million-dollar sale.

Just hours before the auction was to start, a Texas judge issued a Temporary Restraining Order, putting the sale of the Mongolian tyrannosaur on hold, but the auction house went ahead with the sale, receiving a final bid over $1 million, and arguing that the Texas judge had no jurisdiction over a New York auction house. This occurred even as the attorney for the Mongolian government was in the auction room with the Texas judge on his cell phone, trying to make himself heard and to get anyone to listen to the judge. Shortly after the bid of $1 million, the sale was stopped pending further investigation. The wheels of justice began to grind slowly as investigators dug into the data about the fossil, tracked down the anonymous "paleontologist" who had procured it, and supplied the information to our legal system. Just before New Year's Eve 2013, the story broke that the culprit, poacher Eric Prokopi of Gainesville, Florida, had been arrested and pleaded guilty.

The evidence against him was overwhelming. Not only was he caught with multiple dinosaur specimens from Mongolia in his possession in his

Florida home, but *another* dinosaur from Mongolia arrived at his address while he was being investigated! In his possession were illegal specimens from Mongolia and illegal fossils from China, such the "four-winged" dinosaur *Microraptor gui*, which comes from a very specific area of the Liaoning Province of China and cannot be sold or even removed from China under any legal means. Investigators found abundant emails from Prokopi's computer, and from those of his associates, that established he knew the specimen was illegal and was doing his best to cover up its illegality (despite his assurances to the auction house and forged documents proving that the specimen was not from Mongolia). There were even pictures of him in Mongolia collecting the specimens from a well-known locality easily identified in the photo, proving that he not only smuggled specimens in and out but apparently snuck in and out of the country illegally as well. No wonder he pleaded guilty—no competent lawyer could make the case that he was not aware of what he was doing in the face of such damning evidence.

The judge had the option of sentencing this criminal to as much as 17 years in prison. Most paleontologists were in favor of a stiff sentence because this guy was so blatant in his lying and deception and stood to make millions on his poached specimens. In addition, it would send a message to other poachers and smugglers that their long-unsupervised activities just got a lot riskier. Sadly, the judge gave him only a three-month sentence, a mere slap on the wrist, because he agreed to turn informant and cooperate in the arrest of the entire smuggling pipeline. His only other punishment was the loss of all his illegally gained fossils. The entire story is told in detail in Paige Williams' new book, *The Dinosaur Artist*.

Sources tell me that nearly all other Mongolian fossils are off the block for a while because other auction houses are leery of bad publicity or getting in trouble with the law. Maybe it will put a chill in the smuggler's pipeline back to Mongolia and China as well. What hasn't been revealed yet is how such a large specimen was so easily smuggled out of Mongolia and, even more amazing, through U.S. Customs. So far all we've heard is that Prokopi mixed the tarbosaur bones with a crate of "miscellaneous reptile fossils" shipped from a middleman in England. It turned out that the "75 percent complete" skeleton was actually a composite of several different skeletons, not one individual. Such fakery is common among commercial fossil dealers and smugglers. Clearly further investigation of this criminal enterprise is needed, and Prokopi is just the tip of the iceberg.

Some people attempted to excuse Prokopi by comparing him to "Indiana Jones," but they apparently misremember the movies. Indy was dedicated to bringing archeological relics to museums and keeping them in the public trust, not selling them to the highest bidder. (Today most archeologists wouldn't even consider removing artifacts from their country of origin due to the increased sensitivity of most countries to the pillaging of their cultural heritage.) No, Prokopi is like the character Belloq in the first Indiana Jones movie (or the thieves in the beginning of the third Indiana Jones movie) who gleefully sells any object (and their services) to whoever would pay for them. As paleontologist Peter Harries of the University of South Florida pointed out at the end of an article that defended Prokopi: "I don't think Indiana Jones wanted to sell the Ark of the Covenant for $1 million."

FOR FURTHER READING

Carpenter, Kenneth, and Peter Larson. *Tyrannosaurus Rex, the Tyrant King*. Bloomington: Indiana University Press, 2008.

Colbert, Edwin. *Men and Dinosaurs: The Search in the Field and in the Laboratory*. New York: Dutton, 1968.

Erickson, Gregory M., Peter J. Makovicky, Philip J. Currie, Mark A. Norell, Scott A. Yerby, and Christopher A. Brochu. "Gigantism and Life History Parameters of Tyrannosaurid Dinosaurs." *Nature* 430, no. 7001 (2004): 772–775.

Farlow, James, and M. K. Brett-Surman. *The Complete Dinosaur*. Bloomington: Indiana University Press, 1999.

Fastovsky, David, and David Weishampel. *Dinosaurs: A Concise Natural History*, 3rd ed. Cambridge: Cambridge University Press, 2016.

Henderson, M. *The Origin, Systematics, and Paleobiology of Tyrannosauridae*. Dekalb: Northern Illinois University Press, 2005.

Holtz, Thomas R., Jr. *Dinosaurs: The Most Complete, Up-to-Date Encyclopedia for Dinosaur Lovers of All Ages*. New York: Random House, 2011.

——. "The Phylogenetic Position of the Tyrannosauridae: Implications for Theropod Systematics." *Journal of Palaeontology* 68, no. 5 (1994): 1100–1110.

——. "Tyrannosauroidea." In *The Dinosauria*, 2nd ed., ed. David B. Weishampel, Peter Dodson, and Halszka Osmólska, 111–136. Berkeley: University of California Press, 2004.

Hone, David. *The Tyrannosaur Chronicles: The Biology of Tyrant Dinosaurs*. London: Bloomsbury, 2016.

Horner, John R., and Don Lessem. *The Complete T. Rex.* New York: Simon & Schuster, 1993.

Horner, John R., and Kevin Padian. "Age and Growth Dynamics of Tyrannosaurus Rex." *Proceedings of the Royal Society B: Biological Sciences* 271, no. 1551 (2004): 1875–1880.

Molnar, Ralph. *Tyrannosaurid Paleobiology.* Bloomington: Indiana University Press, 2013.

Naish, Darren. *The Great Dinosaur Discoveries.* Berkeley: University of California Press: 2009.

Naish, Darren, and Paul M. Barrett. *Dinosaurs: How They Lived and Evolved.* Washington, D.C.: Smithsonian Books, 2016.

Osborn, Henry Fairfield. "*Tyrannosaurus* and Other Cretaceous Carnivorous Dinosaurs." *Bulletin of the American Museum of Natural History* 21, no. 14 (1905): 259–265.

Wilford, John Noble. *The Riddle of the Dinosaur.* New York: Knopf, 1985.

Williams, Paige. *The Dinosaur Artist: Obsession, Betrayal, and the Quest for the Earth's Ultimate Trophy.* New York: Hachette Books, 2018.

Xing Xu, Mark A. Norell, Xuewen Kuang, Xiaolin Wang, Qi Zhao, and Chengkai Jia. "Basal Tyrannosauroids from China and Evidence for Protofeathers in Tyrannosauroids." *Nature* 431, no. 7009 (2004): 680–684.

GIGANOTOSAURUS

Skeletons of a huge, meat-eating dinosaur that overshadows *Tyrannosaurus rex* have been discovered in Argentina. The newly revealed species is one of the biggest carnivores ever to have walked the Earth, dinosaur experts say. At least seven of the animals were uncovered together in a mass fossil graveyard in western Patagonia, a region famous for giant-dinosaur remains. Living some 100 million years ago, the largest specimen was more than 40 feet (12.5 meters) long.

—JAMES OWEN, *NATIONAL GEOGRAPHIC NEWS*, 2006

WHO'S ON FIRST?

Thanks to a century of publicity and fame, most people think *Tyrannosaurus* is the largest land predator ever known. That was certainly true for most of the twentieth century when it was one of the few large theropod dinosaurs known. In the past 30 years, more remote and less explored parts of the world have yielded spectacular dinosaurs that routinely exceed the early North American finds everyone knows for their weirdness and great size.

Patagotitan, Argentinosaurus, and the other enormous titanosaurs of South America are discussed in chapter 10. Indeed, South America has been one of the most productive areas for finding new dinosaurs in the past 40 years (along with China, which was shut off from the Western world until 30 years ago when the Cultural Revolution ended). But thanks to the hard work of José Bonaparte (see chapter 5) and many of his students and colleagues, more and more of the huge and bizarre dinosaurs of Argentina have been discovered.

One of the richest areas is the Neuquén Province in northern Patagonia, which covers much of the dry foothills of the Andes Mountains. Amateur fossil collector Ruben Carolini was riding his dune buggy through these Middle Cretaceous badlands near Villa Chocon in 1993 when he spotted a large bone. He called specialists from the Natural History Museum of Comahue to come and look at it, and they quickly excavated a large nearly complete skeleton of a theropod. The partial skull was scattered over a wide area of 10 square meters (110 square feet) when it was originally found. Like most dinosaur finds, the rest of the skeleton was scattered around with no two bones attached to each other (disarticulated). However, it was still about 70 percent complete, with the skull and jaws, most of the backbone, pectoral and pelvic girdles, and most of the hind limb bones. The arms and hands were missing, but this is a lot more complete than many other dinosaurs (including most of the huge sauropods discussed in chapter 10).

By 1994, my friend and colleague, theropod specialist Rodolfo Coria of the Muséo Carmen Funes, was working on the fossil with Leonardo Salgado. I vividly remember hearing him speak about the find at the 1994 meeting of the Society of Vertebrate Paleontology, and the enormous size of this predator took my breath away; the entire room filled with the world's best dinosaur paleontologists was equally impressed. Science writer Don Lessem saw pictures of the enormous leg bones and helped the Argentinians fund further excavation and preparation of the fossil.

By 1995, Coria and Salgado published their description of all the bones that had been found. They named it *Giganotosaurus carolinii*. The genus name is Greek *gigas* for "huge," *noto* for "southern," and *sauros* for "lizard," and the species name honors its discoverer. The correct pronunciation, as I've heard it from all the dinosaur experts at the Society of Vertebrate Paleontology meetings, is "GIG-a-NOTE-o-saurus" (just as we say "GIG-a-byte" not "JIG-a-byte"). Unfortunately, most people read the name as "Gigantosaurus" and mispronounce it as well. The name "Gigantosaurus" was originally used for a few tail bones of a huge titanosaur sauropod from England (see chapter 10), and it is no longer considered diagnostic enough to be a valid genus. So not only is the name *not* "Gigantosaurus," as many misread it, but no dinosaur could be given that name because the name was used previously.

The specimen is currently displayed at the Muséo Municipal Ernesto Bachman in Villa Chocon, near where it was found (figure 15.1A). The original fossils are not placed in life position in a mount but lie flat on a bed of sand in articulated position, looking as if they had just been exposed. (Remember, it was not discovered in this state.) In the next room is a mounted replica of those same bones, posed in a lifelike position (figure 15.1B). *Giganotosaurus* has become so famous in recent years that casts of it are on display elsewhere in Argentina, including at the Muséo Municipal Carmen Funes in Plaza Huincul, as well as at the Academy of Natural Sciences in Philadelphia, the Fernbank Museum of Natural History in Atlanta, the Naturmuseum Senckenberg and the Frankfurt Hauptbahnhof in Frankfurt am Mein, Germany, the Hungarian Natural History Museum, and the Australian Museum in Sydney.

What kind of dinosaur was *Giganotosaurus*? Like its close relative *Carcharodontosaurus* (figure 15.2), the skull was lightly built with arches and struts of bone and lots of openings; the skull was not quite as high and arched as *Carcharodontosaurus* but lower and flatter on top. There were roughened areas above the eyes and the top of the snout. The rear of the skull had a forward slant, so the jaw joint hangs behind and beneath the attachment point between the neck and the back of the skull. It was definitely not the robust, heavy crushing skull seen in *Tyrannosaurus*. It also lacked the sagittal crest along the top midline of the skull, where the jaw muscles attached, suggesting its bite was not as powerful as that of some other huge theropods. Some estimates suggest its bite was only a third as powerful as the bite of a *T. rex*. The teeth were like knife blades, compressed side to side and with the point curved backward, and they had serrations on both the front and back cutting edges. Nonetheless, it had a robust neck, so it could cope with struggling prey after it had a bite on them. This suggests it did not try to crush its prey with a bulldog bite, as did a tyrannosaur, but may have slashed it and then allowed it to bleed to death. It also suggests that it may have taken smaller prey than that of the largest dinosaurs.

How big was *Giganotosaurus*? It is very difficult to get a reliable estimate of size for dinosaur skeletons, even when they are complete, like *Tyrannosaurus*, or nearly complete, like *Giganotosaurus*. Most of the back and tail

Figure 15.1 ▶

(A) Original skeleton of *Giganotosaurus* on display at Muséo Municipal Ernesto Bachman in Villa Chocon, Argentina; (B) mounted replica of the original skeleton in the Museo Municipal Carmen Funes in Plaza Huincul, Argentina. ([A] Courtesy of Wikimedia Commons; [B] courtesy of R. Coria)

Figure 15.2 ▲

The skull of *Carcharadontosaurus* from the Kem Kem beds of Morocco. (Courtesy of P. Sereno)

vertebrae of *Giganotosaurus* are known, so we can estimate its length at around 12.5 meters (41 feet), whereas Sue, the largest complete tyrannosaur skeleton, is about 12.3 meters (40 feet) in length. A more direct comparison is the length of complete bones that scale with size, so the thighbone of *Giganotosaurus* is 1.3 meters (4.7 feet) long and more robust than that of a tyrannosaur, and that of Sue is 5 centimeters (2 inches) shorter. Coria has also described a fragmentary jaw for a *Giganotosaurus* that is even larger than the original type specimen. Although the size differences are not dramatic, it seems that the largest *Giganotosaurus* is longer than the largest tyrannosaur.

What about weight? Estimating weight is even less imprecise than direct comparisons of complete bones or spinal columns. Coria and colleagues originally estimated the weight of *Giganotosaurus* around 6–8 metric tonnes (6.7–8.7 tons), although in 1997 Coria upped the estimate to 8–10 metric tonnes (8.8–11 tons) based on the incomplete lower jaw that was about 8 percent larger than the original specimen. Other methods of calculating the weight of dinosaurs have given results for 6.6 metric tonnes (7.3 tons) to 6.5 metric tonnes (7.2 tons) to 8 metric tonnes (8.2 tons) to 8.21 metric tonnes (9.0 tons) to as much as 13.8 metric tonnes (14 tons). In short, the size ranges of tyrannosaurs and *Giganotosaurus* overlap, with the largest specimens of *Giganotosaurus* just slightly larger than the largest *T. rex*.

Giganotosaurus was not the only huge predator in Argentina at that time. From 1997 to 2001, the Argentinian-Canadian Dinosaur Project was digging in the Huincul Formation at Cañadón del Gato, the same beds that produced *Argentinosaurus*. These beds are only slightly younger than those that produced *Giganotosaurus*, dating between 93.5–97 million years old, whereas the *Giganotosaurus* fossils come from beds 97–100 million years in age. They began to find another huge theropod, with fragments of bones representing most parts of several skeletons, and in 2006 Rodolfo Coria and Canadian paleontologist Phil Currie from the Royal Tyrrell Museum in Alberta described these fossils. They gave it the name *Mapusaurus*, from the Mapuche word *mapu* meaning "of the earth" and *saurus*, "lizard." Although it is less complete and far more fragmentary than the extraordinary *Giganotosaurus* skeleton, the similar parts of *Mapusaurus* are almost all the same size as those in *Giganotosaurus*.

The bone bed contained the fossils of at least seven different individuals of *Mapusaurus*, mostly in different growth stages. The accumulation of so many theropods has stimulated a lot of speculation and discussion. Some think that this represents a death assemblage of a family or a pack of theropods, and that *Mapusaurus* was a social pack hunter. Others argue that the death assemblage is the accidental accumulation of a number of skeletons washed into a backwater sandbar. This kind of all-predator assemblage is not unique to *Mapusaurus*. The famous Cleveland-Lloyd Quarry in northern Utah is composed almost entirely of *Allosaurus* and other predators, and there is no consensus on whether this is a natural biological assemblage or something that the water currents arranged and collected.

If those two huge predators were not enough, yet another one was found in 2005 and described by Argentinian paleontologists Fernando Novas, Silvina de Valais, and Australian paleontologists Pat and Tom Rich. They come from the Chubut Province in the southern part of Patagonia and are older than the previous dinosaurs, dating from 112–120 million years in age. Named *Tyrannotitan chubutensis* by Novas and colleagues, it consists mostly of a partial skull, a few backbone elements, and part of the hips and hind limbs. It is too fragmentary to reliably estimate its size, but the original estimates of length were 11.4–13.0 meters (37–44 feet), only slightly shorter than *Mapusaurus* and *Giganotosaurus*, and between 4.9 and 7 metric tonnes (5.4–7.7 tons). As with *Mapusaurus, Giganotosaurus,* and *Tyrannosaurus,* the measurements of the largest individuals overlap, so it's difficult to know which of these four dinosaurs was truly the largest.

BACK TO AFRICA

No claim of "largest this" or "longest that" goes unchallenged in dinosaur paleontology for very long. Soon after *Giganotosaurus* was becoming familiar and established as the "largest predator," another discovery in Africa pulled the spotlight away from South America. In 1995, my classmate, coauthor, and good friend Paul Sereno (see chapter 5) was in northwestern Africa on the fringe of the Sahara Desert in the Kem Kem beds of Morocco. The crew found enormous bones of sauropods in many places, but one of the most striking discoveries was the skull of an enormous predator. When Sereno and his colleagues put the skull together, they realized it was a complete skull of a dinosaur that had been previously named (figure 15.2). In 1924, French paleontologists Deperet and Savornin obtained two enormous shark-like dinosaur teeth from the Middle Cretaceous of Algeria. With only teeth to go on, they referred the specimens to the wastebasket taxon *Megalosaurus,* which was used for nearly every fragmentary theropod for a long time (see chapter 1), but they called it *Megalosaurus saharicus.* Then German paleontologist Ernst Stromer (see chapter 13) obtained a partial skull, teeth, vertebrae, partial hip bones, leg bones, and claws during his excavations at Bahariya Oasis in western Egypt. The original fossils were found in 1914, but they did not reach him until the early 1920s due to World War I and other delays. When he finally studied and published the fossils in 1931, he realized the teeth resembled the ones found in Algeria, but it was a completely different dinosaur than *Megalosaurus.* Based on those fragments and the similarity of the teeth to those of the great white shark *Carcharodon,* he renamed the dinosaur *Carcharodontosaurus saharicus.*

Unfortunately, the original fossils of *Carcharodontosaurus* in the Munich Natural History Museum were destroyed during the same 1944 bombing that wiped out the original material of *Spinosaurus* and nearly all of Stromer's Egyptian fossils. Sereno and his colleagues realized that they had found an even more complete fossil of Stromer's *Carcharodontosaurus*, sufficient to replace the lost specimens destroyed during the air raid. It's not often that we find even better specimens when the originals are lost due to unfortunate circumstances!

Similar to *Giganotosaurus* but unlike *Tyrannosaurus*, the skull of *Carcharodontosaurus* is built of struts and arches of bones with lots of open spaces for reducing weight (figure 15.2). However, the skull of *Carcharodontosaurus* has a much higher arch than the relatively flat-topped, low skull of *Giganotosaurus*. As previously mentioned, its distinctive feature is the narrow, knife-like teeth suitable for slashing, which are not very different from the teeth of the great white shark. Like all of these dinosaurs (*Giganotosaurus, Mapusaurus,* and *Tyrannotitan*), these features of the skull and teeth define a group that is now called the carcharodontosaurs. This group of dinosaurs is, in turn, closely related to *Concavenator* from the Cretaceous of Spain; *Acrocanthosaurus* from Oklahoma, Texas, and Wyoming; and *Shaochilong* from China. Similar to most of the dinosaur groups we have seen, carcharodontosaurs were distributed across most of the Pangea continents (South America, North America, Africa, Europe, and Asia), with only the poor Cretaceous record in Australia and Antarctica preventing them from being completely cosmopolitan—although Australia has now produced a distant relative of the carcharodontosaurs, known as *Australovenator*.

How big was *Carcharodontosaurus*? We have only a skull and fragments of the skeleton to go on. The most common method of establishing body size in this case is the length of the skull, which measured 1.6 meters (5.2 feet), one of the longest predatory dinosaur skulls known. Using typical scaling relationships, this gives a length for *Carcharodontosaurus* of about 12–13 meters (39–43 feet), and a weight between 6 and 15 metric tonnes (6.6–16 tons).

The original skull of *Giganotosaurus* was fragmentary, and there is some question of the reliability of the skull length estimates for a skull reconstructed from pieces. Coria and coauthors believe that it reached 1.8 meters (5.9 feet), with a body length of about 13 meters (42 feet). This is only slightly larger than *Carcharodontosaurus*, but the measurements of the two dinosaurs largely overlap. However, the largest specimen of *Giganotosaurus* is a lower jaw, which suggests a skull 1.95 meters (6.4 feet) long, and a length of 13.2 meters (43 feet). Once again, we are faced with the dilemma of who is biggest. At least four large carcharodontosaurs seem to have an overlapping range of sizes. The largest specimens of *Giganotosaurus* are just marginally larger than the rest, but there is no dramatic difference that makes any one dinosaur stand out as the clear Numero Uno.

Regardless of the outcome of the debate, it's clear that the largest predators that ever lived were the carcharodontosaurs, and the largest ones lived in the Early-Middle Cretaceous of South America and Africa.

THE "FLESH BULL"

Predatory dinosaurs show a lot of regional differences in the Cretaceous (figure 15.3). Tyrannosaurs and ornithomimids (ostrich dinosaurs) ruled North America, China, and Mongolia, but they were not found much outside of these regions. In contrast, South America and Africa not only had carcharodontosaurs and spinosaurs in the Late Cretaceous but also had

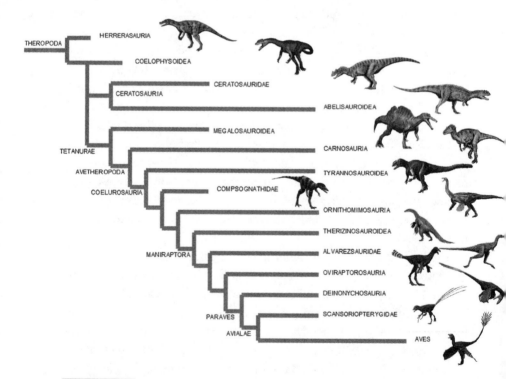

Figure 15.3 ▲

Family tree of the theropods showing the major branches discussed in this book. (Redrawn from several sources)

several predatory dinosaurs not found in North America. This is comparable to the way that titanosaurs dominated the sauropod faunas of South
America and Africa in the Cretaceous but were found only in the southwestern part of North America, and only in the latest Cretaceous.

In addition to the huge carcharodontosaurs and spinosaurs, another
characteristic group of theropods found mainly in South America and other
southern continents, such as India and Madagascar, were the abelisaurs.
Most of these theropods were not as large as the carcharodontosaurs or
spinosaurs, although one specimen from Kenya has been reported at professional meetings but not yet published that may be as long as 12 meters.
Abelisaurs were distinctive, with relatively short and deep snouts and
lots of smaller, simpler conical teeth (rather than the shark-like teeth of
carcharodontosaurs or the banana-shaped teeth of tyrannosaurs). Most of
them have extremely reduced forearms.

The first known example was *Abelisaurus* itself, discovered and described
by (who else?) José Bonaparte and Fernando Novas in 1985. Another was
Aucasaurus, described by Rodolfo Coria, Luis Chiappe, and Lowell Dingus
in 2002 from the Auca Mahuevo locality that produced so many dinosaur
eggs. Unlike *Abelisaurus*, which was known only from a skull, *Aucasaurus*
is based on a nearly complete skeleton that measured 6.1 meters (20 feet)
in length. This is not as long as the carcharodontosaurs but it was still a big
predator. Other large South American abelisaurs include *Skorpiovenator,*
Ekrixinatosaurus, and *Ilokelesia.*

But the weirdest (and most famous) of these strange predators was
Carnotaurus (figure 15.4A–B). Discovered and described by José Bonaparte
in 1985, its name means "flesh bull" in Latin, in reference to its flesh-eating
habits and the strange pair of bull-like horns over its eyes. The horns give
it a wicked or even diabolical appearance, so naturally it was the villain
in Disney's 2000 movie *Dinosaur,* the first film on dinosaurs produced
with computer graphics. According to insider accounts, Disney's animators
originally planned to use *Tyrannosaurus rex* because of its iconic status
as a terrifying giant predator, but they switched to *Carnotaurus* when
they saw its horns. Since then, *Carnotaurus* has appeared in a number of
films and TV shows, including the kid's show *Dinosaur Train* (where it is
friendly and benevolent, not a nasty killer—the show is for preschoolers
after all). It also made an appearance in the 2018 movie *Jurassic World:*
Fallen Kingdom, where it almost kills dinosaur wrangler Owen Grady

Figure 15.4 ▲ ▶

Carnotaurus, the "flesh bull": (*A*) mounted replica of the original skeleton in the Natural History Museum of Los Angeles County; (*B*) reconstruction of the animal in life; (*C*) diagram of the skull; (*D*) close-up of the tiny, nonfunctional arms with almost no fingers. ([*A, D*] Photographs by the author; [*B*] courtesy of N. Tamura; [*C*] courtesy of R. Coria)

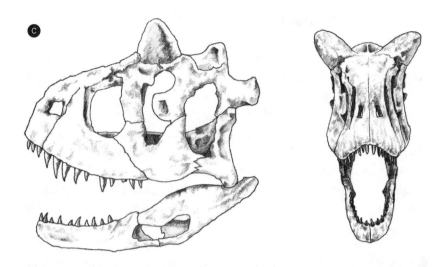

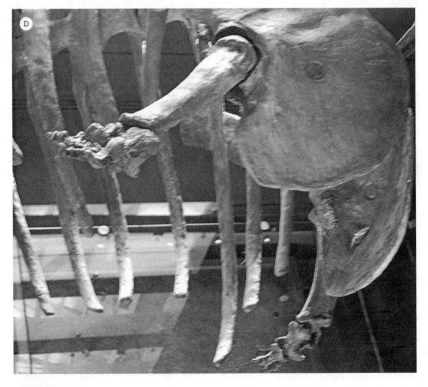

Figure 15.4 ▲
(*continued*)

(played by Chris Pratt) until a *T. rex* kills it—all while the volcano is erupting in the background.

Like other abelisaurs, *Carnotaurus* had a short deep snout (figure 15.4C) with lots of smaller conical teeth, which are suited for a quick slashing or nipping bite on the prey. *Carnotaurus* also had a strong, bull-like neck for grappling with prey once it had a bite on them, or wrestling with other members of its species. The skull had lots of bony struts and was lightly built, so *Carnotaurus* could probably stretch the joints and open its mouth very wide around a smaller prey item. In contrast, the skull of a *T. rex* was solidly built for inflicting a powerful crushing bite, not ripping out chunks of the prey animal until it bled to death. The bite force of *Carnotaurus* was not that strong, but it still was much stronger than that of a crocodile or an alligator.

If people mock tyrannosaurs for having small arms, abelisaurs had ridiculously tiny arms (figure 15.4D). *Aucasaurus* had tiny arm bones with no finger bones in its hands. *Carnotaurus* was almost as bizarre, with lower arm bones much shorter than the upper arm bones, and only two functional finger bones (index and middle finger), which were short and stubby. Its wrist bones never developed but apparently remained as cartilage, as did any other fingers it might have had. Only the middle finger had a claw. The tip of the ring finger is missing, and instead a long bony spur stuck out of the side of the hand. Paleontologists have made the argument that tyrannosaurs might have had some limited function of their tiny arms (see chapter 14), but clearly the limbs of abelisaurs were virtually useless. They are what are known as vestigial organs, which were in the process of being lost but were not completely gone (even if they had no real function).

In addition to the lightweight skull, the overall build of *Carnotaurus* was lighter than the tyrannosaurs or allosaurs or other large predators. Unlike most dinosaurs, *Carnotaurus* is based on a nearly complete articulated skeleton found by José Bonaparte in 1985 in its death pose, so most of the skeleton (except the lower hind leg bones and most of the tail) is present. It measures 9 meters (30 feet) long, with a body weight of 1.35 metric tonnes (1.5 tons). It had relatively long slender hind limbs, suggesting it was one of the fastest runners among the theropods. This is confirmed by the structure of the thighbone, which appears to be able to withstand a lot of bending and twisting. It is hard to reliably estimate speed in the dinosaurs, but studies suggest it was not as fast as the modern ostrich, which can travel at 69 kilometers per hour (43 miles per hour). Contrary to the *Jurassic Park* movies, all the research shows that tyrannosaurs were much slower still and could not outrun the fastest humans, let alone a speeding car.

Not only was original type specimen of *Carnotaurus* preserved in a death pose nearly complete, but there were skin impressions as well. Its scales were shaped like polygons that did not overlap, and they are scored with parallel grooves in many places. The scales of the head are not as regular as those of the body, with differences between the scales on the right and left side of the face. There were also long knob-like bumps on the skin running along the side of the neck, back, and tail in irregular rows. These may be condensed dermal armor, perhaps to protect its flanks against attacks. Like *Coelophysis*, the head of the original specimen of *Carnotaurus* is drawn backward on a curved neck, probably because the nuchal ligament along the back of the neck contracted when the animal died. There is no evidence of feathers in any of the skin impressions, contrary to the evidence we have that tyrannosaurs had feathers—although the preservation of *Carnotaurus* may just not be suitable for showing feathers.

Finally, what about the horns that gave *Carnotaurus* its name? There have been lots of arguments and suggestions about their use. The obvious feature is that they help in species recognition and perhaps in showing dominance in males, as they do in cattle, deer, and antelopes today. Some paleontologists argue that they were used in head butting or wrestling with other members of their own species. Combined with their powerful neck, the head wrestling idea is plausible, but their lightly built skulls were not strong enough to survive high-speed head butting as in rams. The horns have flattened surfaces on top suitable for horn-to-horn wrestling and pushing, and they are also fused to the skull strongly so could withstand the stresses of head wrestling. One suggestion is that the bony horn cores were covered by long outer sheaths made of keratin (like the horns of cattle and antelopes) which would have made good offensive weapons and may have been used to stab or injure prey. However, there are no known instances of animals using horns for predation because nearly all of the horned animals alive today are herbivores.

Carnotaurus comes from the latest Cretaceous of Argentina. As the largest predator of that time, *Carnotaurus* probably preyed on titanosaur sauropods as well as smaller prey, such as duck-billed dinosaurs and ankylosaurs. Indeed, most of the weird abelisaurs come from the Late Cretaceous, where they are known largely from Argentina.

However, there were several poorly known abelisaurs in India, and one of the oddest of all is from Madagascar. Known as *Majungasaurus*, it comes from the latest uppermost Cretaceous beds of that Gondwana remnant. Although *Majungasaurus* was first named and described based on fragments found by French paleontologists in 1896, recent discoveries by Field Museum scientists have produced much better material. Like other abelisaurs, *Majungasaurus* had a short and deep snout with relatively small teeth. However, instead of the paired horns of *Carnotaurus*, it had a dome of bone over the top of the skull, which was originally thought to be from a pachycephalosaur. *Aucasaurus* also had thick ridges of bone on the top if its skull. This suggests that head wrestling was a normal feature of all the abelisaurs, and the horns of *Carnotaurus* were mostly for head wrestling as well.

We have now seen the full spectrum of Gondwana predators (such as abelisaurs and carcharodontosaurs) and prey (especially titanosaurs).Next we look at the weirdest predators of all—so weird that they gave up carnivory altogether and became one of the few theropods to become herbivorous.

FOR FURTHER READING

Bonaparte, José F. "A Horned Cretaceous Carnosaur from Patagonia." *National Geographic Research* 1, no. 1 (1985): 149–151.

Bonaparte, José F., Fernando E. Novas, and Rodolfo A. Coria. "*Carnotaurus sastrei Bonaparte*, the Horned, Lightly Built Carnosaur from the Middle Cretaceous of Patagonia." *Contributions in Science* 416 (1990): 1–41.

Calvo, Jorge Orlando, and Rodolfo A. Coria. "New Specimen of *Giganotosaurus carolinii* (Coria & Salgado, 1995), Supports It as the Largest Theropod Ever Found." *Gaia* 15 (1998): 117–122.

Carrano, Matthew T., and Scott D. Sampson. "The Phylogeny of Ceratosauria (Dinosauria: Theropoda)." *Journal of Systematic Palaeontology* 6, no. 2 (2008): 183–236.

Colbert, Edwin. *Men and Dinosaurs: The Search in the Field and in the Laboratory.* New York: Dutton, 1968.

Currie, Philip J. "Out of Africa: Meat-Eating Dinosaurs That Challenge *Tyrannosaurus rex.*" *Science* 272, no. 5264 (1996): 971–972.

Farlow, James, and M. K. Brett-Surman. *The Complete Dinosaur.* Bloomington: Indiana University Press, 1999.

Fastovsky, David, and David Weishampel. *Dinosaurs: A Concise Natural History*, 3rd ed. Cambridge: Cambridge University Press, 2016.

Holtz, Thomas R., Jr. *Dinosaurs: The Most Complete, Up-to-Date Encyclopedia for Dinosaur Lovers of All Ages*. New York: Random House, 2011.

Mazzetta, Gerardo V., Per Christiansen, and Richard A. Fariña. "Giants and Bizarres: Body Size of Some Southern South American Cretaceous Dinosaurs." *Historical Biology* 16, no. 2 (2004): 71–83.

Mazzetta, Gerardo V., Richard A. Fariña, and Sergio F. Vizcaíno. "On the Palaeobiology of the South American Horned Theropod *Carnotaurus sastrei* Bonaparte." *Gaia* 15 (1998): 185–192.

Naish, Darren. *The Great Dinosaur Discoveries*. Berkeley: University of California Press, 2009.

Naish, Darren, and Paul M. Barrett. *Dinosaurs: How They Lived and Evolved*. Washington, D.C.: Smithsonian Books, 2016.

Novas, Fernando E. *The Age of Dinosaurs in South America*. Bloomington: Indiana University Press, 2009.

Senter, Phil. "Vestigial Skeletal Structures in Dinosaurs." *Journal of Zoology* 280 (2010): 60–71.

Sereno, Paul C., Didier B. Dutheil, M. Iarochene, Hans C. E. Larsson, Gabrielle H. Lyon, Paul M. Magwene, Christian A. Sidor, David J. Varricchio, and Jeffrey A. Wilson. "Predatory Dinosaurs from the Sahara and Late Cretaceous Faunal Differentiation." *Science* 272 (1996): 986–991.

Sereno, Paul C., Jeffrey A. Wilson, and Jack L. Conrad. "New Dinosaurs Link Southern Landmasses in the Mid-Cretaceous." *Proceedings of the Royal Society B: Biological Sciences* 271, no. 1546 (2004): 1325–1330.

Therrien, François, and Donald M. Henderson. "My Theropod Is Bigger Than Yours . . . or Not: Estimating Body Size from Skull Length in Theropods." *Journal of Vertebrate Paleontology* 27, no. 1 (2007): 108–115.

Tykoski, Ronald B., and Timothy Rowe. "Ceratosauria." In *The Dinosauria*, 2nd ed., ed. David B. Weishampel, Peter Dodson, and Halszka Osmólska, 47–71. Berkeley: University of California Press, 2004.

Weishampel, David B. Paul M. Barrett, Rodolfo A. Coria, Jean Le Loeuff, Xu Xing, Zhao Xijin, Ashok Sahni, Elizabeth M. P. Gomani, and Christopher R. Noto. "Dinosaur Distribution." In *The Dinosauria*, 2nd ed., ed. David B. Weishampel, Peter Dodson, and Halszka Osmólska, 517–606. Berkeley: University of California Press, 2004.

Wilford, John Noble. *The Riddle of the Dinosaur*. New York: Knopf, 1985.

DEINOCHEIRUS

The area was remote, relatively unexplored, and access was difficult—both physically and politically: the Mongolian People's Republic was closed to Western scientists for much of the 20th century. But as members of the Eastern Bloc were allowed entry, access was the least of Zofia's worries. The inability to speak Mongolian, organising logistics for thirty other people, the dangers of sandstorms, and the lack of water were constant troubles. It was worth it though: the expeditions were huge successes.

—"Zofia Kielan-Jaworowska," Trowelblazers.com

INTO THE GOBI DESERT

Politics often gets in the way of science, especially paleontology. During much of the twentieth century, it was impossible for Western scientists to visit the Soviet Union or see their amazing fossils thanks to the Cold War. It was equally difficult for Soviet scientists to leave their country and see foreign fossils, or even get copies of paleontological research published outside Russia. The same thing occurred in China when the "Cultural Revolution" (1966–1976) purged China of most of its scholars and intellectuals, and paleontologists had to hide and work quietly in isolation for fear of being arrested (see chapter 18). In addition, both world wars brought paleontological field research to a standstill because paleontologists could not travel or collect fossils during the war. Many served in the armed forces and put their research on hold. Some even died in the wars.

One of the richest but least accessible areas for dinosaur fossils is the famous Gobi Desert in Mongolia. For centuries, fossils poached from there were making their way to the Chinese market for "dragon bones," which were ground up into a powder and turned into "medicine." After World War I ended, the American Museum of Natural History in New York sent the Central Asiatic Expeditions to Mongolia throughout the 1920s (see chapter 17). Their amazing discoveries shook the world (from the first dinosaur eggs to gigantic hornless rhinoceroses as large as five elephants). But the Great Depression ended funding for most paleontological field work during the 1930s, and World War II put research further on the back burner.

When the war ended, paleontologists were eager to return to Mongolia. However, the northern half of Mongolia ("Outer Mongolia") was under Soviet influence, so most Western scientists could not get there due to the Cold War. The southern part of Mongolia ("Inner Mongolia") was under Chinese rule, and when the Communists took over China in 1949, it too was closed to Western scientists until the early 1980s. My graduate advisor, Dr. Malcolm McKenna of the American Museum, spoke Russian and had many powerful contacts and friends, and he was able to visit Moscow several times when no one else could. He was also one of the first Americans to be allowed back into Mongolia, and he helped organize and lead the first American Museum-Mongolian Academy of Sciences expeditions from 1990–1992.

Only the Russians had scientific access to Outer Mongolia during this time, but few expeditions by Soviet paleontologists occurred during the reign of Stalin or Khrushchev. Instead, a group of scientists from Poland stepped into the breach and became the first to collect there extensively since the American Museum's efforts in the 1920s. Poland was an Eastern-bloc country with Soviet influences at that time, so Polish scientists could get permission from Moscow to work in Mongolia when Western scientists were shut out.

The leader of these legendary Polish-Mongolian Palaeontological Expeditions from 1963 to 1971 was Zofia Kielan-Jaworowska (figure 16.1). Born in Sokolow Podlaski, Poland, in 1925, Kielan-Jaworowska was a student in Warsaw when the Germans invaded Poland in 1939. During the war years, she watched as most of Warsaw was destroyed, including the University of Warsaw, where she was enrolled in the Department of Geology. Many Poles

Figure 16.1 ▲

Zofia Kielan-Jaworoska (*left*), Halska Osmólska (*right*), and a young girl, in a yurt in Mongolia in 1968. (Courtesy of Wikimedia Commons)

were executed in the Nazi concentration camps. She described her experience this way:

> During the Second World War, the Germans occupied Poland for more than five years. The official aim announced by Adolf Hitler and his accomplices was the extermination of most Poles, with the rest (those who could "count to 500, sign their name, and not necessarily be able to read") turned into German slaves. However, at the risk of the death penalty, Poles organized a clandestine countrywide system of education at all levels. High school teachers and university professors continued to give lessons and lecture in private homes. Gradually, a secret network of instruction was established, constituting part of the Polish Resistance. . . . I attended a clandestine high school; lessons were conducted in private homes of the pupils, in groups of six to eight persons. In 1943/1944 I attended the clandestine University of Warsaw, studying zoology. These classes were suspended after the Germans put down the Warsaw Uprising of August/September 1944 (in which I took part as a medic). Systematically, block by block, and in cold blood, the Germans blew up and burned down my city. Intense evil guided this job. The result was

more than 200,000 murdered civilians, the total destruction of 85 percent of Warsaw, and the expulsion of all the surviving inhabitants. The ruined and burned-down city was left empty: "annihilated forever" according to Hitler's order. But in December 1945 the university reopened. (Kielan-Jaworowska, *In Pursuit of Early Mammals*, 35)

After the war, her paleontology professor, Roman Koslowski, held classes in his apartment while the slow rebuilding of Warsaw took place. Kielan-Jaworowska earned her master's and doctorate at the rebuilt University of Warsaw, and she became a professor there. Because trilobite fossils were readily available and poorly studied, she began her research career working on Polish trilobites and on polychaete worm jaw fossils.

Kielan-Jaworowska knew of the amazing fossils that had been found in Mongolia in the 1920s. The American Museum had recovered a complete skull of *Zalambdalestes*, one of the few Mesozoic mammals known from more than just jaws and teeth. Through her hard work and the work of her mentors, including Roman Koslowski, she received funding to begin the Polish-Mongolian Palaeontological Expeditions in 1963. She also arranged an exchange agreement between the Polish Academy of Sciences and the Mongolian Academy of Sciences and received the necessary permissions from Moscow and Ulaanbaatar in Mongolia. Her particular interest was in the tiny mammals that lived during the Cretaceous, but other scientists on the team focused on dinosaurs. One of these was Teresa Maryańska, who later became famous for her work on many different kinds of Mongolian dinosaurs, especially ceratopsians, ankylosaurs, and pachycephalosaurs. Halszka Osmólska (figure 16.1) also joined the expedition and later published many papers on Mongolian dinosaurs such as hadrosaurs, dromaeosaurs, oviraptorids, as well as pachycephalosaurs and ceratopsians with Maryánska. The Polish research on dinosaurs was led mostly by women, which was not typical for paleontologists in the 1950s and 1960s. Paleontology would not include many prominent female researchers until the 1980s and later.

The first year of the Polish-Mongolian Palaeontological Expeditions was mostly scouting and reconnaissance, but they began major excavations in 1964. In addition to the three female paleontologists leading the expedition and 12 other Poles in various jobs (drivers, cooks, mechanics, interpreters, and so on), many Mongolians were hired from Ulaanbaatar to handle the heavy labor and to help with the local people. They drove the huge Soviet-made six-wheel-drive military trucks known as the Star 66, along with smaller vehicles for scouting. They had to bring along all the gasoline needed for the entire expedition in 200-liter drums, along with camping equipment and food for 30 people for several months. The entire load was shipped by rail from Warsaw to Ulaanbaatar. Local Mongolian workers then unloaded the trucks and gear and set up a base camp.

The logistics for the excavations were daunting. A single sauropod skeleton collected in 1965 weighed 12 tons and had to be trucked back to the rail line in Ulaanbaatar, more than 1,000 kilometers (620 miles) away, which required four round trips by a three-ton truck. During the first year alone, they found many dinosaurs, including skeletons of the tyrannosaur *Tarbosaurus*, the duckbill *Saurolophus angustirostris*, and the ostrich dinosaur genus *Gallimimus* (featured in the first *Jurassic Park* movie). In the 1965 season, they found the large sauropods *Opisthocoelicaudia* and *Nemegtosaurus*, the pachycephalosaurs *Homalocephale* and *Prenocephale*, and the ceratopsian *Bagaceratops*. Later expeditions recovered many more new dinosaurs, as well as dozens of skulls and even skeletons of Late Cretaceous mammals and many other fossils.

In the 1964 season, they made another incredible find. As Kielan-Jaworowska recounted it:

> One rainy day, very rare in the Gobi Desert during summer, I was walking alone along the gullies of Altan Uul III and found an unusual skeleton consisting only of complete forelimbs and a shoulder girdle of enormous size, along with fragmentary ribs. The limbs . . . were 2.5 meters long, ending in three powerful fingers armed with claws more than 30 cm long. The bones were scattered on the flat surface of a small hill. (Kielan-Jaworowska, *In Pursuit of Early Mammals*, 54)

The crew took two days to remove and jacket the specimens, but there was never sign of any other parts of the skeleton—just the forelimbs plus rib fragments. They were on top of a rapidly eroding hill, so other parts of the skeleton might have been there but were eroded away before the fossils were found. Alternatively, they might have been limbs that were separated from the carcass and left some distance from the rest of the body. Whatever the reason for just finding the arms, their size was unbelievable. When the specimen was cleaned and prepared and mounted in Warsaw (figure 16.2),

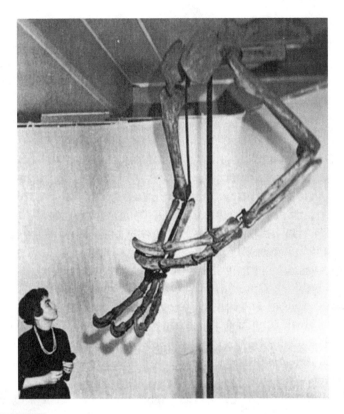

Zofia Kielan-Jaworowska and the original display of the incredible arms of *Deinocheirus* in Warsaw. (Courtesy of Wikimedia Commons)

the implications were staggering! The Polish team could not figure out what kind of theropod dinosaur had such huge hands, but if the hands were proportional to the body in the same ratio as the forelimbs of *Allosaurus* or some other large theropod, it seemed to indicate a giant predator that would dwarf a *T. rex*. In 1970, Halszka Osmolska and Eva Roniewicz officially described it and named it *Deinocheirus mirificus*, Greek for "amazing terrible hands."

For decades, the creatures that bore the "terrible hands" were a mystery. What kind of animal could have had arms and hands as huge as these? Various paleontologists called the creature "enigmatic," "mysterious," and "bizarre." However, close examination of the wrist bones provided a lot of clues, even without the rest of the skeleton. As early as 1971, John Ostrom (see chapter 17) noticed that the wrist bones mostly closely resembled the ostrich dinosaurs, or the ornithomimids. This suggestion was confirmed in 2004 by Peter Makovicky and colleagues, who noted that although the wrist bones looked like those of ornithomimids, it was a very primitive member of the Ornithomimosauria, lacking many of the advanced features found in others including *Ornithomimus, Struthiomimus,* and *Gallimimus.*

MYSTERY SOLVED!

Finally, in 2008 a Korean-Mongolian expedition relocated Kielan-Jaworowska's original *Deinocheirus* locality and found additional fragments, including the gastralia (rib-like bones on the belly). They had scars on them showing they had been bitten by a large tyrannosaur like *Tarbosaurus,* which suggested that the reason the specimen was so incomplete was that it had been scavenged, and only the arm bones had been left behind.

The real break in the case occurred in 2009 when Phil Currie of the Royal Tyrrell Museum in Alberta and his Korean and Mongolian colleagues found two partial skeletons of *Deinocheirus.* Both had been plundered by poachers who had robbed the sites of their skulls, hands, feet, and some other bones, and damaged the rest. Based on clues left behind by the poachers, the crime had occurred in 2002. Nonetheless, Currie and colleagues were able to recover and describe most of the remaining skeleton of the animal. Rumors began flying around that the skull and other bones had been spotted in the European black market. Fossil trader Francois Escuillé saw the specimens in a private European collection and alerted Belgian paleontologist Pascal Godefroit in 2011. The specimens apparently had been poached from Mongolia, sold to a Japanese middleman, then resold to a German buyer. Currie and colleagues were finally able to recover most of the stolen specimens and were able to confirm that they belonged to the poached skeleton from Mongolia because the toe bones from the two specimens fit together perfectly. The paleontologists then assembled and prepared the fossils at the Royal Belgian Institute of Natural Sciences, where replicas were made before the originals were sent back to Mongolia for permanent safekeeping.

When Currie and colleagues finally had all the bones prepared and studied, they presented their results at the 2013 meeting of the Society of Vertebrate Paleontology in Los Angeles, hosted by my very own institution,

Figure 16.3 ▲
Reconstruction of the entire skeleton of *Deinocheirus* in life. (Courtesy of N. Tamura)

the Natural History Museum of Los Angeles County, in the fancy Westin Bonaventure Hotel downtown. I vividly remember the session in which Currie presented the bizarre beast that was *Deinocheirus* to the dinosaur specialists for the first time (figure 16.3). The room was buzzing with amazement at the weird features of the beast. When they projected a virtual model of the animal in life, the whole audience burst out in applause. Steve Brusatte of the University of Edinburgh said he had never been more surprised by a new dinosaur in all his life (new discoveries are reported each year at the meetings). Thomas Holtz of the University of Maryland, a theropod specialist, said it looked like the "product of a secret love affair between a hadrosaur and *Gallimimus*." Many surprising new dinosaur finds have been presented in the past few decades, but *Deinocheirus* was one of the weirdest and most surprising I've ever seen.

Just how weird was *Deinocheirus*, and why did it look like a cross between a duckbill and *Gallimimus*? They were huge animals all right, but they were not predators. Despite having the claws of a theropod, they were herbivores or omnivores! The largest known specimen was 11 meters (36 feet) long, and may have weighed 6.4–12 metric tonnes (7–13 tons), but the smaller specimens were only about 75 percent the size of the largest. Even though it had a huge bulky body, the bones were hollow, which made it lighter and caused less strain on its relatively short legs and toes with blunt claws, which bore all the weight, because the hands were for grasping, not walking. The oddest feature is the long spines on the top of its backbone from the lower back to the base of the tail, which may have given it a tall "sail" similar to that of *Spinosaurus* or *Ouranosaurus*, or possibly supported a large fleshy "hump." The tail ended with a fusion of most of the tail vertebrae into a pygostyle (like the "parson's nose" in birds), which apparently supported a fan-like array of tail feathers as in birds. The huge skull was long and narrow and over a meter (3.36 feet) long. Yet it was nothing like a typical theropod predator; it was more like the toothless ostrich-like heads of ornithomimids. It had a wide bill and deep lower jaw, resembling a duck-billed dinosaur snout rather than a predator (but was toothless, unlike duckbills or most theropods). The eyes were relatively small, with a ring of bone around the pupil (sclerotic ring). Because they were herbivores, it is thought that they were mostly daytime feeders. *Deinocheirus* had a relatively small brain, with a ratio of brain size to body mass more like the huge sauropods than others in its own group, the more intelligent theropods.

So what did this weird creature eat? The beak suggests a herbivorous or omnivorous diet of plants, but fish scales were found in one specimen, so they ate at least some fish, if not meat. One specimen had hundreds of gastroliths (gizzard stones) in its gizzard to grind down its plant diet. They had enormous bellies, which would be expected for a secondary plant eater that needed a big gut and a long digestive tract to process and ferment large volumes of vegetation. The huge clawed hands were apparently for grasping and pulling down branches, not attacking prey, and possibly were used for digging for roots and tubers or to defend against predators.

As bizarre as *Deinocheirus* seems, it was not the only big-handed heavy-bodied herbivorous ornithomimosaur from the Cretaceous of Asia. There is also *Garudimimus*, named after the Garuda bird in Hindu and Buddhist mythology, which was about 2.5 meters (8.2 feet) long. Larger still is *Beishanlong* from the Early Cretaceous Ghost Castle site in the White Mountains of Gansu Province, China. Its name means "White Mountain dragon" in Mandarin. It reached about 7 meters (23 feet) long and about 550 kilograms (1,212 pounds) in weight. Deinocheirids were apparently widespread in Asia throughout most of the Cretaceous.

And let us not forget the ornithomimids themselves. Thanks to *Jurassic Park*, almost everyone had heard of *Gallimimus* (whose name in Greek means "chicken mimic"), but it was a relatively late addition to the list of ostrich-like dinosaurs with toothless beaks. The first one discovered was an incomplete set of legs and foot bones from the Upper Cretaceous beds of the Denver Basin, called *Ornithomimus* (bird mimic) because of its bird-like feet and legs. It was described by O. C. Marsh in 1890 and was about 3.8 meters (12 feet) in length. The first nearly complete skeleton that showed the ostrich-like neck and legs and general build was *Struthiomimus* (ostrich mimic) from the Upper Cretaceous beds of Alberta (figure 16.4). Described by Henry Fairfield Osborn in 1917, it was about 4.3 meters (14 feet) long. These are the oldest named genera and some of the best known, but there are at least 10 other genera in the family Ornithomimidae, and another 9 genera of ornithomimosaurs that are not members of the family Deinocheiridae or the family Ornithomimidae. The ostrich mimics were a diverse group of dinosaurs found all over Eurasia and North America during most of the Cretaceous—and possibly Africa, if *Ngwebasaurus* is an ornithomimosaur.

Figure 16.4 ▲

Struthiomimus (ostrich mimic) on the right had many similarities to a modern ostrich (genus *Struthio*) on the left, but it was much larger, and it had a long bony tail and long arms with strong, clawed hands. (Photograph by the author)

WEIRD HERBIVOROUS CARNOSAURS

By the 1980s, it seemed that most of the major groups of dinosaurs had been discovered, and few paleontologists thought that there would be a whole new group with a previously unexpected body plan. In addition, dinosaurs seemed to be very stereotyped: all the theropods were predators, and all the other groups were herbivores. No one imagined that we'd find fossils that required a completely new group of dinosaurs, let alone imagining that the predatory theropods could ever revert to becoming herbivorous. But they were in for some big surprises during the mid-1980s, and all those rules were broken.

If *Deinocheirus* seemed like a strange animal with its huge hands and secondary development of herbivory or omnivory, this was not the first time this strange adaptation had appeared in the normally carnivorous theropods. An eerie parallel to the *Deinocheirus* discovery was the solution to the puzzle of the therizinosaurs. They were another mainly Asian group of theropods that became herbivorous and had large hands and a toothless beak. Just like *Deinocheirus*, their initial fragmentary specimens were badly misinterpreted for a long time until more complete specimens were found. In fact, for a long time paleontologists thought that even *Deinocheirus* was a therizinosaur before the complete skeletons found in 2012 revealed it was an ornithomimid.

So what are these weird monsters called therizinosaurs? The first to be discovered was *Therizinosaurus cheloniformis* itself (figure 16.5A), found during the 1948 Soviet-Mongolian expedition, and formally named by Evgeny Maleev in 1954. But all that was found was enormous hands with long sickle-shaped claws. The name *Therizinosaurus* means "sickle lizard" in Greek. With nothing else to go on, Maleev imagined that they came from some large turtle-like reptile (hence the species name "*cheloniformis*") that used its claws to harvest seaweed. Not until 1970 did Anatoly Rozhdestvensky realize that the fossils came from a theropod dinosaur. In 1976, Mongolian paleontologist Rinchen Barsbold described the shoulder girdle, arms and hands, a few ribs, and the hind legs. They clearly came from an enormous beast with giant hand claws (parallel to the discovery of *Deinocheirus*), but there was simply not enough of the skeleton to imagine what kind of animal it was. We now know it came from a huge dinosaur about 10 meters (33 feet) long and weighing about 5 metric tonnes (5.5 tons), one of the largest theropods known (figure 16.5B).

Figure 16.5 ▲

(A) Profiles of different therizinosaurs to scale, showing the bones (in white) that are known for the different genera; (B) reconstruction of *Therizinosaurus*. ([A] Courtesy of Wikimedia Commons; [B] courtesy of N. Tamura)

For another 25 years, no further clues were uncovered to explain what kind of creature *Therizinosaurus* was. The next therizinosaur to be discovered was *Segnosaurus* (figure 16.5A), found by another Soviet-Mongolian expedition in 1973 and named by Mongolian paleontologist Altangerel Perle in 1979. This fossil had parts of the arms preserved, plus the hips, hind legs, and the spine from the hips to the tail. The fragment of lower jaw has a downturned snout and leaf-shaped teeth, suggestive of a herbivore. *Segnosaurus* had powerful forelimbs with long claws that were strongly curved and flattened, like the blade of a sickle. The pelvis was very broad, the hind limbs robust and still retaining four toes pointing forward (most theropods had only three, plus a tiny "big toe" or hallux that points backward), suggesting it was a very heavy, slow-moving biped with a huge gut. The feet are odd in that the first toe is large and contacts both the ankle joint and the ground, whereas the first toe of most theropods is extremely reduced and doesn't touch the ankle joint. Even more odd, the pubic bone of the hip points backward, but in a different way than it does in ornithischian hips, or in the hips of true birds—yet all its other features were indicative of a nonavian theropod dinosaur that had a saurischian pelvis.

The crucial discovery was made in the Gobi Desert in Chinese Inner Mongolia and was formally described by Dale Russell and Dong Zhiming in 1994. Named *Alxasaurus* (after the Alxa Desert, also spelled Alshan Desert, where it was found), it consisted of five partial skeletons that together gave us the first look at the entire anatomy of therizinosaurs (figure 16.5A). The only part missing was the skull, which is still not well known for most therizinosaurs although *Alxasaurus* is represented by teeth and a lower jaw. *Alxasaurus* was about 3.8 meters (12 feet) long, with a long neck, long jaw with simple leaf-shaped teeth, and long arms with the same characteristic long scythe-like claws. Yet it has a short tail and the typical broad pelvis and short four-toed feet, again indicating a large digestive tract for eating plants.

More and more of these strange creatures have been found, and there are now 13 genera of therizinosaurs—and counting. One of the smallest was *Beipiaosaurus* (figure 16.6A), at 2.2 meters (7.3 feet) in length. It was found in the Lower Cretaceous beds of northeastern China in 1999. *Beipiaosaurus* came from the famous Liaoning lake shale beds that preserved fine feather impressions, confirming that therizosaurs were feathered (as were most dinosaurs, which at least had some kind of downy covering or filaments; see chapter 19).

Figure 16.6 ▲

(A) Reconstruction of the Chinese therizinosaur *Beipiaosaurus*. (B) The skeleton of the American therizinosaur *Nothronychus* from New Mexico (*left*) and the primitive therizino-saur *Falcarius* (*right*). ([A] Courtesy of M. Martinyuk/Wikimedia Commons; [B] photograph by the author)

Along with all the Asian species, a few therizinosaurs have now been found in North America. *Nothronychus* (sloth claw) came from the mid-Cretaceous of the Zuni Basin in southwestern New Mexico (figure 16.6B). Described by my friend Jim Kirkland and Doug Wolfe in 2001, it consists of two species. The larger species was up to 6 meters (20 feet) long and had the typical long neck, large hands with scythe-like claws, broad pelvis and pot-bellied build, and short tail.

In 2005, the news media were all abuzz with the discovery of *Falcarius utahensis* (sickle cutter of Utah) from the Early Cretaceous Cedar Mountain Formation in east-central Utah near Grand Junction, Colorado (figure 16.6B). Described by Jim Kirkland, Lindsay Zanno, Scott Sampson ("Dr. Scott" of *Dinosaur Train* fame), Jim Clark, and Don DeBlieux, they found an entire bone bed of fossils with more than 2,700 bones representing a minimum of about 300 individuals, with many juveniles as well as adults. Several other localities with *Falcarius* in them also have been found. As one would expect from such an early member of the therizinosaurs, *Falcarius* had many primitive features and was not as specialized as the Late Cretaceous forms. At 4 meters (13 feet) long, it was small by theropod standards, but it had the long neck typical of the therizinosaurs, a small head adapted for eating plant material (but still no complete skull is known), robust arms, and the sickle-like claws on the hands. Yet it retained some of the primitive features found in theropods, including long lower legs, an oval-shaped torso without the pot belly found in later therizinosaurs, and a long tail. In other words, *Falcarius* is a mosaic of primitive and advanced features typical of most transitional species in the fossil record, showing advanced anatomy in some parts of the body but retaining their ancestral anatomy in others. Sampson said that *Falcarius* "is the missing link between predatory dinosaurs and the bizarre plant-eating therizinosaurs." Lindsay Zanno described *Falcarius* as "the ultimate in bizarre: a cross between an ostrich, a gorilla, and Edward Scissorhands."

We have now learned that the theropods were not always meat eaters, as was long assumed. Discoveries in the past 20 years have revealed at least two groups of dinosaurs (one previously unknown) that switched from a carnivorous lifestyle to herbivory. (Herbivory may have evolved in oviraptorids as well, and, of course, it occurs in birds.) In the process, these animals developed similar heads with long beaks and small herbivore teeth, long necks, huge hands with long claws for pulling down branches, and eventually pot-bellied bodies for containing the long digestive tract necessary to ferment and digest plant material. Yet both groups developed these adaptations independently; they are distantly related branches of the theropods that independently evolved from different predatory ancestors. Even when we think we have the groups of dinosaurs all figured out, nature is still full of surprises.

FOR FURTHER READING

Clark, James M., Teresa Maryańska, and Rinchen Barsbold. "Therizinosauroidea." In *The Dinosauria*, 2nd ed., ed. David B. Weishampel, Peter Dodson, and Halszka Osmólska, 151–164. Berkeley: University of California Press, 2004.

Colbert, Edwin. *Men and Dinosaurs: The Search in the Field and in the Laboratory*. New York: Dutton, 1968.

Farlow, James, and M. K. Brett-Surman. *The Complete Dinosaur*. Bloomington: Indiana University Press, 1999.

Fastovsky, David, and David Weishampel. *Dinosaurs: A Concise Natural History*, 3rd ed. Cambridge: Cambridge University Press, 2016.

Holtz, Thomas R., Jr. *Dinosaurs: The Most Complete, Up-to-Date Encyclopedia for Dinosaur Lovers of All Ages*. New York: Random House, 2011.

Kielan-Jaworowska, Zofia. *Hunting for Dinosaurs*. Cambridge, Mass.: MIT Press, 1969.

——. *In Pursuit of Early Mammals*. Bloomington: Indiana University Press, 2013.

Kirkland, James I., Lindsey E. Zanno, Scott D. Sampson, James M. Clark, and Donald D. DeBlieux. "A Primitive Therizinosauroid Dinosaur from the Early Cretaceous of Utah." *Nature* 435, no. 7038 (2005): 84–87.

Lee, Yuong-Nam, Rinchen Barsbold, Philip J. Currie, Yoshitsugu Kobayashi, Hang-Jae Lee, Pascal Godefroit, François Escuillié, and Tsogtbaatar Chinzorig. "Resolving the Long-Standing Enigmas of a Giant Ornithomimosaur *Deinocheirus mirificus*." *Nature* 515 (2014): 257–260.

Makovicky, Peter J., Yoshitsugu Kobayashi, and Philip J. Currie. "Ornithomimosauria." In *The Dinosauria*, 2nd ed., ed. David B. Weishampel, Peter Dodson, and Halszka Osmólska, 137–150. Berkeley: University of California Press, 2004.

Naish, Darren. *The Great Dinosaur Discoveries*. Berkeley: University of California Press, 2009.

Naish, Darren, and Paul M. Barrett. *Dinosaurs: How They Lived and Evolved*. Washington, D.C.: Smithsonian Books, 2016.

Wilford, John Noble. *The Riddle of the Dinosaur*. New York: Knopf, 1985.

Zanno, Lindsay E. "A Taxonomic and Phylogenetic Re-evaluation of Therizino-sauria (Dinosauria: Maniraptora)." *Journal of Systematic Palaeontology* 8, no. 4 (2010): 503–543.

Zhiming, Dong, Y. Hasegawa, and Y. Azuma. *The Age of Dinosaurs in Japan and China*. Fukui, Japan: Fukui Prefectural Museum, 1990.

VELOCIRAPTOR

In the public imagination then, dinosaurs were plodding, thunderous monsters, cold-blooded and stupid. Even paleontologists had lost interest in these "symbols of obsolescence and hulking inefficiency," Ostrom's student Robert T. Bakker later wrote. "They did not appear to merit much serious study because they did not seem to go anywhere: no modern vertebrate groups were descended from them."

—RICHARD CONNIFF, *THE MAN WHO SAVED THE DINOSAURS*, 2014

THE NEW CONQUEST OF CENTRAL ASIA

The ambitious paleontologist Henry Fairfield Osborn, head of the American Museum of Natural History, wanted to explore the long-neglected Gobi Desert region for fossils of early humans (see chapter 14). Like many scientists of his time, he believed that humans first evolved in Eurasia, not in Africa as Darwin had suggested (because our closest relatives, chimps and gorillas, live there). Anthropologists were convinced that modern humans evolved in the rigorous climates of Eurasia, and their deeply ingrained racism prevented them from considering that humans might have originated in Africa, where the "inferior" races were found. Fossils like "Java man" and "Peking man" (now both considered *Homo erectus*), plus the hoax known as "Piltdown man" ("discovered" in 1912 but not exposed until 1953), only whetted his appetite for reaching this untapped resource of fossils.

Once World War I had ended and the economy was booming, Osborn set about raising funds among his rich relatives and friends, swaying them with promises of amazing fossil finds of early humans. Finally, in 1922, the American Museum sponsored one of the most ambitious scientific expeditions ever attempted (figure 17.1). Led by the legendary explorer Roy Chapman Andrews, the expedition traveled to China and Mongolia with a huge caravan of 75 camels (each carrying 180 kilograms [400 pounds] of gasoline and other supplies), three Dodge touring cars and two Fulton trucks, and a large party of scientists, guides, and helpers. The party included not only Andrews but also paleontologist Walter Granger, a veteran of many fossil-hunting expeditions in the United States and elsewhere, who was experienced in hunting fossils in China, as well as two geologists (Charles P. Berkey and Frederick K. Morris). Many other assistants were needed to drive the trucks and cars and camels, cook the food and set up the camp, and act as guides and interpreters. Osborn told Andrews, "The fossils are there. I know they are. Go and find them."

The "biggest scientific expedition ever to leave the United States" passed through a gate in the Great Wall of China on April 21, 1922, and headed into

Figure 17.1 ▲

The Flaming Cliffs of Shabarakh Usu, Mongolia, with the American Museum camel caravan in the foreground. (Image #410767, courtesy of the American Museum of Natural History Library)

Mongolia. Aided by their cars, they traveled 426 kilometers (265 miles) in just four days, which was much more efficient than any previous expeditions mounted on horse or camel. Over the course of the many expeditions, they faced bandits, blinding sandstorms, blistering heat and freezing cold, and lots of dangerous vipers, but the trip went off with relatively few problems.

THE REAL "INDIANA JONES"?

Roy Chapman Andrews (figure 17.2) was a flamboyant and colorful character. One of the last classical "scientific explorers" who was not a scientist with a doctorate in a particular specialty, Andrews's gift was in raising funds, leading and organizing the trips, and in conveying the

Figure 17.2 ▲
Roy Chapman Andrews. (Courtesy of Wikimedia Commons)

excitement of his many exploits to the general public through his popular books and lectures (which further aided in fundraising). Despite his image, Andrews was not trained as a paleontologist, nor was he good at collecting fossils. In the Mongolian expeditions, the real paleontologists urged him not to try to excavate the fossils before they got there to do the job properly because he often damaged the fragile specimens. When someone botched the collection of a fossil on the trip, or butchered it trying to get it out of the ground, they would say it had been "RCA'd" (after the initials for Roy Chapman Andrews).

Despite his limitations, Andrews was a bold and fearless leader. Several times Andrews scared off Mongolian bandits by shooting before they could draw their weapons (and occasionally using guns to intimidate corrupt border guards or greedy officials). In one incident, he charged the bandits with his car, shooting as he approached, and the bandits fled as their horses were spooked. Many people consider him the model for the "Indiana Jones" character played in the popular movie series by actor Harrison Ford, but George Lucas, Steven Spielberg, nor anyone else connected with the films has ever confirmed this. In fact, they have said that Indiana Jones is based loosely on a number of heroic movie explorers of the Silver Screen that they grew up with (although some of those may have been inspired by Andrews).

Born in Beloit, Wisconsin, in 1884, Andrews taught himself marksmanship and taxidermy, and he earned a degree from Beloit College (paid for by his earnings from taxidermy). When he talked his way into the director's office and asked to work at the American Museum as a taxidermist, he was told there were no openings, so he started as a janitor. While mopping floors, he earned a master's degree in mammalogy at Columbia University. In 1909 and 1910, he joined an expedition on the USS *Albatross* in the East Indies, where he collected lizards and snakes and studied mammals. In 1913, he was on the crew of the schooner *Adventuress* in the Arctic, where they hoped to obtain a bowhead whale specimen. That effort failed, but they filmed some of the best footage of seals ever seen. In 1916 and 1917, Andrews led the American Museum's Asiatic Zoological Expedition through western and southern Yunnan Province in China, where he collected many specimens and developed valuable skills and contacts in Asia.

By 1920, Andrews was planning the first of several American Museum expeditions to Mongolia. The first, in 1922, was a short exploratory trip to find out whether there were any fossils at all. They were so successful that

that they returned in 1923, 1925, 1928, 1929, and 1930. Almost immediately after arriving in Mongolia, fossils were found, and their second expedition found spectacular dinosaur bones and the first dinosaur eggs. This made the expeditions world famous and helped fundraising for three additional expeditions. They found not only dinosaurs but also the first good specimens of tiny mammals from the Age of Dinosaurs. But by the time of the last expedition, the political situation in Mongolia had deteriorated so badly that no further expeditions were possible.

By the 1930s, Andrews's ability to mount great expeditions to Asia had ended. The Great Depression worldwide made it impossible to raise funds to mount another trip. Many of the formerly rich museum donors had lost their fortunes in 1929 and 1930, and some of the museum's investments had become nearly worthless as well. By 1932, the museum was so strapped for funds that it canceled all fieldwork entirely and cut its staff to the bone. In addition, tensions between China and Japan were rising as the Japanese prepared to invade China and other parts of Asia.

Andrews spent much of his time in the 1930s writing books about his exploits; he was also designated one of the first "Honorary Boy Scouts" and served as president of the Explorers' Club (1931–1934). In 1935, Andrews was appointed director of the American Museum of Natural History, replacing Osborn, but he was unable to do much to help the museum during the depths of the Depression. Despite his great skills in organizing expeditions and raising money for them, he proved to be so inept at running the museum that the trustees replaced him in 1941. It took two more directors, plus the end of the Depression and World War II, for the museum to recover its strength. Andrews retired to Carmel-by-the-Sea, where he lived out the rest of his life writing popular books until his death in 1960 at the age of 76.

WALTER GRANGER, PALEONTOLOGIST

The other key figures in the American Museum's Central Asiatic Expeditions of the 1920s were Walter Granger (figure 17.3) and Henry Fairfield Osborn (see figure 14.1). Granger was the main paleontologist on all of the expeditions, and there could not have been a more competent person assigned to the task. Born in Vermont in 1872, he was one of five children of Civil War veteran and insurance agent Charles H. Granger. Like Andrews, he developed an early talent for taxidermy, and by 1890, when he was only

Figure 17.3 ▲
Walter Granger, the chief paleontologist on the expedition. (Courtesy of Wikimedia Commons)

17, he got a job doing taxidermy at the American Museum. Within a few years, he went on field expeditions to the American West searching for vertebrate fossils with the American Museum paleontologists. After two field seasons (1894 and 1895), his fossil-hunting talents were better appreciated, and he was transferred to the Department of Vertebrate Paleontology in 1896. Granger discovered the legendary Bone Cabin Quarry near Laramie, Wyoming, in 1897, which he worked for the next eight field seasons. The site yielded thousands of bones representing 64 species of dinosaur, including the mounted skeletons of the big *Apatosaurus* and *Stegosaurus* on display at the museum today.

After working Bone Cabin Quarry from 1897–1906, Granger accompanied Osborn on an expedition to the Eocene-Oligocene Fayûm beds of Egypt in 1907. This was the first American Museum fossil trip outside North America. They made many important discoveries that complemented what had been described by British Museum paleontologist Charles W. Andrews just a few years earlier. By this point, Granger was promoted to assistant curator, and his flexible schedule allowed him at least five months in the field every year to find more fossils. He continued to write two or three scientific papers each year as well.

In 1921, Granger began explorations in China, which eventually led to the discovery of "Peking man" in the Zhoukoudian caves near Beijing. This work in China laid the foundation for negotiations to go through China to Mongolia that allowed the 1922 American Museum expedition to succeed. When he finished the expeditions, he was promoted to curator of fossil mammals, allowing him to continue to work on his many amazing discoveries until his death at age 68 of heart failure in 1941. Entirely self-taught, Granger never earned a formal academic degree; this was rectified in 1932 when he received an honorary doctorate from Middlebury College.

Andrews, Granger, and American Museum preparator Peter Kaisen were in the badlands of Outer Mongolia on August 11, 1923, late in the first field season, when Kaisen spotted bone weathering out of the rock. He and Granger collected it, protected it with a plaster jacket, and shipped it to New York at the end of the season. When it was prepared, the specimen turned out to be a nearly complete skull of a small predatory dinosaur (badly crushed), plus a large sickle-shaped claw (figure 17.4A). In 1924, Osborn described the specimen as *Velociraptor mongoliensis* (fast robber from Mongolia). Nothing else was known of the fossil, so it was impossible to tell much about the rest of the dinosaur. In fact, nothing more would be known about this head and claw until the Polish-Mongolian expeditions in the early 1960s. When those specimens were found, it turned out that the huge sickle-shaped claw was on the second toe of each hind foot. The rest of the dinosaur was built rather lightly, with a long thin neck, a long narrow tail, and strong forearms and hands with sharp curved claws. Complete specimens showed that the biggest *Velociraptor* was the size of a large turkey (figures 17.4 and 17.5), with a total length of only 2 meters (7 feet) and weighing about 15-20 kilograms (33-44 pounds).

The "fighting dinosaurs" was most famous specimen the Poles found in Mongolia (figure 17.4B). A *Velociraptor* skeleton was buried in a collapsed sand dune and fossilized in a pose attacking a *Protoceratops*. The specimen vividly shows that *Velociraptor* had gripped the edge of the frill of the prey with its hands, and its sickle-clawed feet were slashing the throat of the *Protoceratops* and that the *Protoceratops* had a bite on the arm of *Velociraptor*.

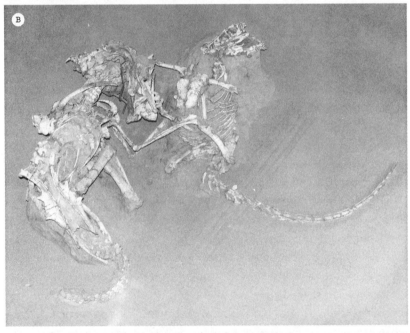

Comparison of the size of different dromaeosaurs. On the man's arm is the four-winged *Microraptor gui*. At his feet is the Canadian Cretaceous taxon *Dromaeosaurus albertensis*, which gives the group its name. To the right is the large *Austroraptor cabazai*. Below it to the right is the turkey-sized *Velociraptor mongoliensis*, followed by the largest dromaeosaur of all, *Utahraptor ostromi*. On the extreme right is *Deinonychus antirrhopus*. (Courtesy of Wikimedia Commons)

Eventually, over a dozen different genera of extinct animals like *Velociraptor* would be discovered and placed in a group known as the dromaeosaurs ("running lizards" in Greek). The group got its name after the first genus to be discovered, *Dromaeosaurus albertensis*, found in 1914 by Barnum Brown in the Upper Cretaceous beds of the Red Deer River of Alberta (figure 17.5).

Meanwhile, another important discovery, the second dromaeosaur to be found, would put *Velociraptor* and the dromaeosaurs into context.

"TERRIBLE CLAW"

The discovery of the fossil that changed our understanding of dromaeosaurs like *Velociraptor* can be traced to one man: John Ostrom (figure 17.6). Born in New York in 1928, Ostrom planned on being a doctor like his father until he read George Gaylord Simpson's 1944 book, *Tempo and Mode of Evolution*. This changed his life plans completely, and he set out to make a career in paleontology. He got his BA degree in 1951 at Union College, but in 1950 he was a field assistant to his idol George Gaylord Simpson. This helped

(A) The original type skull of *Velociraptor mongoliensis*. (B) The famous "fighting dinosaurs" specimen found by the Poles in Mongolia, which entombed a *Velociraptor* attacking the head of *Protoceratops*. (C) The mounted skeleton of *Velociraptor*. ([A] Courtesy of Wikimedia Commons; [B] courtesy of D. Fowler; [C] photograph by the author)

Figure 17.6 ▲
John Ostrom with a sculpture of his most famous discovery, *Deinonychus*. (Courtesy of Wikimedia Commons)

him become a student at the American Museum under Ned Colbert, where he began working on dinosaurs. After getting his PhD from Columbia University, Ostrom taught at Brooklyn College (1955–1956) and Beloit College (1956–1961) before becoming the curator of fossil reptiles at the Peabody Museum of Natural History of Yale University in 1961. There he inherited the Marsh Collection, and he spent the rest of his career there, from 1961 until he retired for health reasons at age 65 in 1993. From the 1960s to the 1990s, he mentored a lot of PhD students at Yale, including many American dinosaur paleontologists. If they were not Ostrom's students directly, they were students of Peter Dodson of the University of Pennsylvania, one of Ostrom's first acolytes. Ostrom passed away at age 77 in 2005 after years of suffering from Alzheimer's disease.

I remember John well, both as a friend and as an inspiration. In 1994, I invited him to give a talk on birds and dinosaurs for a short course on vertebrate evolution that his former student Robert Schoch and I had organized, and his presentation was the highlight of the session. I can still see him giving talks with the excitement in his voice, a twinkle in his eye, looking down at his audience through the half-rimmed glasses perched on his nose, and giving us all a sly grin. In fact, nearly all of my generation of vertebrate paleontologists knew him well as a jovial, provocative, and inquisitive man who never accepted dogmatic answers and handled his forays in major controversies without losing his calm or the smile on his face. He was famous not only for his discoveries and insight but for his unsparing search for the right answer, and his unflinching honesty and integrity.

Early in his career, John embarked on a series of field seasons working in the Lower Cretaceous Cloverly Formation of the Bighorn Basin in Montana. Barnum Brown (chapter 21) had shown Ostrom what turned out to be an early specimen of a dromaeosaur that he never got a chance to describe before he died. Brown found it in the Cloverly Formation in Montana, but it sat in the American Museum unstudied for decades. Ostrom was looking for Early Cretaceous dinosaurs, which were virtually unknown in North America (in contrast to the abundant Wealden dinosaurs of Britain, or the huge diversity of Lower Cretaceous dinosaurs in Asia). As recounted by Richard Conniff in "The Man Who Saved the Dinosaurs":

But dinosaurs had begun to look a lot more interesting one afternoon in late August, 1964, near Bridger, Montana. Ostrom and his assistant Grant E. Meyer had been walking a landscape of prairie punctuated with rocky, eroding outcrops, considering sites for the following summer's fieldwork, when they spotted a large, clawed dinosaur hand protruding from the earth on a slope just below them. They scrambled down, dropped to their knees beside it, and because they hadn't brought their toolkit, began digging excitedly with their hands, and then with their jackknives, turning up a scattering of the serrated teeth of a predator. Next day, returning with proper tools, they unearthed an astonishing foot. Two of three toes had ordinary claws. But from the innermost toe, a sharp sickle-shaped claw curved murderously up and out. It had a slashing arc, Ostrom later calculated, of 180 degrees. Hence the eventual name *Deinonychus*, or "terrible claw." Ostrom and his crew spent two full field seasons digging at the site and three years in study and reconstruction at the Peabody, working with more than a thousand bones from at least four individuals of the same species. Then in 1969, Ostrom announced what he called a "grandiose" conclusion: that foot was "perhaps the most revealing bit of anatomical evidence" in decades about how dinosaurs really behaved. In place of the plodding, cold-blooded dinosaur stereotype, *Deinonychus* "must have been a fleet-footed, highly predaceous, extremely agile, and very active animal, sensitive to many stimuli and quick in its responses," Ostrom wrote.

Deinonychus was not only an amazing creature (figures 17.5 and 17.7), but its anatomy completely forced a rethinking of the "slow sluggish" dinosaur image. Its tail was long, straight, and pointed, and was held rigid by a truss of crisscrossing struts of bone from the vertebrate (now turned to stone). With such a rigid structure, the tail could not have dragged on the ground,

Figure 17.7 ▲

Deinonychus: (A) mounted skeleton in a dynamic pose; (B) reconstruction showing the feathers that would have covered its body. ([A] Photograph by the author; [B] courtesy of N. Tamura)

instead serving like a tightrope walker's balancing pole. *Deinonychus* was completely bipedal, yet to use the huge slashing claws on its feet it would have to leap up and strike with its entire foot. This simply was impossible for a sluggish reptile that was slow and inactive.

This is the animal that thrilled movie audiences watching the *Jurassic Park* movies—except instead of calling it by the proper name, *Deinonychus*, author Michael Crichton and director Steven Spielberg opted to call the dinosaur *Velociraptor*. According to some accounts, Crichton was misled by a 1988 book by dinosaur artist Greg Paul, who falsely argued that *Velociraptor* and *Deinonychus* were the same dinosaur, making *Velociraptor* the first valid name. Other accounts suggest that Crichton just thought *Velociraptor* was easier to read, spell, and pronounce or that it sounded cooler than the correct name.

Unfortunately, the movies got the science completely wrong. First of all, the actual *Velociraptor* was the size of a large turkey (figure 17.5). In addition, *Velociraptor* is only known from Mongolia, yet the expedition finds it in "Snakewater, Montana" (which was actually filmed in Red Rock Canyon State Park, California, where the beds yield Miocene mammals, not dinosaurs). Third, *Velociraptor* and most small predatory dinosaurs had feathers. There are even specimens of *Velociraptor* from Mongolia with quill knobs on their arm bones showing where their largest feathers attached. Thanks to Crichton and the movies, the general public now has a slightly more accurate image of what dinosaurs (especially the "raptors") looked and acted like, but everyone attaches the wrong name to the animal that has become so famous. For example, the Toronto NBA team is called the "Raptors" but shows images of the large dromaeosaurs like *Deinonychus*—even though the name they use is that of the turkey-sized *Velociraptor*.

This small but bad choice by Crichton and the moviemakers still drives paleontologists crazy! The other annoying mistake is the fact that *Velociraptor/Deinonychus* had feathers (figures 17.5 and 17.7B), something we've known since 1996. The moviemakers refuse to put feathers on their dinosaurs, so the science is not up to date in the last three *Jurassic Park* movies. Only the first movie was relatively accurate for its time.

HOT AND COLD RUNNING DINOSAURS?

The beginning of the Dinosaur Renaissance, or the change in the way we thought of dinosaurs—from slow sluggish lizards dragging around in swamps to active animals—can be traced to Ostrom's work. In 1963, he wrote a paper arguing that duck-billed dinosaurs were not slow, stupid swamp dwellers but active land-based herbivores that could be compared to buffalo. His discovery of *Deinonychus* in 1964 accelerated his general rethinking of how dinosaurs were built. If *Deinonychus* was an active, hopping, fast-running, jumping, and slashing predator, then maybe the rest of the dinosaurs were more active and agile than once thought. Ostrom also pointed out that dinosaurs all had fully upright posture, with their limbs beneath their bodies, so they were never slow, sprawling, sluggish creatures like crocodilians or lizards. Instead, they seem to have been active fast runners, more like elephants and mammals, and especially birds. Soon other discoveries, such as additional dinosaurs with rigid tails and trackways that showed how fast dinosaurs could run and how they never left tail drag marks, confirmed the "running dinosaur" model.

Once Ostrom began to think about dinosaurs as active creatures, the next obvious question was, "What was dinosaur physiology like?" If *Deinonychus* and many other dinosaurs were fast and agile and active, wouldn't they also be warm blooded? Ostrom cautiously suggested these ideas in a famous paper given at the first North American Paleontological Convention in Chicago in 1969, and he talked about it occasionally afterward. But his former student at Yale (and later Harvard PhD) Bob Bakker aggressively marketed the idea of "hot-blooded dinosaurs" until he was world-famous for it, even publishing cover articles in *National Geographic* and *Scientific American*. During the entire 1970s and early 1980s, the battle over "hot and cold running dinosaurs" raged in scientific meetings and in publications and even in popular books. Eventually, however, the topic reached a limit of how much we could really know about ancient extinct animal physiology. Richard Conniff described it this way:

> Bakker had latched onto many of Ostrom's ideas as an undergraduate and, to Ostrom's occasional chagrin, he ran with them. Bakker—"the infamous Bob Bakker," as Peter Dodson, another former Ostrom student, says—became the outspoken advocate of dinosaurs as active, warm-blooded, and even "superior" animals. "Where John was cautious, Bob was evangelical," Dodson and Philip Gingerich later wrote. "Each deserves considerable credit for revolutionizing our concept of dinosaurs." In his book *The Riddle of the Dinosaur*, science writer John Noble Wilford added that Bakker "was the young Turk whose views could be dismissed by established paleontologists. Ostrom, however,

could not be ignored." Late in 1969, Ostrom took the challenge directly to the North American Paleontological Convention in Chicago, declaring in a speech that there was "impressive, if not compelling" evidence "that many different kinds of ancient reptiles were characterized by mammalian or avian levels of metabolism." Traditionalists in the audience responded, Bakker later recalled, with "shrieks of horror." Their dusty museum pieces were threatening to come to life as real animals.

The problem with talking about dinosaur metabolism is that vertebrate physiology is more complex and not amenable to oversimplifications such as "cold-blooded" animals and "warm-blooded animals." There are actually two components to physiology: the *source* of the heat and whether the heat is *regulated*. Animals that get their body heat from the environment are called *ectotherms*, and those that burn food to create body heat through metabolism are called *endotherms*. Animals that let their body temperatures change with the surrounding temperature are called *poilkilotherms*, and animals that try to hold their body temperature constant are called *homeotherms*.

In the modern world, the boundaries seem pretty clear: all living birds and mammals are homeotherms and endotherms, and the rest of the animals are all poikilotherms and ectotherms. Homeothermic endotherms can live in almost any environment, no matter how hot or cold, but they pay a heavy price in that they burn most of the food they consume for metabolism. Poikilothermic ectotherms regulate their body temperature by moving in and out of hot and cold areas, and if it gets too cold or too hot, they die. For example, a desert lizard typically has a higher body temperature than a "warm-blooded" mammal like you when it is running or active—but it regulates its temperature by shuttling between sun and shade, or burrowing down into the cool sand, not by burning food.

But even with these broad generalizations, there are exceptions that are informative. For example, ectotherms like pythons can generate body heat by shivering when they are incubating their eggs, and sea turtles, tuna, sharks and even some insects are capable of some endothermy. Many homeotherms (such as the platypus, sloths, and certain rodents, shrews, and small birds) let their body temperature fluctuate tremendously, as do animals that go into torpor when they hibernate. These animals allow their body temperature to drop as they go into their suspended animation state. At the other extreme of body size, camels are famous for letting their body temperature cool down during the cold desert night. With their large body mass and small surface area, it takes a long time for the heat of the desert to warm them up. They can even let their bodies reach unusually high temperatures at the end of the day because they are about to cool down in the cold desert night.

During the peak of the controversy, there were lots of arguments back and forth about the possible evidence for dinosaur endothermy. French paleophysiologist Armand de Ricqles pointed out that dinosaur bones have large canals for blood vessels inside them, called Haversian canals, a feature found in mammals but not in reptiles. But it turns out that the presence of these canals is also affected by body size and rate of growth. Some large ectotherms (sea turtles, tortoises, and crocodilians) also have them, but some small birds, bats, shrews, and rodents may not have them. The presence of Haversian canals seems to be more an indicator of rapid growth to large body size than an explanation for endothermy. Dinosaurs are now known to have had extremely rapid growth after they hatched, which better explains the Haversian canals.

Bob Bakker's favorite argument was talking about ratios of biomass of predators in a food pyramid to the biomass of prey. When the predator is an endotherm (say, a lion), the biomass of prey species needs to be about 10 times that of the biomass of lions, because most of the lion's food goes to body heat. In other words, the predator/prey biomass ratio is about 1:10. If the predator is an ectotherm (say, a crocodile), it eats rarely and doesn't use its food for body heat but for activity and growth, so the biomass of predatory crocodiles to the biomass of prey species can be almost equal (the predator/prey biomass ratio is 1:1). When you look at the predator/prey ratios in Early Permian assemblages from Seymour, Texas (preyed upon by the fin-backed protomammal *Dimetrodon*), Bakker claimed that the ratio was about 1:1, but for most dinosaur faunas, the prey biomass is about 10 times that of predators.

This sounds convincing at first, but on closer examination it breaks down. Too many factors bias which animals are fossilized and which are not, so you cannot interpret fossil collections as perfect reflections of what was originally living. Museum collections tend to be highly biased because the collectors are after only the spectacular skulls and other diagnostic parts, and they don't take an unbiased sample of what was present in the collecting area. Lots of things just don't fossilize well, or are overabundant or rare, and these factors may have nothing to do with biology. For example, there are numerous examples of dinosaur quarries that are nothing but predators (such as the *Falcarius* quarry in Utah or the Cleveland-Lloyd Quarry in Utah, which is full of allosaurs). The famous La Brea tar pits in Los Angeles have far more predators and scavengers (primarily dire wolves and saber-toothed cats, plus vultures and predatory birds) than then do prey species. If you took this overabundance of predators at face value, it would suggest a world entirely full of carnivores who were mostly cannibals, with almost no prey species to eat.

Many other arguments were debated back and forth, and the "hot-blooded" dinosaur controversy raged for several decades, but it now seems to be resolved. So were dinosaurs endotherms or ectotherms? The answer is, "It's complicated." Certainly the smaller predatory dinosaurs (like the "raptors" of *Jurassic Park* fame) were endotherms. With their small body size and high levels of activity, they would need a high metabolism to be successful. Indeed, there is good evidence that "raptors" and most predatory dinosaurs (including even *T. rex*) were covered by a downy coat of feathers for insulation; these animals were not slow and stupid but active and smart and warm-blooded.

But for huge dinosaurs like the sauropods, size presents a different problem. At such large body sizes, they have a relatively small surface area compared to their huge volume (see chapter 9). Remember, area only increases as a square, but volume increases as a cube, so the volume increases much faster than the surface area in larger sauropods. They had no obvious ways of rapidly gaining or losing heat from their bodies.

The living elephants are a good example of this physiological dilemma. At its huge size, an elephant must spend much of its time in mud or water or resting in the shade to dump excess body heat. Its huge ears are primarily used as radiators to shed heat from its body. Most sauropods would have had even greater difficulties if they were endotherms and generating body heat from their metabolism of food. Instead, such large beasts kept warm because of the warm climates around them. With their large size, they would have gained or lost body heat only very slowly, so they could obtain a stable warm body temperature by sheer size alone. This strategy is known as "inertial homeothermy," or "gigantothermy," and it probably characterized all of the larger nonpredatory dinosaurs, including sauropods, and possibly stegosaurs, horned dinosaurs, duckbills, and many others.

It is instructive to consider the big mammals, the largest living endotherms, as a reality check. The largest living endotherm is the elephant, which is nearly at the thermal limit for an endotherm. Most of its day is spent finding places to cool down, and it feeds mostly in the cool of the night. Some of the largest extinct mammoths and the gigantic hornless rhinoceros known as *Paraceratherium* are about twice as large as the living elephants, suggesting that there is a size limit for a land-dwelling endothermic animal. No land mammalian endotherm has ever gotten bigger. Yet some of the huge sauropods are 10 times the mass of elephants or *Paraceratherium*, and there is even less convincing evidence that they had efficient heat-regulating surfaces that would allow endothermy to work for an animal with such an unfavorable surface area to mass ratio.

However, the idea of inertial homeothermy in these animals has been challenged recently by Martin Sander and colleagues. They argue that the long necks and tails of sauropods gave them much more surface area than has been suggested, all of which could have helped radiate some of their excess body heat. They also point to the enormous air sacs and passages that connect to the lungs, which could have allowed for passing the air from their bodies in a one-way flow system similar to modern birds, ventilating a lot of body heat from their core body temperature.

At the moment, it's hard to decide which system makes more sense. No one has done the rigorous physiological calculations and modeling to test Sander's model and establish whether sauropods had enough surface area in their skins and respiratory system to ventilate their huge furnaces of bodies running as endotherms. For that matter, no rigorous studies to see whether inertial homeothermy is a viable option have been done either. For now, we'll just have to reserve judgment until rigorous studies have been done to show which model is more plausible.

FOR FURTHER READING

Alexander, R. McNeill. *Dynamics of Dinosaurs and Other Extinct Giants*. New York: Columbia University Press, 1989.

Bakker, Robert T. "Anatomical and Ecological Evidence of Endothermy in Dinosaurs." *Nature* 238, no. 5359 (1972): 81–85.

——. *The Dinosaur Heresies*. New York: Morrow, 1986.

Carpenter, Kenneth. *The Carnivorous Dinosaurs*. Bloomington: Indiana University Press, 2005.

Chinsamy, Anusuya, and William J. Hillenius. "Physiology of Nonavian Dinosaurs." In *The Dinosauria*, 2nd ed., ed. David B. Weishampel, Peter Dodson, and Halszka Osmólska, 643–659. Berkeley: University of California Press, 2004.

Colbert, Edwin. *Men and Dinosaurs: The Search in the Field and in the Laboratory.* New York: Dutton, 1968.

Conniff, Richard. "The Man Who Saved the Dinosaurs." *Yale Alumni Magazine*, July/August 2014. https://yalealumnimagazine.com/articles/3921-the-man-who -saved-the-dinosaurs.

Desmond, Adrian. *Hot-Blooded Dinosaurs: A Revolution in Paleontology.* New York: Dial Press, 1976.

Farlow, James. "Predator/Prey Biomass Ratios, Community Food Webs and Dinosaur Physiology." In *A Cold Look at the Warm Blooded Dinosaurs*, ed. Roger D. K. Thomas and Everett C. Olson, 55–83. Boulder, Colo.: American Association for the Advancement of Science, 1980.

Farlow, James, and M. K. Brett-Surman. *The Complete Dinosaur.* Bloomington: Indiana University Press, 1999.

Fastovsky, David, and David Weishampel. *Dinosaurs: A Concise Natural History*, 3rd ed. Cambridge: Cambridge University Press, 2016.

Holtz, Thomas R., Jr. *Dinosaurs: The Most Complete, Up-to-Date Encyclopedia for Dinosaur Lovers of All Ages.* New York: Random House, 2011.

Lessem, Don. *Kings of Creation: How a New Generation of Scientists Is Revolutionizing Our Understanding of Dinosaurs.* New York: Simon & Schuster, 1992.

Long, John A., and Peter Schouten. *Feathered Dinosaurs: The Origin of Birds.* Oxford: Oxford University Press, 2008.

McGowan, Christopher. *Dinosaurs, Spitfires, and Sea Dragons.* Cambridge, Mass.: Harvard University Press, 1991.

Naish, Darren. *The Great Dinosaur Discoveries.* Berkeley: University of California Press, 2009.

Naish, Darren, and Paul M. Barrett. *Dinosaurs: How They Lived and Evolved.* Washington, D.C.: Smithsonian Books, 2016.

Norell, Mark A., and Peter J. Makovicky. "Dromaeosauridae." In *The Dinosauria*, 2nd ed., ed. David B. Weishampel, Peter Dodson, and Halszka Osmólska, 196–210. Berkeley: University of California Press, 2004.

Ostrom, John H. "The Evidence of Endothermy in Dinosaurs." In *A Cold Look at the Warm Blooded Dinosaurs*, ed. Roger D. K. Thomas and Everett C. Olson, 82–105. Boulder, Colo.: American Association for the Advancement of Science, 1980.

——. "A New Theropod Dinosaur from the Lower Cretaceous of Montana." *Postilla* 128 (1969): 1–17.

———. "Osteology of *Deinonychus antirrhopus*, an Unusual Theropod from the Lower Cretaceous of Montana." *Peabody Museum of Natural History Bulletin* 30 (1969): 1–165.

Sander, P. M., A. Christian, M. Clauss, R. Fechner, C. Gee, E.-M. Griebeler, H.-C. Gunga, J. Hummel, H. Mallison, S. Perry, H. Preuschoft, O. Rauhut, K. Remes, T. Tütken, O. Wings, and U. Witzel. "Biology of the Sauropod Dinosaurs: The Evolution of Gigantism." *Biological Reviews of the Cambridge Philosophical Society* 86, no. 1 (2011): 117–155.

Thomas, Roger D. K., and Everett C. Olson, eds. *A Cold Look at the Warm Blooded Dinosaurs*. Boulder, Colo.: American Association for the Advancement of Science/Westview, 1980.

Turner, Alan H., Peter J. Makovicky, and Mark A. Norell. "Feather Quill Knobs in the Dinosaur *Velociraptor*." *Science* 317, no. 5845 (2007): 1721.

Wilford, John Noble. *The Riddle of the Dinosaur*. New York: Knopf, 1985.

SINOSAUROPTERYX

And if the whole hindquarters, from the ilium to the toes, of a half-hatched chick could be suddenly enlarged, ossified, and fossilised as they are, they would furnish us with the last step of the transition between Birds and Reptiles; for there would be nothing in their characters to prevent us from referring them to the Dinosauria.

—THOMAS HENRY HUXLEY, *FURTHER EVIDENCE OF THE AFFINITY BETWEEN DINOSAURIAN REPTILES AND BIRDS*, 1870

THE BIG CHILL IN CHINA

During the Chinese Cultural Revolution from 1966 to 1976, scholars and educated people were purged from most institutions, and many were killed or forced to work in labor camps and undergo indoctrination for not being good communists. Chinese scientists at the Institute for Vertebrate Paleontology and Paleoanthropology (IVPP) were severely affected, and most of them tried to lay low and hope that the authorities would not arrest them. Even if their lives were not threatened, they had almost no contact with the outside world. Almost all research ceased, and Chinese scientists had no access to the scientific work of the rest of the world, nor could they leave China to see other fossils and make comparisons with their own specimens.

Some of the most important Chinese localities include the famous Lower Cretaceous lake shales of the Yixian Formation in Liaoning Province of northeastern China. These shales were deposited at the bottom of a stagnant lake with no oxygen at the bottom. When animals died, they sank but did not decay until they were covered with many fine layers of silt and mud. The lack of oxygen allowed even soft tissues to be preserved. As a result, the Jehol biota from these shales includes many incredible specimens and thousands of leaves, insects, and small animals. More famous are the birds and nonavian dinosaurs with their original feathery coating (some with the original pigment cells preserved) and stomach contents, plus some of the earliest complete fossils (including fur and body outline) of marsupial and placental mammals and other more archaic groups. There are just a handful of these extraordinary localities known anywhere in the world fossil record. Paleontologists call them *Lagerstätten*, German for "mother lode."

There was very limited collecting in Liaoning Province in the early twentieth century, and after 1949 China was closed to most Western scientists. This remained true during the dark days of the Cultural Revolution and well into the early 1980s. Specimens kept accumulating in the collections and were studied but not published. In the late 1970s and early 1980s, the political winds in China changed; Mao Zedong died in 1976, and the "Gang of the Four" were overthrown. A more open and practical regime began under Zhou Enlai.

In the late 1980s and 1990s, the floodgates burst, and more and more scientists found and studied these amazing specimens. Chinese scientists were once again allowed to travel and publish, and Western scientists were allowed into China for the first time since the 1940s. In addition, a few Chinese scientists were allowed to travel abroad, study fossils in other museums, attend professional meetings, and present their work. They described their fossils in international journals, and everyone began to see their incredible discoveries.

The Liaoning fossils included hundreds of exquisitely preserved Early Cretaceous birds, some of which were published but many are still being studied. (My former student Dr. Jingmai O'Connor is now a curator at the IVPP, working on enantiornithine birds and publishing many papers each year.) Each fall at the annual Society of Vertebrate Paleontology (SVP) meeting during the 1980s and 1990s, paleontologists were stunned by presentations showing these incredible fossils. Some seeing the light of the scientific world for the first time in 50 years.

In 1996, a local farmer and part-time fossil hunter named Li Yumin found an extraordinary specimen preserved on a slab with nearly all the bones present in a death pose (figure 18.1). He knew his fossil had great

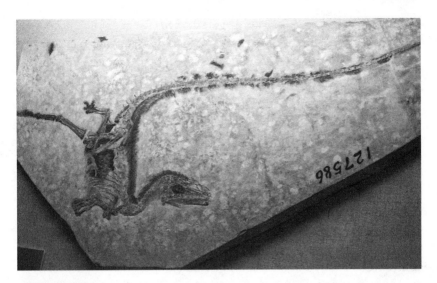

Figure 18.1 ▲

The original specimen of *Sinosauropteryx* from the Liaoning beds, showing the complete animal with soft tissues preserved and even the feathers along its back. (Courtesy of Wikimedia Commons)

value, so he sold the slab and its counterpart (the other half of the split fossil) to two different museums. Ji Qiang, director of the Beijing National Geological Museum, realized it was an important specimen. As luck would have it, Phil Currie of the Tyrrell Museum in Alberta was visiting at the time, having just led a fossil tour of Mongolia. He recognized its great value immediately. As he said later to the *New York Times*, "When I saw this slab of siltstone mixed with volcanic ash in which the creature is embedded, I was bowled over." The Chinese scientists immediately began to work on a paper about the specimen and banned anyone from publishing the images until the paper appeared.

But Currie and another Chinese scientist, Chen Pei Ji, had photographs of the specimen, which they brought with them to the SVP meeting in the American Museum of Natural History in the fall of 1996. They were not on the program to make an official presentation about the fossil, but rumors about the specimen spread like wildfire through the meeting. Whenever they were standing in the hallways or at receptions, they were mobbed by paleontologists who wanted to see for themselves. I remember this vividly, and it took me several tries before I got my own chance to see the images. This is not unusual for the SVP meetings. Lots of people bring photographs or casts or even original specimens with them to get input and reactions from the qualified specialists at the meeting. This is especially true if they want this feedback before they write up the specimen for publication and presentation at a future meeting. I can't even count how many times someone has come up to me at SVP, slipped a specimen out of his or her pocket, and asked me for my opinion.

Why were these photographs so shocking and surprising? After more than 25 years of debate about whether birds evolved from dinosaurs, here was the first fossil of a nonbird dinosaur with clear imprints of some kind of feathers, especially the ridge of fluffy down along the spine. All the paleontologists who fought hard to deny any connection between birds and dinosaurs were suddenly and completely discredited. John Ostrom, who had led the battle to accept the fact that birds evolved from dinosaurs, was overjoyed and even in a state of shock. A year after that SVP meeting, Ostrom, Alan Brush (an expert on feather evolution), Peter Wellnhofer (an expert on *Archaeopteryx*), and Larry Martin (one of the last deniers of the evidence that birds are dinosaurs) met in Beijing to look closely at the specimen and argue about it. By then several additional specimens with even clearer imprints had been found, and over the next 20 years, hundreds of dinosaurs not even close to the branch that led to birds showed evidence of feathers. Published later in 1996 in a small Chinese museum journal, Ji Qiang and Ji Shuan named it *Sinosauropteryx*, "Chinese lizard wing." It was the final proof for an idea that had been simmering since a chance discovery in 1861.

"ANCIENT WING"

During the early days of printing, to create a print of a drawing in a book or newspaper, one had to cut an image into a wooden block or etch an image into a block of fine-grained limestone, using wax drawings on a smooth rock surface to prevent the acid from etching the wax-covered part of the limestone. This technique is called lithography ("stone writing" in Greek). In Bavaria in southern Germany, there was a naturally occurring extremely fine-grained rock known as the Solnhofen Limestone, which the stonecutters had been quarrying for centuries. Every once in a while they would find an extraordinarily complete and well-preserved fossil of something that had died and sunk to the bottom of this stagnant, shallow lagoon surrounding an archipelago of islands formed when Late Jurassic seas drowned much of Europe. These creatures were buried without being disturbed by scavengers. All their soft tissues, including skin, were intact, and they were articulated in a death pose. Fossils of marine creatures such as shrimp and horseshoe crabs were found more frequently, but occasionally land animals that had washed out into the lagoon, such as the first pterodactyl specimens, were recovered. Because of their great value on the commercial market, the quarrymen saved these fossil slabs from becoming a lithographic plate or a building stone. In 1860, they found the fossil of a feather, and in 1861, a complete but partially disarticulated specimen of what appeared to be a bird (figure 18.2A). The idea of a bird fossil with feathers was sensational. Before the German authorities could coordinate their funding and make a bid, the British Museum in London had purchased the specimen, and it still resides there today (now known as the "London" specimen).

Naturally, it fell to Richard Owen, the leading paleontologist in England, to describe it (see chapter 3). The discovery was particularly timely as Darwin's *On the Origin of Species* had been published a year earlier and was a national sensation. Darwin was overjoyed at the news of a primitive bird that provided a transition between two major groups. The original feather fossil had already been named *Archaeopteryx* ("ancient wing" in Greek) by Hermann von Meyer. Owen dutifully described the anatomy of the complete but somewhat jumbled "London" specimen as a Jurassic bird. But as an opponent and critic of Darwin, he made no mention of all the archaic dinosaur-like features of this bird, which made it exactly the kind of transitional fossil that Darwin had predicted.

Darwin's great friend and supporter, Thomas Henry Huxley, had no such reservations. Often nicknamed "Darwin's Bulldog," Huxley was already a rising star in paleontology and biology, and after 1859 he spent a lot of his time defending Darwin's ideas in print and in debates. (Darwin was notoriously shy and retiring and sick much of the time, so he never got into the battle directly.) When Huxley looked at the *Archaeopteryx* fossil, he could see that most of its bony features suggested that it was a small dinosaur. At a famous presentation in front of the Royal Society in 1863, he proposed that birds were descended from dinosaurs and listed 35 features shared only by nonavian dinosaurs and birds (17 of these are still used by modern paleontologists).

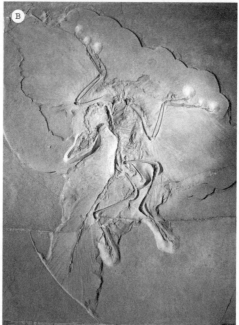

Figure 18.2 ▲

Archaeopteryx lithographica: (*A*) the "London" specimen, now in the Natural History Museum in London; (*B*) the "Berlin" specimen, now in the Museum für Naturkunde in Berlin. (Courtesy of Wikimedia Commons)

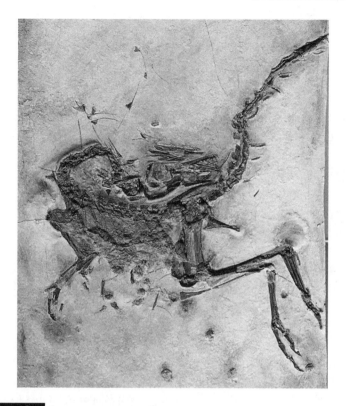

Figure 18.3 ▲
Articulated skeleton of *Compsognathus* from the Upper Jurassic Solnhofen Limestone. (Courtesy of Wikimedia Commons)

In 1868, he studied another fossil from the Solnhofen Limestone, the first known specimen of the small dinosaur *Compsognathus* (figure 18.3). This was the basis for the dinosaurs nicknamed "compys" in the *Jurassic Park* books and movies. It had been found in 1859 and briefly described and named by Johann Wagner, but now it was more important than ever. Huxley immediately noticed how similar its bones were to the bones of *Archaeopteryx*, convincing him that birds evolved from small dinosaurs. (In 1973, F. X. Mayr found a specimen that had been misidentified as *Compsognathus* until he saw feather impressions and realized it was *Archaeopteryx*.) Finally, in 1877 the most famous of the 12 known specimens of *Archaeopteryx* was found in a perfect death pose with the neck drawn back (figure 18.2B). This fossil is known as the "Berlin" specimen because the German authorities got the money from Wilhelm Siemens to buy it and prevent it from "flying the coop." Today it is featured in a vault behind bulletproof glass at the Museum für Naturkunde in Berlin.

In the late 1800s, the "birds are dinosaurs" idea became less popular, and paleontologist Harry Govier Seeley argued strongly against it. Then, in 1926, paleoartist Gerhard Heilmann proposed that birds arose from some primitive unspecified group of archosaurs (then called "thecodonts"; see chapter 5). His main evidence against the dinosaur origin of birds was that birds have a wishbone formed of two fused collarbones or clavicles, and he claimed that no dinosaur had collarbones. How could birds with collarbones evolve from animals that had already lost their collarbones? It turns out that most dinosaurs did have collarbones (this has been demonstrated in numerous specimens now), but collarbones are small and fragile and are rarely preserved except in the most complete specimens. In 1926, this argument seemed conclusive however, and the "birds are dinosaurs" idea died for decades.

The "birds are dinosaurs" idea was nearly forgotten until the 1970s, when John Ostrom was reexamining and redescribing all the known specimens of *Archaeopteryx*. He was already struck by the idea that dinosaurs like *Deinonychus* were active, intelligent predators (see chapter 17). As he looked closer at *Archaeopteryx*, he realized he was seeing almost the same bony anatomy as in *Deinonychus*. Not only were the bones almost the same shape, but some strikingly detailed features in the anatomy were shared by *Deinonychus* and *Archaeopteryx*. These features were only seen in birds and dromaeosaur dinosaurs such as *Deinonychus*. For example, some of the wrist bones in these animals are fused into a crescent-moon-shaped bone called the semilunate carpal (figure 18.4). This wrist bone allows dromaeosaurus like *Deinonychus* to snap their hand in a quick motion forward and down to grab prey—the same motion of the wrist seen in the downward flight stroke of a bird.

Like all other dinosaurs and pterosaurs, *Archaeopteryx* and all birds have a mesotarsal joint. Unlike almost all other land animals, the joint in their foot does not hinge between their shinbone and first row of ankle bones (tarsals), but between the first and second row of ankle bones (see figure 5.4B). You can see this any time you are eating a chicken or turkey drumstick. The drumstick bone itself is the shinbone of the bird, and the tiny cap of cartilage on the "handle" of the drumstick is the remnant of the

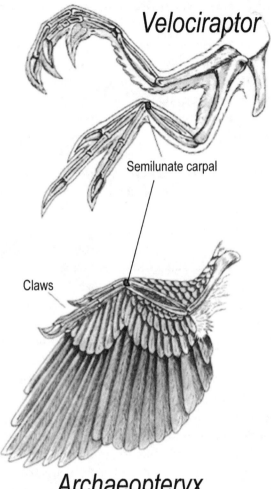

Velociraptor

Semilunate carpal

Claws

Archaeopteryx

Figure 18.4 ▲
Anatomy of the wrist bones, including the semilunate carpal (dark bone) found in dromae-osaurs like *Velociraptor* and in birds like *Archaeopteryx*. (Redrawn from several sources)

first row of ankle bones. Only birds, dinosaurs, and pterosaurs have this unique configuration of the ankle. Finally, the astragalus bone in the first row of ankle bones (fused to the end of the shinbone) has a spur of bone (called the "ascending process") that overlaps the front of the lower end of the shin. This feature only occurs in dinosaurs and birds, yet another distinctive evolutionary novelty that can be explained only if birds evolved from dinosaurs.

As Ostrom revived Huxley's idea that birds are dinosaurs during the 1970s and 1980s, he faced resistance from scientists who were used to the old way of thinking. But along with the Dinosaur Renaissance, largely started by Ostrom, the momentum for "birds are dinosaurs" quickly built as more and more evidence accumulated to support it. By the time of Jacques Gauthier's pioneering analysis of archosaurian relationships in 1986, hundreds of unique anatomical specializations supported the idea that birds evolved from dromaeosaurs such as *Velociraptor* and *Deinonychus*. The discovery of *Sinosauropteryx* in 1996, and the many other feathered nonbird dinosaurs found since then, should have been the final nail in the coffin of the "birds are not dinosaurs" (BAND) idea, but even this was not enough.

Entrenched beliefs don't die easily, especially for something as radical as arguing that birds are just modified dinosaurs. Some of the deniers simply didn't like or understand cladistics or realize the importance of the evidence of unique evolutionary specializations. Others tried to nitpick one or two of the anatomical lines of evidence, ignoring the fact that there were hundreds of features supporting the argument that birds are dinosaurs. They were stuck on rigid older ideas such as "birds have feathers, reptiles don't" and "feathers evolved for flight." More and more feathered nonflying nonbird dinosaurs were discovered, showing that feathers first evolved for insulation—and most of the down feathers on a bird still function that way today. The development of certain feathers into airfoils that create lift was a late event in feather evolution, taking advantage of a feature that was already present on the bodies of birds and of the dinosaurs.

Many of the deniers were obsessed with the idea that birds turned into flyers by gliding down from trees, and they could not imagine birds developing flight from the ground up. Ostrom showed that many ground birds propel themselves by flapping as they run across the ground, and this is still seen in ground birds such as the chicken, turkeys, pheasants, and quail. These birds fly just well enough to escape a ground predator, but they cannot fly very fast or very far. A study by Ken Dial in 2002 on chukar partridges found that feathered wings were very good for assisting in scrambling up steep inclines, even though chukars rarely actually fly. It is now thought that birds probably evolved flight feathers primarily for short escape flights and for climbing up surfaces, eventually finding a way to use them to glide and fly longer distances.

The dino deniers often used extreme language in their battle against the wave of evidence against them. Ornithologist Alan Feduccia wrote in *Science* that "the theropod origin of birds will be the greatest embarrassment of paleontology in the 20th century." In his book, *Unearthing the Dragon*, American Museum paleontologist Mark Norell goes over the weird debate in detail (226–235). As he writes, the "birds are not dinosaurs," and the gang "have never quite gotten it that the origin of flight and feathers is decoupled from the origin of birds. To them, if it has feathers it is an avian. They have odd ideas about how all this relates to flight. As Feduccia claims, 'It is biophysically impossible to evolve fight from such large bipeds with foreshortened limbs and heavy, balancing tails.' Before more discussion about feathers, the point of about large bipeds incapable of climbing trees is clearly falsified by the presence of so many nonavian dinosaurs in the Jehol fauna" (227).

As the debate dragged on, the BAND band kept "moving the goalposts." Whenever one piece of evidence showed the dino deniers were wrong, they'd quickly abandon their old argument and grasp at straws to find another attack that maintained their position. As Kristopher Kripchak posted, "As soon as one theory on why birds cannot be dinosaurs is demolished by a new discovery, the BAND crowd comes up with a new theory that is even less plausible than their previous one. Over the past few years, these folks have adopted more positions than the Kama Sutra" (quoted in Norell, 235).

The debate was more intense than most scientific arguments for another reason—it made a good spectacle in the media. In 2005, David Fastovsky and David Weishampel wrote: "The debate has been unnaturally prolonged by media attention. The origin of birds has been a topic of great public interest for the past twenty years, so much so that the leading proponents are frequently interviewed for newspaper articles and TV specials. The rules of journalism require that 'equal time' is given to representatives of each viewpoint. So the supports of the basal diapsid origin of birds often have as much airtime as the supporters of birds as dinosaurs, even though the latter represent probably more than 99 percent of working vertebrate paleontologists."

As Norell put it, "When more evidence is garnered, whether through the analysis of additional characters, through the discovery of new specimens, or by pointing out errors and problems with the original data sets, new trees can be calculated. If these new trees better explain the data (taking fewer evolutionary transformations), they supplant the previous trees. You might not always like what comes out, but you have to accept it. Any real systematist (or scientist in general) has to be ready to heave all that he or she has believed in, consider it crap, and move on, in the face of new evidence. That is how we differ from clerics."

Today the debate has largely ended within the paleontological community. The late Larry Martin was the last major denier among vertebrate paleontologists, so no one at SVP is still beating this dead horse. There are still some deniers among ornithologists, who are not familiar with dinosaurs or the evidence or still object to cladistic methods, but they have no influence on the future direction of bird and dinosaur research. Some arguments in science (the debate over what caused the extinction of nonavian dinosaurs at the end of the Cretaceous, or whether dinosaurs were endotherms) are stuck forever in the unresolved, "It's complicated" category. Fortunately, the debate about whether birds are dinosaurs is over.

FEATHERS, FEATHERS, EVERYWHERE

The discovery of *Sinosauropteryx* was the first of hundreds of additional specimens from the Liaoning Province, and other places in China, that produced evidence of feathers in nonbird dinosaurs. In fact, *Sinosauropteryx* is not that closely related to birds at all. It turns out that it is a close relative of *Compsognathus*, which Huxley first featured in 1868 (figure 18.3) and is a very primitive branch of theropods related to allosaurs, megalosaurs, and spinosaurs. As mentioned previously, none of these nonbird dinosaurs have flight feathers. Instead their feathers are simple shafts (like the pin feathers of modern birds) or downy coverings, or possibly larger feathers that can be used for coloration, display, and courtship.

Prum and Brush published an analysis of the evolution of feathers in 2003 (figure 18.5). First of all, they showed that feathers were not (as dogma had taught) modified reptilian scales but come from a different embryonic primordium with a different developmental pathway. What they call Type 1 feathers are simple, hollow pointed shafts. These occur in the most primitive dinosaurs and appeared in the Ornithischia as well. In other words, all dinosaurs have some kind of feathery covering on some parts of their bodies, even if they are just simple quills. What Prum and Brush called Type 2 feathers are simple down with no vanes in the center. This is what excited paleontologists at SVP in 1996 when they first saw photos of *Sinosauropteryx*. Type 3 feathers have a vane and a shaft, but the vanes are not

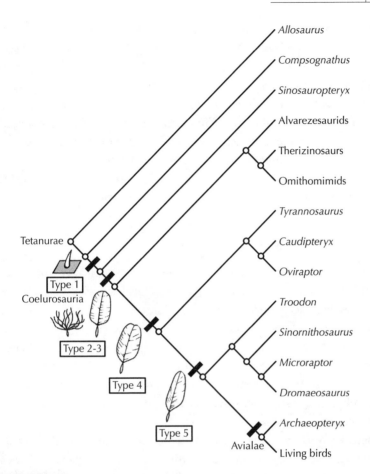

Figure 18.5 ▲

The evolution of feather types from simple pinshafts to down plumes to complex flight feathers with asymmetric vanes and shaft. On the basis of their appearance in various feathered nonflying dinosaurs from Liaoning, we can demonstrate that most predatory dinosaurs (including *T. rex*) probably had feathers of some sort.

linked by hook-like barbules sticking them together like Velcro. This is the kind of feathers found in the Chinese tyrannosaur *Yutyrannus* and its close relatives, so that would mean *Tyrannosaurus rex* had feathers of the Type 3 category. Both Type 2 and 3 feathers were found in the large therizinosaur *Beipiaosaurus*. This shows that most of the advanced theropods related to the therizinosaurs, such as ornithomimids and dromaeosaurs, must have had such feathers.

According to Prum and Brush, their Type 4 feathers have barbules that link the vanes of the feather into a continuous surface, but the shaft is symmetrical down the middle of the feather. We find feathers like this in the advanced theropod *Caudipteryx*, which suggests that these occurred in all the advanced theropods related to it, such as the oviraptorids and the dromaeosaurs. Finally, the most familiar type of feather is the asymmetric flight feather with the shaft near the leading edge of the vane, making an airfoil. Feathers like this first occur in *Archaeopteryx* and in a few of its close relatives, and many paleontologists think *Archaeopteryx* was one of the first dinosaurs to actually be able to truly fly and not just glide.

THE "ARCHAEORAPTOR" HOAX

The market for specimens from the Liaoning beds is so lucrative that many farmers spend most of their time quarrying local outcrops of the Yixian Formation in hopes of finding a great specimen that will pay far more than farming. Making the specimen more attractive and preparing it to show all of its good features increases the pressure—and pressure to "enhance" the specimen so it will fetch a better price is also felt. Paleontologists who work with these fossils, particularly those not quarried by scientists and that come from questionable private sources, have learned to watch out for these problems.

There are also pressures on the media to find flashier and flashier stories to grab readers' attention and sell magazines or newspapers (or these days to get more clicks on a story). This is even more true now that most media are commercial ventures owned by big corporations, driven to make a profit, not to publish a well-checked truth. Together these two forces combined to make the most famous hoax since Piltdown man. This contrasts with most scientific journals, which require careful fact-checking and critiques by scientific peers to sniff out problems and weed out bad research.

In 1997, a farmer in Xiasanjiazi, China, found a specimen of a toothed bird, but the slab was broken into several pieces. In a nearby pit, he found another fossil with a feathered tail and legs that complemented the front end of the animal found in the first slab. He cemented the pieces together, thinking they were parts of the same animal, and sold them to a dealer in June 1998. The dealer then smuggled the contraband specimen to the United States. (It's illegal to export *any* fossils from China. Once they are out of China, however, they are freely marketed with no worries about their illegal source.) Note that the original finder of the fossil was not deliberately trying to hoax the scientific world or the public; he was simply "enhancing" his "art object" to fetch a better price.

When I attended the SVP meeting in October 1998 in Snowbird, Utah, the gossip was flying about an extraordinary specimen with a toothed bill and feathered body and feathered tail that was being circulated among the commercial dealers in a national mineral and gem show. (This was just after *Sinosauropteryx* was first published, and very few other feathered dinosaurs had been found.) The word reached paleoartists Steve and Sylvia Czerkas, who got their patrons to spend $80,000 to buy the fossil for their museum so it would not become another specimen lost to rich collectors. Meanwhile a writer from *National Geographic* magazine became interested and was anxious to get the scoop on the scientific journals and write a cover story about this amazing feathered dinosaur.

Various parties told Steve and Sylvia Czerkas that they needed to have experts look at it, so they asked Phil Currie of the Royal Tyrrell Museum in Drumheller, Alberta, and Xu Xing from IVPP to come and study it on March 6, 1999. Right away Currie noticed that the left and right halves of the top part of the animal were mirror images, so the fossil had been made by gluing the slab and its counterpart together side by side. He also noticed that the tail and hind legs did not match the top of the specimen, and there was no evidence they had ever belonged together. They got Tim Rowe of the University of Texas to CAT scan it, and the scan showed the pieces didn't belong together. Nevertheless, Steve and Sylvia Czerkas refused to listen to the paleontologists, and *National Geographic* didn't want its scoop in the next issue to be delayed while waiting for careful analysis and publication in a peer-reviewed scientific journal. They ran their article on October 15, 1999. *National Geographic* held a big press conference to announce their discovery, and it appeared in the November 1999 issue with the name "Archaeoraptor."

When the article appeared, it sparked outrage among paleontologists. First of all, publishing a new scientific name in a popular magazine rather than in a scientific journal is a strict no-no. (In retrospect this was lucky. After the hoax was discovered, the naming mistake meant that scientists didn't have to formally deal with the name—it is invalid.) More important, other people began to agree with Currie and Xu that the specimen appeared to be a composite of more than one animal. Xu returned to the collecting area, interviewed many of the Chinese collectors, and soon located the site where the tail and hind legs had been found. Eventually, he found the counterpart slab that matched the lower half of the "Archaeoraptor" fossil; it even had a matching yellow iron oxide stain. Yet the rest of the slab did not have parts that matched the top of "Archaeoraptor." Soon several people followed Xu and Currie in announcing that it was a composite of several specimens. *National Geographic* launched their own investigation, and in October 2000 it was finally concluded that "Archaeoraptor" was a hoax.

What are we to learn from this? The BAND band always points to "Archaeoraptor" when they refuse to accept new specimens that prove them wrong. Even creationists bring up "Archaeoraptor" as proof than *no* fossil is real—they are all hoaxes cooked up by paleontologists to prove evolution. It is important to note that the hoax was discovered as soon as qualified paleontologists looked at it. Amateur paleoartists Steve and Sylvia Czerkas did not have the years of experience and anatomical training required to know what is real and what is not. Moreover, the hoax would never have gotten so far if *National Geographic* hadn't tried to "scoop" everyone and rushed to publication before proper scientific analysis and review were completed.

This episode once again reminds us of the serious threat to paleontology represented by the commercial market for fossils. Yes, many of these specimens are authentic, and sometimes these fossils end up in the hands of scientific institutions where they can be studied by qualified people instead of decorating someone's living room. But as we saw with the skull of *Irritator* (chapter 13), many of the fossils are "enhanced" or even faked to improve their selling price. Even worse, *no* fossil from China or Mongolia should ever leave its home country except on a short-term loan. But smuggling and poaching operations are so out control that the market is flooded with these illegal fossils. Any fossil from a commercial dealer is automatically suspect, not only because it lacks the data describing where the fossil was collected and what rock layer it came from but because it may be a fake. In addition, there's a good chance that the fossil was poached (see chapter 14 about the Mongolian tarbosaur poachers or chapter 16 about the therizinosaurs that were poached). Even when good-quality authentic specimens (such as the primate fossil called *Darwinius*, which came from a private collector) are announced, paleontologists are wary. Not only might the locality data be faked or lost, but they are also encouraging poaching by buying it. Worst of all, the fossil might be illegal or even a hoax.

WHAT COLOR WERE DINOSAURS?

One of the major problems with the public concept of dinosaurs is that they take the images presented in the media and in merchandising too seriously. Kids often ask me questions like "What color were dinosaurs?"For years, I could only tell them "No one knows." The general public doesn't realize that we usually have only the skeletons, so most of the muscles, tendons, skin textures, and coloration are pure guesswork supplied by the artist who reconstructs the dinosaur and is not based on direct scientific evidence. (A handful of dinosaurs also have skin impressions, but they don't begin to represent all the dinosaurs that are reconstructed.) Most of the sounds and behaviors animators provide for their subjects are pure guesswork, often informed by very little scientific evidence.

Dinosaur reconstructions come in a complete range of colors, from the various drab greens and browns, based on most animals in nature to today, to brightly colored patterns that seem almost psychedelic. The possible colors for dinosaurs are limited only by the artist's imagination because there was no fossil evidence one way or another. Skin only rarely preserves in dinosaurs, muscles even less often, and color and texture of the skin surfaces seemed to have no fossilization potential.

At least that was the story until 2010. Then a startling article came out in *Nature,* the leading British science journal, which described a specimen of a small feathered dinosaur, *Anchiornis,* from the Liaoning beds, which is a close relative of the dromaeosaurs. This beautiful specimen had feather impressions that were so well preserved that they also included some of the melanosomes, the color-determining pigments that give feathers their hues. From this specimen, we can now say with confidence that *Anchiornis* had a red head, black and white stripes and blotches on its body and wings, and a black tail.

Sinosauropteryx (now represented by many more specimens than the original found in 1996) was recently analyzed. It was even more gaudy (figure 18.6). Even though it's a small ground runner related to *Compsognathus,* its body was covered by orange feathers, its tail had black rings around it, and the head had white, orange, and red blotches. Now there are specimens of dromaeosaurs including *Caudipteryx, Microraptor,* and *Sinornithosaurus* with color impressions, along with a bunch of birds, such as *Confuciusornis,* and a number of Early Cretaceous birds that preserve their color patterns. Even the colors of *Archaeopteryx* have been deciphered.

Figure 18.6 ▲
Color patterns of *Sinosauropteryx* based on melanosomes in the specimen. (Courtesy of N. Tamura)

It is no longer guesswork to say that dinosaurs had feathers and to guess what color they were. I never imagined that this day would ever come to paleontology.

FOR FURTHER READING

Chiappe, Luis, and Meng Qingin. *Birds of Stone: Chinese Avian Fossils from the Age of Dinosaurs*. Baltimore, Md.: Johns Hopkins University Press, 2016.

Long, John A., and Peter Schouten. *Feathered Dinosaurs: The Origin of Birds*. Oxford: Oxford University Press, 2008.

Martinyuk, Matthew A. *A Field Guide to Mesozoic Birds and Other Winged Dinosaurs*. New York: Pan Aves, 2012.

Mayr, Gerald. *Avian Evolution: The Fossil Record of Birds and Its Paleobiological Significance*. New York: Wiley-Blackwell, 2016.

Norell, Mark. *Unearthing the Dragon: The Great Feathered Dinosaur Discovery*. New York: Pi Press, 2005.

Pickrell, John. *Flying Dinosaurs: How Fearsome Reptiles Became Birds*. New York: Columbia University Press, 2014.

Shipman, Pat. *Taking Wing: Archaeopteryx and the Evolution of Bird Flight*. New York: Simon & Schuster, 1999.

Wellnhofer, Peter. *Archaeopteryx: The Icon of Evolution*. Berlin: Dr. Friedrich Pfeil, 2009.

PART IV

HORNS AND SPIKES AND ARMOR AND DUCK BEAKS
THE ORNITHISCHIANS

Oh how unlike Iguanodon next me

In dignity, yet moving at my nod.

The Mega-Plesi-Hylae-Saurian tribes-

Ranked next along the grand descending scale:

Testudo next below the Nautilus

The curious Ammonite and kindred forms,

All giants to the puny races here,

Scarce seen except by Ichthyosaurian eye,

Gone too the noble palms, the lofty ferns,

The Calamite, Stigmaria, Voltzia all:

And Oh! what dwarfs, unworthy of a name,

Iguanodon could scarce find here a meal!

Grow on their graves! Here, too, where ocean rolled,

Where coral groves the bright green waters graced,

Which glorious monsters made their frolic haunts,

Where strange Fucoides, strewed its very bed,

And fish of splendid forms and hues, ranged free,

A shallow brook troop, where only creatures live

Which in my day were Sauroscopic called,

Scarce visible, now creeps along the waste.

—EDWARD HITCHCOCK, "THE SANDSTONE BIRD," 1836

HETERODONTOSAURUS

The pubes also present two types. First there are the genera in which the bones are directed anteriorly and meet by a median symphysis and have no posterior extension except for the proximal symphysis with the ischium. This type is represented by *Cetiosaurus*, *Ornithopsis*, *Megalosaurus*, and many genera figured by Professor Marsh. The second form of pubis has one limb which is directed backward parallel to the ischium, and another limb directed forward. It is typically seen in *Omosaurus* and *Iguanodon*. There are many variations in stoutness and details of form of the bones, but so far as I am aware these two plans comprise all the Dinosaurian genera.

—HARRY GOVIER SEELEY, *ON THE CLASSIFICATION OF ANIMALS COMMONLY NAMED DINOSAURIA*, 1887

"BIRD HIPS"

In the 1880s, the number of known dinosaurs was beginning to multiply as Cope and Marsh rapidly added to the list of species already known from Europe. All of these scientists tried to put together a scheme to classify all these groups of individual genera, such as *Iguanodon* and *Megalosaurus* and *Cetiosaurus* plus the new American dinosaurs. In 1866, Cope created the names "Ornithopoda" for iguanodonts and scelidosaurs, "Goniopoda" for *Megalosaurus*, and "Symphopoda" for *Compsognathus*. In 1870, Huxley only recognized the families Scelidosauridae, Iguanodontidaea, Megalosauridae, and Compsognathidae; and Harry Govier Seeley added the Cetiosauridae to the list (after Owen had originally described it but not realized it was a dinosaur; see chapter 3). Marsh used a different classification scheme from 1878 to 1884, recognizing orders for Stegosauria, Ornithopoda, Sauropoda, and Theropoda. All four of those groups are still in use today in some capacity. Cope, naturally, refused to recognize most of Marsh's ideas or names, so in 1883 he wrote about groups he called the Ornithopoda, plus the "Opisthocoelia" for sauropods, "Goniopoda" for theropods, and Hallopoda for a group now known to be pseudosuchian archosaurs, not dinosaurs.

The chaos of names was a product not only of the Cope-Marsh feud but of the choice of anatomical parts that were emphasized, from the shape of the foot to the shape of the skull. Each scientist recognized lower-level groups for predators (Theropoda or "Goniopoda"), sauropods or "Opisthocoelia," and nearly everyone used Ornithopoda for iguanodonts and their kin. But no one could cluster these many orders of dinosaurs into two or three larger groups.

In 1887, Seeley again tried to make sense of dinosaur classification. He noticed that the structure of the hips seemed to have only two basic forms. Many dinosaurs, such as *Megalosaurus, Cetiosaurus, Compsognathus,* and a few others known at that time, have a simple pelvis structure, with the pubic bone pointing only in the forward direction (see figure 5.3C). These were the "lizard-hipped" dinosaurs, or "Saurischia."

In the hips of other dinosaurs, such as *Iguanodont, Scelidosaurus, Stegosaurus,* and a few others known at that time, at least part (if not all) of the pubic bone projected backward, running parallel to the ischium in the hip (see figure 5.3A, B). Seeley called these "bird-hipped" dinosaurs, or Ornithischia, because living birds also have a backward-pointing pubic bone. Within a few years, most paleontologists of the late 1800s and 1900s had adopted these two categories because it seemed that every new dinosaur found fit into one or the other.

As a group, the Saurischia was always problematic. They were defined by having the primitive reptilian hip structure with a forward-pointing pubic bone, a feature seen in crocodiles and other reptiles. Seeley noticed they had a few unique characteristics, such as hollow, pneumatic bones, but he never emphasized these features. In 1986, Jacques Gauthier proposed a few anatomical features in addition to the hip that seemed to unite the Saurischia. But the latest work by Barron, Norman, and Barrett in 2017 argued that theropods and sauropods were not that closely related after all because few unique evolutionary specializations are found in their skeletons (see figure 5.2)—although many paleontologists beg to differ with them. Various scientists have argued either that sauropods are closer to ornithischians (sometimes called the "Phytodinosauria" or "plant dinosaurs"), clustering nearly all the herbivores, or that theropods are closer to ornithischians (the "Ornithoscelida" hypothesis). Thus whether the Saurischia is a natural group is now under question.

The same cannot be said about Ornithischia. The more people look at the group and its members, the stronger the evidence becomes for it being a natural monophyletic group. In addition to their hip condition, ornithischians have another unique specialization: a bone in the tip of their lower jaw forms a beak and is made of a bone called the predentary (because it is in front of the dentary bone that makes up the tooth-bearing part of the jaw). In fact, the presence of this bone is so distinctive and consistent that Marsh called the group the "Predentata" in 1894. Fortunately, Seeley's name Ornithischia already had priority, or we might be mistaking a group of dinosaurs for a grouping of anteaters, sloths, and armadillos that were long called the "Edentata."

Beyond the hip structure and the predentary bone, a long list of features confirms the reality of the grouping of ornithischians. The bones at the tip of the upper jaw and snout are usually toothless, and a horny beak probably occluded against the beak on the toothless predentary bone in the lower jaw. In the "eyebrow" area of the eye socket, ornithischians develop a bone called the palpebral, which not found in any other dinosaur. The jaw joint is below the line of the tooth row, which was helpful for the leverage of the jaw needed to chew up plants. Nearly all ornithischians have simple "leaf-shaped" teeth, suitable for cropping vegetation. The tooth rows of most ornithischians are inset deep in the skull, creating a region where fleshy cheeks might have covered the sides of their mouths. This would help keep the food in their mouths as they chewed. Finally, there are ossified tendons all through the backbone, hips, and tail, and in advanced ornithischians, five of the hip vertebrae were fused to the pelvis (primitively, only three were fused in most dinosaurs).

Ironically, the group's name, Ornithischia, means "bird-hipped" dinosaurs—yet birds are descended from the Saurischia. The earliest relatives of birds have a forward-pointing pubis similar to that of other saurischians, but later in bird evolution the pubis rotated backward, the same condition found in all modern birds. However, birds did not do it in the exact same way as Ornithischia. In addition, recent discoveries have found that weird herbivorous saurischians such as the therizinosaurs and the deinocherids (see chapter 16) also rotated their pelvis backward, but not exactly in the ornithischian manner either. Thus the backward-pointing pubic bone evolved at least three or four times. The ornithischians and therizinosaurs apparently rotated the pubic bone back to allow for a large gut and digestive tract. In birds, the position of the pubic bone is related to the way their skeletons are adapted for flight.

WHENCE ORNITHISCHIANS?

Once the grouping was recognized, paleontologists sorted them into lots of different types of advanced ornithischians: iguanodonts, stegosaurs, ankylosaurs, hadrosaurs, and eventually ceratopsians. But there was no fossil record showing how they were related. Some paleontologists hoped to find fossils that showed how all these diverse groups evolved from a common primitive ornithischian ancestor, and the search was on.

Paleontologists had examples of highly specialized ornithischians, such as stegosaurs, scelidosaurs, hadrosaurs and iguanodonts, but they needed to find primitive dinosaurs with ornithischian features without the specialized armor or jaw features. Naturally, they looked at the relatively generalized ornithischians known as "ornithopods" first. But ornithopods such as *Iguanodon* had weird specializations, including the thumb spike and the shape of their skull and teeth, so it was not close to the ancestral condition.

One of the early candidates for the most primitive ornithischian was a dinosaur called *Hypsilophodon*. The first specimens of this creature were found in 1849 on the Isle of Wight on the southern English coast. A group of workers found a block with some vertebrae and limb bones embedded in them (figure 19.1A). To maximize their profits, they sold one half to Gideon Mantell and the other to James Scott Bowerbank. The specimen was so incomplete, however, that its small size led Mantell to identify it as a young *Iguanodon* when he first saw it. Owen reached the same conclusion in 1855 when he published his own description of the fossil after Mantell had died and it was in the collections of the British Museum.

In January 1868, the Reverend William Fox found more specimens from the Isle of Wight locality, including a complete skull and other postcranial bones (figure 19.1B). These were sent to Thomas Henry Huxley, who immediately recognized that the vertebrae matched the Mantell-Bower-bank specimen, but the skull was very different. It was clearly a small adult, not a juvenile, with a relatively short snout and was a much smaller size than an adult *Iguanodon*. In a presentation in 1869, and then a full paper published in 1870, Huxley called it *Hypsilophodon foxii*, named after the genus of iguana known as *Hypsilophus* (now a junior synonym of *Iguana*) because its teeth resembled that living lizard. The species name honored its discoverer, Reverend Fox. In that same 1870 paper, Huxley was the first to describe the backward-pointing pubic bone, a feature Seeley would use 17 years later to define the Ornithischia. Huxley thought this pubic bone was proof that *Hypsilophodon* was even closer to living birds than *Archaeopteryx*, although we now know that the birds are not related to the ornithischians and their backward pubic bones are due to convergent evolution.

More specimens of *Hypsilophodon* were found over the years: some scholars thought it was quadrupedal, but others argued that it was bipedal (figure 19.1C). The fact that it still had four fingers and four toes (compared to three in *Iguanodon* and most other known dinosaurs) suggested that it was more primitive, and some thought of *Hypsilophodon* as being close to the ancestry of ornithischians. One misconception about *Hypsilophodon* propagated by Hulke in 1874 stated that it had armor on its neck and back, but these bones turned out not to be armor at all. Another mistaken recon-struction of its foot by Othenio Abel in 1912 suggested that it had an oppos-able big toe like that of a bird, capable of grasping branches. This led to the famous reconstruction by British paleoartist Neave Parker of *Hypsilopho-don* perched in a branch like a bipedal lizard trying to fly (figure 19.1D).

Later research showed this was incorrect, and in most modern recon-structions of *Hypsilophodon* it appears as a small, fast-running biped with strong hind limbs and relatively small forelimbs (figure 19.1E). It was about 1.8 meters (6 feet) long and weighed about 20 kilograms (44 pounds). It clearly had a beak on the upper and lower jaws, but it also had the recessed tooth row, so most artists give *Hypsilophodon* cheeks as well. Further analy-sis in the past decade has found many more specialized features in *Hypsilo-phodon* that firmly place it within the Ornithopoda and not ancestral to the entire Ornithischia. It comes from the Early Cretaceous, so its appearance is too late to be an ancestor. The same goes for some of the candidates pro-posed by others, such as the Early Cretaceous (Purbeck beds of England) jaws called *Echinodon* by Richard Owen in 1861, or the Upper Jurassic

Hypsilophodon foxii: (*A*) the first specimen, composed of two blocks sold to Gideon Mantell and James Scott Bowerbank, was misidentified as a juvenile *Iguanodon*; (*B*) skeleton of *Hypsilophodon* as it is known today; (*C*) an 1894 reconstruction by Joseph Smit, making it look both like a quadrupedal lizard and a biped; (*D*) the Neave Parker reconstruction, based on the mistaken notion that it had a grasping big toe, so it was drawn as a tree-dweller; (*E*) a modern reconstruction. ([*A–D*] Courtesy of Wikimedia Commons; [*E*] courtesy of N. Tamura)

(Morrison Formation) ornithopods *Dryosaurus* or *Camptosaurus* described by Marsh in the 1880s. For something more primitive than the Jurassic ornithischians like the primitive scelidosaurs and stegosaurs, we need to go back to the Early Jurassic or even the Late Triassic.

OUT OF AFRICA (AGAIN)

The problem with so many of the potentially ancestral ornithischians is that they were found to be too specialized once we obtained better skeletal material, and nearly all were too late in time. But Triassic beds with the preservation potential for the earliest ornithischians are not that common around the world, and for almost a century after *Hypsilophodon* was found, they produced nothing that resembled a primitive ornithischian. Other Triassic dinosaurs, prosauropods such as *Plateosaurus* (chapter 6), primitive theropods such as *Coelophysis* (chapter 11), and even primitive saurischians such as *Herrerasaurus* and *Eoraptor* (chapter 5), were found, but surprisingly few fossils could be called ornithischian. In 1964, French paleontologist Leonard Ginsburg published a description of fragmentary lower jaws and teeth and some other bones from Basutoland (now Lesotho) in southern Africa. From the scraps he had, it looked like a primitive ornithopod jaw with the characteristic leaf-shaped teeth, so it was certainly an ornithischian. Ginsburg named it *Fabrosaurus*, after Jean Fabre, the geologist who led the French expedition with Ginsburg that found the fossil. Soon every primitive ornithischian in the 1960s was being thrown into a taxonomic wastebasket group called the "fabrosaurids," even though the original *Fabrosaurus* fossils were too fragmentary to be identified as one animal. By 1978, paleontologist Peter Galton (who had proposed Fabrosauridae in 1972) realized that the *Fabrosaurus* fossils were insufficient to diagnose a genus; the Fabrosauridae were not a real group, just a wastebasket for primitive ornithischians. Galton then proposed a new name, *Lesothosaurus* ("lizard of Lesotho"), for the best new fossils that had once been called *Fabrosaurus*.

Lesothosaurus is only known from partial skeletal material, but it is clearly a relatively small and very primitive Triassic ornithischian. The skeleton is lightly built and completely unspecialized, with delicate arms and legs and a slender tail. It reached a total length of about 2 meters (6.6 feet), counting the long tail. The skull was not completely primitive, having leaf-shaped teeth, the predentary bone, and evidence of a horny beak on the upper and lower jaws as seen in all ornithischians. It also had the palpebral bone in the eyebrow but does not have the recessed cheek tooth row seen in more advanced ornithischians, suggesting that it didn't have cheeks.

The specimen is still pretty fragmentary and poorly preserved in key areas, so its relationships are controversial. It was originally considered to be a more specialized ornithopod. Paul Sereno considered it one of the most primitive ornithischians known, but in 2005 Richard J. Butler argued that it was a much more advanced creature, close to the ancestry of the ornitho-pod-ceratopsian branch. A 2008 study by Butler, Upchurch, and Norman placed it near the base of the stegosaur-ankylosaur branch of ornithischi-ans. This is not surprising for such an incomplete and confusingly primitive fossil in which just a few anatomical features can radically change its posi-tion on the family tree.

So far, all the possible candidate fossils were too young, too special-ized, or too fragmentary to tell us much about the origin of ornithischians. But one fossil does not suffer from these handicaps. During the winter of 1961–1962, a group of British paleontologists combined with South African colleagues to lead an expedition to the interior of South Africa, especially Lesotho (where *Lesothosaurus* was found much later). Paleontologists Alan Charig of the British Museum and A. W. "Fuzz" Crompton of the South African Museum found a well-preserved fossil skull that was clearly from a very primitive ornithischian. It had the predentary bone and toothless lower beak in front, the palpebral bone in the eyebrow, and columnar chisel-like plant-eating cheek teeth that were slightly inset, suggesting a set of cheeks. However, it had its own specializations in the teeth, including a set of fang-like canine teeth in the snout just behind the toothless beak, plus incisor-like teeth in front of the tusk. These three different types of teeth are extremely unusual for any dinosaur, so Charig and Crompton named it *Heterodontosaurus* (different toothed lizard) in 1962. In 1974, Crompton, Charig, and their field crew also found a beautiful articulated nearly com-plete skeleton in its death pose (figure 19.2), also from Lesotho. It clearly showed that *Heterodontosaurus* had long arms with five-fingered hands with curved claws, and even longer leg bones. The hind foot had four toes, primitive for all dinosaurs, and not the three seen in more advanced ornith-ischians. The thighbone was relatively short, but the shinbone was long, as were the ankle and toe bones. This showed *Heterodontosaurus* was a fast, bipedal runner, running with its body horizontal and balanced by its long

Figure 19.2 ▲

Heterodontosaurus: (*A*) complete articulated skeleton from the Triassic of Lesotho; (*B*) reconstruction of the head in life. (Courtesy of Wikimedia Commons)

tail. The complete skeleton and other specimens show that *Heterodontosaurus* was 1.18 meters (3.9 feet) to 1.7 meters (5.7 feet) long and weighed 1.8 kilograms (4 pounds) to 10 kilograms (22 pounds).

In nearly all respects, *Heterodontosaurus* provides a good image of what the earliest ornithischians were like. About the only specialization unique to *Heterodontosaurus* and not found on later dinosaurs were the weird tusks, or fangs, in the mouth. Other skulls have been found with no tusks, so the presence or absence of tusks may be an example of sexual dimorphism, a difference between males and females, or these tuskless specimens may not be *Heterodontosaurus* at all. In addition, even juvenile specimens seem to have large tusks, so it is a feature of the genus.

The specimen was crucial in many ways. Not only was it the best primitive ornithischian fossil ever found, but it showed the kind of anatomy that was the starting point for the evolution of all other groups of ornithischians, without the problems that plagued incomplete specimens like *Lesothosaurus, Fabrosaurus*, and other primitive forms.

Its discovery also came at a crucial time during the Dinosaur Renaissance. Until then, most paleontologists doubted that Saurischia and Ornithischia were closely related, or that Dinosauria was a real, natural group of organisms. They thought that saurischians and ornithischians had independently evolved from different lineages within the thecodonts. The cladistic revolution that established the common ancestry of all dinosaurs was still on the horizon, but a 1974 paper by Peter Galton and Bob Bakker pointed out just how many uniquely dinosaurian features could be seen in a good fossil such as the complete *Heterodontosaurus*, especially in the wrist and ankle. (John Ostrom was coming to the same conclusion independently.) They argued that the Dinosauria was clearly a natural monophyletic group, not an artificial assemblage of saurischians and ornithischians that evolved independently from the thecodonts and were arbitrarily called dinosaurs. That conclusion was supported by John Ostrom in the 1970s and 1980s and confirmed by the cladistic analysis published by Jacques Gauthier in 1986.

HETERODONTOSAURS EVERYWHERE

Even though the genus *Heterodontosaurus* is known from just a few specimens, and primitive ornithischians are rare in Triassic and Jurassic beds worldwide, their diversity has increased as more and more fossils have been discovered around the world. Numerous other genera of heterodonto-saurines are found in the Late Triassic or Early Jurassic of southern Africa, including *Lycorhinus, Pegomastax, Abrictosaurus*, plus *Manidens* from Chubut Province in Argentina (figure 19.3). In addition to these members of the subfamily Heterodontosaurinae, more primitive members of the family Heterodontosauridae include *Echinodon* from the Lower Cretaceous beds of England (first named and studied by Richard Owen in 1861); *Fruitadens* from the Upper Jurassic Morrison Formation in the Fruita area near Grand Junction, Colorado; and *Geranosaurus* from the Early Jurassic of South Africa.

The most interesting new heterodontosaur is *Tianyulong* from China, which is preserved in fine-grained lake shales, so the soft tissues are intact (figure 19.4). This fossil shows that some heterodontosaurs had a covering of long filamentous fibers similar to bristles (and possibly homologous with the primitive Type 1 feathers of Prum and Brush), especially along their back and sides. They were arranged almost like the spines of a porcupine along the back. Sereno actually described *Tianyulong* as a "nimble two-legged porcupine." This specimen, plus a number of others including *Psittacosaurus*, show that feathers were found across the entire Ornithischia,

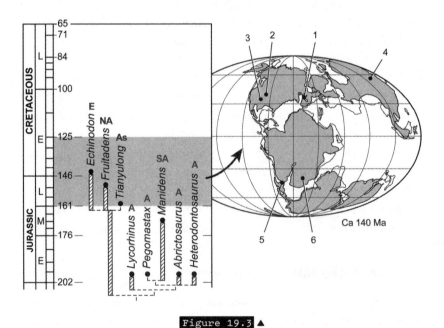

Figure 19.3 ▲

The family tree and geographic distribution of heterodontosaurs. (Courtesy of P. C. Sereno)

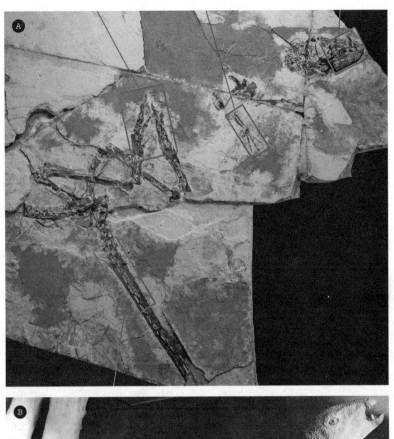

Figure 19.4 ▲
(A) Complete specimen of *Tianyulong* from the Lower Cretaceous Liaoning beds of China. Boxes frame the muzzle, hand, feet, and tail. (B) Life-sized model of *Tianyulong*. ([A] Courtesy of Wikimedia Commons; [B] photograph by the author)

and most dinosaurs in both branches were probably feathered in some way. Added to the other primitive ornithischians, such as *Lesothosaurus*, plus the primitive relative of ankylosaurs and stegosaurs known as *Emausaurus*, the outlines of the diversification of the major groups of ornithischians by the Early Jurassic is becoming better and better understood.

Finally, there is an even more primitive fossil from the lower Upper Triassic Ischigualasto Formation of Argentina, source of *Herrerasaurus* and *Eoraptor*. Discovered in 1962 by Galileo Juan Scaglia, it was named *Pisanosaurus* (figure 19.5) by R. M. Casamiquela in 1967, in honor of

Figure 19.5 ▲

Pisanosaurus: (*A*) reconstructed skeleton; (*B*) artistic reconstruction of the dinosaur. (Courtesy of Wikimedia Commons)

Argentine paleontologist Juan Arnaldo Pisano of the La Plata Museum. It was later redescribed by José Bonaparte in 1976. Although it is missing its tail and the pelvis is too broken to determine whether the pubic bone pointed backward, several features of the skull suggest that *Pisanosaurus* might be the most primitive ornithischian ever found. It is probably a dinosaur, and not a more primitive archosaur, because the hip socket is perforated. The lower jaw has a predentary bone, an inset row of herbivorous teeth, and it has a very low jaw joint, all ornithischian features. The top of the skull is missing, so we do not know if it had a palpebral bone in its eyebrows. These missing features leave it open to lots of confusion and to different interpretations: it has been called a heterodontosaurid, a hypsilophodontid, a fabrosaurid, or the earliest known ornithischian. Some would put it just outside the Dinosauria and within the silesaurids (see chapter 5). We will probably never be able to resolve this with the known specimens to date.

We now have a solid and improving fossil record of the earliest evolution of ornithischians. The remaining chapters focus on the major groups of ornithischians that evolved by the later Jurassic and came to dominate herbivore faunas through the Cretaceous.

FOR FURTHER READING

Baron, Matthew G., David B. Norman, and Paul M. Barrett. "A New Hypothesis of Dinosaur Relationships and Early Dinosaur Evolution." *Nature* 543 (2017): 501–506.

Bonaparte, José F. "*Pisanosaurus mertii* Casamiquela and the Origin of the Ornithischia." *Journal of Paleontology* 50, no. 5 (1976): 808–820.

Butler, Richard, Paul Upchurch, and David Norman. "The Phylogeny of Ornithischian Dinosaurs." *Journal of Systematic Palaeontology* 6, no. 1 (2008): 1–40.

Colbert, Edwin. *Men and Dinosaurs: The Search in the Field and in the Laboratory.* New York: Dutton, 1968.

Farlow, James, and M. K. Brett-Surman. *The Complete Dinosaur.* Bloomington: Indiana University Press, 1999.

Fastovsky, David, and David Weishampel. *Dinosaurs: A Concise Natural History*, 3rd ed. Cambridge: Cambridge University Press, 2016.

Holtz, Thomas R., Jr. *Dinosaurs: The Most Complete, Up-to-Date Encyclopedia for Dinosaur Lovers of All Ages.* New York: Random House, 2011.

Naish, Darren. *The Great Dinosaur Discoveries*. Berkeley: University of California Press, 2009.

Naish, Darren, and Paul M. Barrett. *Dinosaurs: How They Lived and Evolved*. Washington, D.C.: Smithsonian Books, 2016.

Norman, David B., Hans-Dieter Sues, Lawrence M. Witmer, and Rodolfo A. Coria. "Basal Ornithopoda." In *The Dinosauria*, 2nd ed., ed. David B. Weishampel, Peter Dodson, and Halszka Osmólska, 393–412. Berkeley: University of California Press, 2004.

Seeley, Harry Govier. "On the Classification of the Fossil Animals Commonly Named Dinosauria." *Proceedings of the Royal Society of London* 43 (1888): 165–171.

Sereno, Paul C. "Phylogeny of the Bird-Hipped Dinosaurs (Order Ornithischia)." *National Geographic Research* 2, no. 2 (1986): 234–256.

——. "Taxonomy, Morphology, Masticatory Function, and Phylogeny of Heterodontosaurid Dinosaurs." *ZooKeys* 226 (2012): 1–225.

Weishampel, David B., and Lawrence M. Witmer. "Heterodontosauridae." In *The Dinosauria*, ed. David B. Weishampel, Peter Dodson, and Halszka Osmólska, 486–497. Berkeley: University of California Press, 1990.

Zheng, Xiao-Ting, Hai-Lu You, Xing Xu, and Zhi-Ming Dong. "An Early Cretaceous Heterodontosaurid Dinosaur with Filamentous Integumentary Structures." *Nature* 458, no. 7236 (2009): 333–336.

STEGOSAURUS

Behold the mighty dinosaur.

Famous in prehistoric lore,

Not only for his power and strength

But for his intellectual length.

You will observe by these remains

The creature had two sets of brains—

One in his head (the usual place),

The other at his spinal base.

Thus he could reason a priori

As well as a posteriori.

No problem bothered him a bit

He made both head and tail of it.

So wise was he, so wise and solemn,

Each thought filled just a spinal column.

If one brain found the pressure strong

It passed a few ideas along.

If something slipped his forward mind

'Twas rescued by the one behind.

And if in error he was caught

He had a saving afterthought.

As he thought twice before he spoke

He had no judgment to revoke.

Thus he could think without congestion

Upon both sides of every question.

Oh, gaze upon this model beast,

Defunct ten million years at least.

—BERT LESTON TAYLOR, "THE DINOSAUR," 1912

"ROOFED LIZARD"

In 1877, at the height of the Bone Wars, O. C. Marsh received several crates of bones from just north of Morrison, Colorado, collected by Arthur Lakes. When the crates were opened, there were vertebrae of the back and tail, parts of the hip bones, partial limb bones, and a single piece of a flat plate, over 1 meter (3.1 feet) long. The specimen was too incomplete to accurately judge the shape of the entire animal because the front half was missing as well as most of the tail. Even so, the large flat bony plate caught Marsh's attention. He named the creature *Stegosaurus armatus* ("armored roofed lizard" in Latin and Greek) in that same year and set up an order Stegosauria to include it. He imagined the creature was an aquatic turtle-like animal, with plates arranged like shingles providing a "roof" over the back. (Scientists later realized that some parts of this specimen were teeth of diplodocine sauropods and didn't belong to a stegosaur at all.)

Even though the bones were too incomplete for any kind of reconstruction, this did not prevent some artists from imagining a version of *Stegosaurus* looking a bit like a bipedal lizard with spikes and plates along its back (figure 20.1). Marsh originally thought the thick dense limb bones indicated that stegosaur was aquatic. Then he decided it was a bipedal dinosaur because its front legs were so short and its hind limbs so long.

Marsh began to receive more and more bones of this mysterious creature. In 1879, he named *Stegosaurus ungulatus* (hoofed roofed lizard) based on specimens collected by William H. Reed from Como Bluff Quarry 12 near Robber's Roost. Quarry 13 at Como Bluff yielded the largest concentration of stegosaur bones Marsh ever received.

In 1887, he received the best specimen of the genus, which he named *Stegosaurus stenops* (narrow-faced roofed lizard). He heard about the bones from a newspaper account, then paid the landowner, Marshal P. Felch and his family near Garden Park, Colorado, to let Samuel Williston quarry the bones and send them to Yale. *S. stenops* is now known from more than 50 specimens, including the original type specimen, which is nearly complete

Figure 20.1 ▲
An 1884 restoration of *Stegosaurus* as a bipedal long-necked dinosaur with a weird arrangement of spikes and plates on its back. (From A. Tobin, 1884)

(figure 20.2A–B). It was found and preserved flattened on its side, so it has been nicknamed the "Road Kill" specimen. This specimen clearly showed the plates arranged in rows along the back, the short front limbs and long hind limbs, and some of the spikes near the tip of the tail.

Marsh had so much material from so many different places, and he was such a taxonomic "splitter," that in 1881 he named two additional species: *Stegosaurus affinis* (known from a single pubic bone that was never adequately described and since has been lost, so the name is invalid) and *Diracodon laticeps* (named from a few jaw fragments, also invalid). He topped that in 1887 by naming three new species. In addition to *Stegosaurus stenops*, they were *S. sulcatus*, and *S. duplex*. The latter two names have been

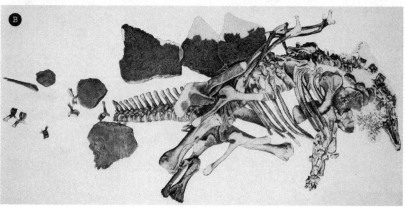

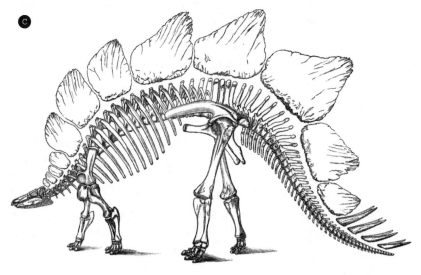

Figure 20.2 ◀ ▲

The nearly complete skeleton of Marsh's type specimen of *Stegosaurus stenops* in the Smithsonian, originally from Felch Quarry in Garden Park, Colorado, nicknamed the "Road Kill." (*A*) Photo of the specimen; (*B*) diagram of the bones showing their position and orientation; (*C*) Marsh's 1891 reconstruction, showing the plates in a single row; (*D*) modern skeletal mount, showing the staggered alternating rows of plates and the throat armor. ([*A–C*] Courtesy of Wikimedia Commons; [*D*] photograph by the author)

abandoned because they were based on nondiagnostic specimens or trivial differences in the skeleton.

Unfortunately, Marsh's original specimen of *Stegosaurus armatus* is so incomplete and nondiagnostic that his original type species is now considered a doubtful name (*"nomen dubium"*) and is no longer used. In 2013, the International Commission on Zoological Nomenclature remedied this by ruling that *S. armatus* was no longer valid, and the designation of the type species of *Stegosaurus* was transferred to *S. stenops*.

By 1891, Marsh finally had enough material to accurately reconstruct *Stegosaurus* in a way that closely resembles what we know today: a small-headed creature with many armored plates (although he thought they were lined up in a single row) standing up along the back, spikes on the tail, short front legs, and a high arched back supported by long hind legs (figure 20.2C).

Meanwhile, Cope was trying his best to compete with the flood of bones Marsh was receiving and the torrent of new names he was publishing. In 1878, Cope received a few back and tail vertebrae and a piece of rib from his collector O. W. Lucas in Garden Park Quarry 3, nicknamed "Cope's Nipple" (in reference to the shape of the hill). Cope gave these nondiagnostic fragments their own new genus and species names, *Hypsirophus discurus*. Marsh and nearly everyone else recognized that it was the same as *Stegosaurus*. Only Richard Lydekker used this genus in 1893 when he mistakenly used Marsh's diagram with Cope's name, *Hypsirophus*. But Cope persisted in using only his own name and never acknowledged Marsh's name, *Stegosaurus*.

STEGOSAURS AROUND THE WORLD

The discovery of *Stegosaurus* in 1877 in Colorado was not the first stegosaur to be found, however. That honor goes to some fragmentary pieces of jaw named *Regnosaurus* that were collected in the Lower Cretaceous Wealden beds. They were in the British Museum collections when Gideon Mantell first saw them in 1839. At first he thought it was the as-yet-unknown lower jaw of *Iguanodon*, and he gave a presentation to that effect to the Royal Society on February 8, 1841. Richard Owen disputed this because the real lower jaws of *Iguanodon* had appeared, so Mantell decided to give the fossil a new name in 1848. *Regnosaurus northamptoni* gets its name from the Regni, the Roman name for the British tribe that once lived in that part of Sussex, and Spencer Compton, the second Marquess of Northampton, who was then president of the Royal Society. All that is known of this dinosaur is a lower jaw fragment and possibly some other bones from the Isle of Wight, which indicate a rather small stegosaur. Because it was so incomplete, it has been assigned to all sorts of different groups of dinosaurs, including iguanodonts, scelidosaurs, stegosaurs, hylaeosaurs, and even sauropods. In 1993, amateur paleontologist George Olshevsky revived the idea it was a stegosaur, and this was confirmed by Paul Barrett and Paul Upchurch in 1995. They found it most closely resembled the Chinese stegosaur *Huayangosaurus*. However, other scholars think the fossils are too incomplete to use the name at all.

In 1845, a fossil skull found in the Jurassic beds of South Africa was sent to London and described by Richard Owen. At first, he thought it belonged to the hippo-like reptiles known as parieasaurs found in the Permian of South Africa, and he assigned it to the pareiasaur genus *Anthodon*. It was forgotten for decades, until a South African paleontologist, Robert Broom, looked at it again in 1912 and realized it was not a parieasaur but a dinosaur, possibly an ankylosaur. He assigned it to the problematic genus *Palaeoscincus* (ancient skink lizard). *Palaeoscincus* was named by Joseph Leidy based on some nondiagnostic ankylosaur teeth, and this name has been completely abandoned as invalid in recent years. In 1929, Baron Franz Nopcsa realized that the skull belonged to a stegosaur. Unaware of Broom's renaming of it, he renamed it *Paranthodon*, which turns out to be the valid name for the specimen. No additional specimens have been found beyond the original snout fragment, so the genus *Paranthodon* is just as enigmatic as it was when first found. All that can be said is that it is some kind of stegosaur.

Another stegosaur found before *Stegosaurus* itself was *Dacentrurus*. In May 1874, James Shopland of Swindon Brick and Tile Company found a specimen in their clay pit near Swindon, Wiltshire, from the Upper Jurassic Kimmeridge Clay. He contacted Richard Owen at the British Museum, who sent William Davies to collect it. The fossil was encased in a nodule that was 8 feet high. When they tried to lift it, the clay crumbled into pieces, so they had to crate each piece separately. When the pieces were finally shipped to London, they weighed 3 metric tonnes (7 tons), and it took two years for Owen's preparator, Caleb Barlow, to extract it and put it all back together. Additional fragments of a stegosaur similar to this were found in other British Jurassic beds throughout the rest of the 1800s. Owen finally described it in 1875 and named it *Omosaurus armatus*. Unfortunately, that name had already been used for a phytosaur, *Omosaurus perplexus*, by Joseph Leidy in 1856, so it was unavailable. In 1902, American paleontologist Frederick A. Lucas realized the problem and gave it a new generic name, *Dacentrurus* ("very pointed tail" in Greek). Since then, stegosaur fossils all over England, France, Spain, and Portugal have been referred to *Dacentrurus*, although they are mostly fragments and the complete skeleton is still unknown. Modern analysis has determined that *Dacentrurus* is a very advanced stegosaur, related to the Portuguese genus *Miragaia* and to the American *Stegosaurus* and not to the dozens of other more primitive stegosaurs.

THE MAN WHO WOULD BE KING

In the early days of paleontology, most of the fragmentary European specimens of stegosaurs were by default assigned to the well-known American genus *Stegosaurus*. Some of these fragments from Middle Jurassic rocks in England and France were named *Stegosaurus priscus* in 1911 by one of the most colorful characters in the history of paleontology, Baron Franz Nopcsa (pronounced "nop-cha"). That material was eventually moved from *Stegosaurus*, then assigned to *Lexovisaurus*, and finally renamed *Loricatosaurus* by Suzanne Maidment and colleagues in 2008. But Nopcsa deserves more than a passing mention here.

Born in Transylvania (then part of Austria-Hungary) on May 3, 1877, to an aristocratic Magyar and Romanian family, Baron Franz Nopcsa von Felsö-Szilvás was easily one of the most bizarre and interesting characters, not just in paleontology but in all of science.

> Baron Nopcsa was a notorious figure in his day. A wild genius with a flair for the dandyish and the dramatic, he was an explorer, spy, polyglot and master of disguise. He crossed the Albanian Alps on foot and befriended local mountain men, sometimes involving himself in their tribal feuds. Once, he was nearly crowned King of Albania. It was said that he would disappear for months at a time only to arrive for polite tea at posh European hotels dressed as a peasant. Along with a younger man whom he called his secretary, he traversed swaths of the Balkans on motorcycle. He kept up years-long correspondences with famous and learned men all across Europe. Later in his life, he was known for chasing villagers from his estate with a pistol. (Vanessa Veselka)

Steve Brusatte adds to this description:

> He seems like the invention of a mad novelist, a character so outlandish, so ridiculous, that he must be a trick of fiction. But he was very real—a flamboyant dandy and a tragic genius, whose exploits hunting dinosaurs in Transylvania were brief respites from the insanity of the rest of his life. Dracula, in all seriousness, had nothing on the Dinosaur Baron.

Nopcsa got an early start when his younger sister, Ilona, found dinosaur bones on their family estate in Transylvania.

> Nopcsa was born to a wealthy noble family, the eldest of three children raised at Sacel. He had a typical upbringing for an aristocrat in a provincial

backwater of an aging empire. At home he spoke Hungarian and learned Romanian, English, German and French. His father, Alexius, had fought in Mexico against Benito Juárez, in 1867, as a hussar in the army of Maximilian, Archduke of Austria and Emperor of Mexico. Later Alexius became a vice-director at the Hungarian Royal Opera, in Budapest. Nopcsa's mother, Matilde, came from an aristocratic family from the nearby city of Arad.

In 1895, Nopcsa's sister Ilona was walking along a riverbank near the family home when she found an unusual-looking skull, and she brought it to her teenaged brother. It soon became his obsession. The skull belonged to a previously undiscovered duck-billed herbivore from the dusk of the Mesozoic. . . . Crushed by geological forces, the skull was in terrible shape. In the fall, Nopcsa entered the University of Vienna and took the skull with him. Like a cat with a gift rat, he presented it to his professor, a famous geologist, expecting him to take it from there. But the professor sent Nopcsa back to Transylvania and told him to figure it out for himself. Whether it was lack of interest or funding or the cunning strategy of a master teacher, it was the making of a great scientist. (Vanessa Veselka)

Inspired to study fossil bones, Nopcsa was largely self-taught, but he received his formal education at the University of Vienna, where he was already giving lectures by age 22. He spent much of his life studying and publishing on the geology and the dinosaurs of Europe, particularly of his home area in Transylvania. His writings on the geology of the Balkans were full of theoretical ideas in tectonics that fit the soon-to-be-proposed theory of continental drift of Alfred Wegener.

His dinosaurian discoveries were impressive. These included the pygmy sauropod *Magyarosaurus* (only 6 meters long rather than 30 meters like elsewhere), the duckbill *Telmatosaurus transylvanicus*, the primitive ornithopod *Zalmoxes robustus*, another species of indeterminate theropod referred to as *Megalosaurus*, the small nodosaurid ankylosaur *Struthiosaurus transylvanicus*, and the English ankylosaur *Polacanthus ponderosus*, which he named. He also described some of the first dinosaur eggs found anywhere, long before the American Museum found dinosaur eggs in Mongolia in 1923. The small body size of many of his Hungarian and Romanian dinosaurs from the Hateg Basin suggested to him that they had once lived on islands in a Cretaceous seaway. Similar to Ice Age pygmy elephants and hippos on Mediterranean islands, these dinosaurs apparently become dwarfed due to the small food resource base of the island and because they no longer needed large body size to cope with predators. Nopcsa was one of the first to develop the theoretical basis for island dwarfism.

Nopcsa's most important pioneering contribution to paleontology was what he called "paleophysiology": the idea of thinking about extinct fossils as complete living breathing animals rather than just a collection of dead bones, as most other scientists did at that time. Nopcsa was one of the first to argue that birds evolved from ground-dwelling dinosaurs and that they developed feathers to aid in running, two ideas that wouldn't be popular until the 1970s. His research on jaw mechanics in dinosaurs was 60 years ahead of his time. He also looked at the great diversity of crests on duck-billed dinosaurs and imagined that the ones with crests were males, and the ones without were females of the same species (as is typical in horns of antelopes or antlers of deer today). He thought *Kritosaurus* was a female of *Parasaurolophus, Prosaurolophus* the female of *Saurolophus*, and many other combinations. Most of these dinosaurs have since been proven to be distinct genera, but sexual dimorphism has been claimed in duckbills like *Lambeosaurus*.

Nopcsa also had another obsession: independence for the tiny province of Albania, which was on the border between the Austro-Hungarian Empire and the Turkish Ottoman Empire. Even though he came from Transylvania, Nopcsa taught himself the Albanian language, learned their folklore and history, and published over 50 scholarly studies on the region. He spent a lot of time in Albania learning about the people and developing good relations with the local leaders. Stephanie Paine wrote this about his exploits:

> At the crack of dawn, Franz Nopcsa mounted his horse and set off for Shkodra, one of the most ancient towns in Albania. It was 1903 and the young aristocrat from Transylvania had just finished his PhD, a meticulous study of dinosaur remains dug up on his family estate. Now he was exploring a place he had dreamed about since childhood—the wild, romantic mountains of northern Albania, a land peopled by lawless tribes who were well-armed and very dangerous. As Nopcsa approached a bend in the road, there was a shot. "The bullet went right through my hat and grazed my head, but did not injure me," he wrote later. "I leapt off my horse and sought shelter." But when he tried to fire back, his assailant had vanished. The rest of the journey, he wrote with some disappointment, "passed without event." Not much of Nopcsa's life passed without event.

Eventually, Nopcsa became a revolutionary, giving fiery speeches and smuggling weapons into the country to aid the revolt against the Turks. In 1912, the entire Balkan region rose up in rebellion and drove out the Turks, and Albania became independent. With his background in Albanian culture and aristocratic roots, Nopcsa thought he should be considered as a candidate to be the first king of Albania, an idea he continually tried to promote. This idea was not so far-fetched. The Austro-Hungarian government could support any suitable nobleman, but they ended up picking a German aristocrat instead of Nopcsa.

There are several dramatic portraits of Nopcsa in Albanian warrior garb (figure 20.3). In his diary, he wrote: "Once a reigning European monarch, I would have no difficulty coming up with the further funds needed by marrying a wealthy American heiress aspiring to royalty, a step which under other circumstances I would have been loath to take."

Figure 20.3 ▲
Baron Franz Nopcsa in the Albanian shqiptar warrior costume, about 1913. (Courtesy of Wikimedia Commons)

Unfortunately, his plans were derailed by the outbreak of World War I. As a rich, educated, multilingual Austro-Hungarian aristocrat, he served as a spy for their empire under the guise of traveling to do scientific research. He even organized and led his own corps of Albanian volunteers in the war. But when the war ended and the Central Powers (Germany, Turkey, and Austria-Hungary) lost, the Austro-Hungarian Empire was broken up into smaller countries. Nopcsa's Transylvania became part of Romania, and he lost all of his ancestral estates and most of his wealth. He took a job at the Hungarian Geological Institute to pay his bills, but this didn't work due to his abrasive personality. Eventually he moved to Vienna to study the fossils he had collected. Accompanying him was his young Albanian secretary and lover for almost 30 years, Bajazid Elma Doda. Many people knew that Nopcsa was gay (a dangerous thing to admit back then). Together they would ride his motorcycle all over Europe, Doda riding in the sidecar.

In Vienna, however, his financial difficulties distracted him from his work, and he slipped into a deep depression. Finally, he was forced to sell all his fossil collections to the British Museum to cover his debts, which depressed him even further. By 1933, at the bottom of this manic-depressive roller coaster, he drugged Doda's tea with a sedative, took a gun and shot his lover in the head, then killed himself.

As Gareth Dyke wrote, Nopcsa's "theories about dinosaur evolution turn out to have been decades ahead of their time.... Only in the past few years, with new fossil discoveries, have scientists begun to appreciate how right he was." Reading through his writings, he was clearly a brilliant scientist and Albanologist, but he was also insensitive and unable to understand the motives of others and had a sociopathic personality. Whether this was due to some sort of psychological problem (many people think he was manic-depressive, and possibly autistic) or was mainly caused by his aristocratic arrogance, is still debated by scholars. Nopcsa was a brilliant paleontologist who pioneered paleobiology, almost became the king of Albania, and was openly gay. He was a very complex, interesting man, to say the least.

THE PALEOBIOLOGY OF STEGOSAURS

Even with a century of familiarity, *Stegosaurus* is still a very weirdly constructed animal. The head was very low to the ground, forcing stegosaurs to eat mostly low-growing brushy vegetation. In the Jurassic and Early

Cretaceous, this would have been mostly ferns and cycads (sego palms), along with mosses, horsetails, and short conifers; grasses and abundant flowering plants did not appear until much later in the Cretaceous. The long narrow skull had a pointed snout without teeth, probably covered by a horny beak like that of a turtle. The cheek teeth were small, flat, and triangular, with wear facets showing some evidence of grinding their food. The most complete specimens show that the throat region was protected by a "chain mail" of tiny bony plates called osteoderms.

Famously, *Stegosaurus* had a small brain, about 80 grams (20.8 ounces), about the size of the brain of a dog, which is tiny considering their huge body mass of 4.5 metric tonnes (5 tons). Scaling brain to body mass, *Stegosaurus* has one of the smallest brains proportional to its size of any dinosaur known. One of Marsh's fossil skulls had a well-preserved brain cavity, allowing him to make a cast of the cavity and describe the brain features in the 1880s. This led to the famous myth that its brain was so small that *Stegosaurus* needed a second brain in its hips just to function (satirized in the poem at the beginning of this chapter). In reality, the "second brain" was just a slightly enlarged ganglion of the nervous system, which would have controlled the muscles in the back of the body; it was not a true brain. It's also likely that most of the space housed a glycogen body (also found in sauropods). This feature is typical in living birds and supplements the supply of glycogen (a sugar) to the nervous system. *Stegosaurus* did not need much intelligence to continually munch away at ferns and low-growing vegetation. Its spiked tail and other defenses and huge size seem to have been sufficient for its needs; stegosaurs were very successful for millions of years and spread worldwide.

The body of stegosaurs was weirdly proportioned, with short forelimbs and long hind limbs. This forced the spine into a big arch that flexed upward over the hips but sloped down steeply to the head and tail (see figure 20.2C). Each hand and foot had three short toes, each of which bore a hoof. In most stegosaurs, the hands had only two finger bones in each finger, and two toe bones in each toe.

The most famous feature of *Stegosaurus* was the huge flat plates over their entire back. The plates were not attached to the spine by a direct bony connection but held in place with cartilage, tendons, and muscles. Originally, Marsh thought that the plates laid flat on the side of *Stegosaurus*, like shingles or tiles, but later specimens proved that idea was wrong. In his 1891 reconstruction (see figure 20.2C), Marsh thought the plates formed a single line down the middle of the back, but this idea was discarded once more plates were found. For many years, a number of reconstructions showed the plates paired with each other down the middle of the back. Today you can still find illustrations and toys with this configuration. (It was the version used in the movie *King Kong*.) But the famous "road kill" type specimen of *S. stenops* (see figure 20.2A–B) clearly showed that the plates were alternating in two rows down the back, and several other recent discoveries have confirmed that arrangement.

What the plates were used for has long been debated. Originally, Marsh and other early paleontologists thought they were protective, although the plates didn't do much of job of shielding their sides or flanks from attack by a theropod such as *Allosaurus* (both are found together in many Morrison localities). Some people thought the plates were not adaptive at all. For example, Frederic Loomis argued that the plates adorning the backs of stegosaurs were maladaptive traits that sapped their vigor and signaled their impending extinction.

More recently, a consensus has formed that they were probably for species recognition and advertising their age and status. Most scientists think that males and females of *Stegosaurus* both appear to have the same sized plates, so it's not a sexually dimorphic feature in that genus. But a study published in 2015 claimed that the plates were different in males and females, with wider plates in males and taller plates in females. The questionable Morrison genus *Hesperosaurus* might have evidence of different male and female plates.

In the 1970s, John Ostrom's former student Jim Farlow did a series of slices through the plates and found they had large cavities and big canals for a lot of blood vessels. Coupled with the other ideas brewing during the Dinosaur Renaissance and the warm-blooded dinosaur debate, this suggested that the plates were for picking up or shedding excess body heat. However, no other group of dinosaurs seemed to need these structures to regulate body temperature. Most other stegosaurs simply have conical spikes or deeply embedded armor plates, so the function of heat regulation would be unique to *Stegosaurus* and not found in any of its close relatives.

In addition, some paleontologists have argued that the bony plates were covered with keratinous horny sheaths to increase their size—but this also would have reduced any heat transport through the outside of the plate to the blood beneath. The surface of the plates, however, were covered by bony grooves for blood vessels, so any horny sheath would have covered and protected these. The horny sheath would have reinforced not only a defensive function but also improved their use as display structures. Like most arguments over the function of unusual structures in extinct animals, we may never know the truth. In addition, there is often no simple "right" answer; it's likely that these structures performed more than one function.

The other distinctive feature of *Stegosaurus* is its spiky tail. Some paleontologists argued that these were just for display, although most have regarded them as defensive weapons. Many of the early reconstructions showed *Stegosaurus* with six to eight spikes, but a more careful analysis shows they had only four. Any model or reconstruction with more than four is in error. Contrary to many reconstructions, the four tail spikes did not point upward. They stuck out upward and sideways away from the tail axis, making them much more effective as a weapon with a side-to-side striking motion. Their tails were not held rigid like most dinosaur tails, so they could swing it around. However, the rows of plates on the upper part of the tail restricted movement to some degree. The most important evidence about the tail as a weapon was published in 2001 by McWhinney and colleagues and showed that the spikes had a very high incidence of damage (9.8 percent of specimens examined), suggesting they were used in defense to strike hard objects. In addition, an *Allosaurus* tail vertebra had puncture marks that fit the tail spikes of *Stegosaurus* perfectly.

The tail spikes had no formal anatomical name until cartoonist Gary Larson published a "Far Side" cartoon showing cave men watching a slide show. The lecturer points to the tail of a *Stegosaurus* and says, "Now this end is called the thagomizer . . . after the late Thag Simmons." "Far Side" cartoons were always hugely popular with scientists because they often talked about scientific topics or were based on scientific in jokes. The term *thagomizer* entered the scientific lexicon when Ken Carpenter used it in a lecture at the 1993 Society of Vertebrate Paleontology meeting. Since then, it has been picked up in numerous dinosaur books, used in the displays at Dinosaur National Monument, and in the BBC series *Planet Dinosaur*. Although there is no formal procedure for making popular nicknames into official anatomical terms, thagomizer is widely used in paleontology, usually with a smile and a chuckle the first time it is mentioned.

Of course, Larson knowingly committed a scientific boo-boo when he showed "cave men" living with dinosaurs, but Larson was fully aware of this, and it was necessary for the joke. Larson has written that "there should be cartoon confessionals where we could go and say things like, 'Father, I have sinned—I have drawn dinosaurs and hominids in the same cartoon.' " A similar anachronism that is usually overlooked is the common pairing of *Stegosaurus* with *Tyrannosaurus rex*, found in many books and cartoons and in the animatronic dinosaurs of "Primeval World" in Disneyland. In reality, *Stegosaurus* vanished about 140 million years ago (Late Jurassic), yet *T. rex* did not appear until 68 million years ago (latest Cretaceous). It's staggering to think about it, but *T. rex* is closer in time to humans than it is to *Stegosaurus*.

THE RISE AND FALL OF STEGOSAURS

Two of the earliest stegosaurs to be described were *Regnosaurus* from Britain and *Dacentrurus* from western Europe—both before *Stegosaurus* itself was named in 1877. As the years went by, more and more different kinds of stegosaurs were found, nearly all with completely different configurations of armor, plates, and spikes.

One of the first stegosaurs to be discovered was the spiky African genus *Kentrosaurus aethiopicus* (figure 20.4) from the Tendaguru bone beds in what is now Tanzania (see chapter 9). It was named by Edwin Hennig in 1915, and its name means "sharp point lizard" in Greek. *Kentrosaurus* is known from hundreds of bones found in multiple quarries between 1910 and 1912 (although many were lost during the bombing of German museums in World War II). It was about 4.5 meters (15 feet) long and weighed about 1 metric tonne (1.1 tons), considerably smaller than some *Stegosaurus*, which reached up to 9 meters (30 feet) in length, and 5.3-7 metric tonnes (6-7.5 tons) in weight. In most respects, *Kentrosaurus* is much like *Stegosaurus*, with a small but long and narrow head, toothless beak, short front limbs and long hind limbs, and a relatively long tail. Unlike *Stegosaurus*, however, it had small plates only on the front half of its backbone, and most of the rest of the spine was covered with paired spikes that clearly served a defensive function.

For a while, the name *Kentrosaurus* was also questioned because the International Code of Zoological Nomenclature forbids names that sound the same (homonyms), and there is also a ceratopsian named

Figure 20.4 ▲
The Late Jurassic spiky stegosaur *Kentrosaurus*, from the Tendaguru beds of Tanzania. (Courtesy of Wikimedia Commons)

Centrosaurus. Some scientists recommended that Hennig's 1916 replacement name, *Kentrurosaurus*, be resurrected, and the name *Doryphorosaurus* was contributed by Nopcsa in 1916. However, the *Kentrosaurus* and *Centrosaurus* are not really homonyms because *Kentrosaurus* is pronounced with the hard "K" sound whereas *Centrosaurus* is pronounced with the soft "C" (as in "center"), so there is no real confusion and no need to drop the name *Kentrosaurus*.

We have discussed the abundant American, European, and African stegosaurs, but their range also extended to China. These amazing fossils had been unknown to Westerners during the political turmoil of the mid-twentieth century, but they finally began to be available for study after the war. The first of these was *Chialingosaurus*, from the Middle Jurassic of China, found during the war years in the 1930s and 1940s and finally named in 1959 by Yang Zhongjian (also written C.C. Young), the "Father of Chinese Paleontology." *Chialingosaurus* is based on a partial skeleton, and some do not consider it to be a valid genus for that reason. However, it apparently had small plates in pairs along its neck and backbone along the shoulders, and paired spikes down the rest of its back and tail, like *Kentrosaurus*.

In 1973, Dong Zhiming (currently the dean of Chinese dinosaur specialists) named *Wuerhosaurus* from the Early Cretaceous of China and Mongolia, one of the very last stegosaurs known. It also consists of a fragmentary skeleton, plus parts of a few more individuals. Its body was much fatter and broader than other stegosaurs, based on the broad pelvis. At one time, it was argued that it had very rounded plates in rows on its back, but this has been dismissed as an artifact of breakage of the few plates found. Dong and others described *Huayangosaurus* in 1982, based on a partial skeleton and some other specimens from Middle Jurassic beds of China (figure 20.5). Unlike other stegosaurs, the plates down its back are tall narrow triangles rather than broad polygons. It also had a Thagomizer of four spikes at the tip of its tail.

Its close relative is the Upper Jurassic stegosaur *Chungkingosaurus*, named and described by Dong and others in 1983. *Chungkingosaurus* had an arrangement of tall, narrow plates on its back and spikes on its tail similar to that of *Huayangosaurus*. Another Late Jurassic stegosaur with similar armor is *Tuojiangosaurus*, described by Dong and colleagues in 1977. It may be closely related to *Paranthodon* from Africa. There is also *Gigantospinosaurus* from the Late Jurassic of Sichuan, which had huge spikes protruding from it shoulders, and *Jiangjunosaurus*, based on a fragmentary skeleton from the Late Jurassic of Inner Mongolia. That makes at least six Middle Jurassic to Early Cretaceous stegosaurs from China and Mongolia, giving it the highest stegosaur diversity in the world, with almost two dozen additional genera in Eurasia, Africa, and North America.

But what about the rest of the Pangea continents: Australia, India, Antarctica, Madagascar, and South America? So far none of them have produced unquestioned stegosaurs, although the fossil record in Australia, Madagascar, India, and, of course, Antarctica is relatively poor during their heyday in the Middle and Late Jurassic. *Dravidosaurus* from the Late Cretaceous of India turned out not to be a stegosaur. In one interpretation, it was based on a weathered set of plesiosaur hip bones and hind limbs. Later authors, however, ruled out the plesiosaur interpretation, concluding that the specimens are too incomplete to tell what they really are. Trackways have been found in Australia that are claimed to be stegosaurian, but so far no bones have come to light.

Figure 20.5 ▲

(*A*) The Middle Jurassic Chinese stegosaur *Huayangosaurus*, with a combination of spikes and narrow triangular plates on its back. (*B*) Restoration of the dinosaur in life. ([*A*] Courtesy of Wikimedia Commons; [*B*] courtesy of N. Tamura)

In 2017, Leonardo Salgado and colleagues described a skull fragment and partial skeleton from the Early Jurassic of Patagonia. The specimen even had gut contents showing that it ate cycads. Named *Isaberrysaura mollensis*, it is definitely an advanced ornithischian, and that is the only commitment Salgado and coauthors would make. However, based on the stegosaur-like characteristics in what is known of the skull, another analysis of it suggested that it might be a very primitive bipedal relative of the stegosaurs.

We have an amazing record of stegosaurs from most of the Jurassic and Early Cretaceous. Along with the possibility that *Isaberrysaura* may be an Early Jurassic stegosaur, stegosaurs are definitely known from the Middle Jurassic, when they evolved from scelidosaurs, their common ancestor with the ankylosaurs. Together, the stegosaurs, scelidosaurus, and ankylosaurs form a group now called the Thyreophora ("armor bearing" in Greek).

The earliest known undoubted stegosaur is *Huayangosaurus* from the early Middle Jurassic of China; followed by late Middle Jurassic stegosaurs such as *Chungkingosaurus, Chialingosaurus, Tuojiangosaurus*, and *Gigantospinosaurus* from China; and *Lexovisaurus* and *Loricatosaurus* from England and France. In the Late Jurassic, stegosaurs were in their heyday in abundance and size, if not diversity, with *Kentrosaurus* in Africa, *Dacentrurus* and *Miragaia* in Europe, *Jiangjunosaurus* in China, and *Stegosaurus* and *Hesperosaurus* in North America.

By the Early Cretaceous, stegosaurs experienced their last phase of evolution, with *Wuerhosaurus* in China, *Paranthodon* in Africa, and *Craterosaurus* from England, plus some undescribed fragments from Russia. Paleontologists have long speculated on what caused the decline and extinction of stegosaurs. Certainly the vegetation was changing, with the decline of cycads (possibly their main food source) paralleling the decline in stegosaurs. In addition, by the Early Cretaceous there was a tremendous bloom of flowering plants, including many types of water plants and primitive trees such as magnolias. Many paleontologists have suggested that the rapidly reproducing flowering plants may have stimulated the evolution of duck-billed dinosaurs with their complex "dental batteries" of hundreds of prismatic teeth fused together. They were clearly more specialized and efficient plant eaters than the almost toothless stegosaurs, and it is possible that they co-evolved with flowering plants to dominate the Cretaceous landscape. The stegosaurs were Jurassic relics and apparently did not do well when facing new competition from Herbivores, changing plants in their diet, and possibly new predators as well. For whatever reason, stegosaurs vanished by the end of the Early Cretaceous.

FOR FURTHER READING

Brinkman, Paul D. *The Second Jurassic Dinosaur Rush: Museums and Paleontology in America at the Turn of the Twentieth Century.* Chicago: University of Chicago Press, 2010.

Carpenter, Kenneth, ed. *Armored Dinosaurs.* Bloomington: Indiana University Press, 2001.

Colbert, Edwin. *Men and Dinosaurs: The Search in the Field and in the Laboratory.* New York: Dutton, 1968.

Farlow, James, and M. K. Brett-Surman. *The Complete Dinosaur.* Bloomington: Indiana University Press, 1999.

Fastovsky, David, and David Weishampel. *Dinosaurs: A Concise Natural History*, 3rd ed. Cambridge: Cambridge University Press, 2016.

Galton, Peter M. "Stegosauria." In *Encyclopedia of Dinosaurs*, ed. Philip J. Currie and Kevin Padian, 701–703. San Diego: Academic Press, 1997.

Galton, Peter M., and Paul Upchurch. "Stegosauria." In *The Dinosauria*, 2nd ed., ed. David B. Weishampel, Peter Dodson, and Halszka Osmólska, 343–362. Berkeley: University of California Press, 2004.

Holtz, Thomas R., Jr. *Dinosaurs: The Most Complete, Up-to-Date Encyclopedia for Dinosaur Lovers of All Ages.* New York: Random House, 2011.

Naish, Darren. *The Great Dinosaur Discoveries.* Berkeley: University of California Press, 2009.

Naish, Darren, and Paul M. Barrett. *Dinosaurs: How They Lived and Evolved.* Washington, D.C.: Smithsonian Books, 2016.

Spaulding, David A. E. *Dinosaur Hunters: Eccentric Amateurs and Obsessed Professionals.* Rocklin, Calif.: Prima, 1993.

Veselka, Vanessa. "History Forgot This Rogue Aristocrat Who Discovered Dinosaurs and Died Penniless." *Smithsonian Magazine*, July 2016. https://www.smithsonianmag.com/history/history-forgot-rogue-aristocrat-discovered-dinosaurs-died-penniless-180959504/.

ANKYLOSAURUS

Ankylosaurs lived at a time when the largest land predators in Earth's history including *T. rex* roamed the landscape, dismembering other dinosaurs with powerful jaws and serrated teeth. In an arms race, some plant-eaters developed defensive weaponry. "A tail club was definitely an effective weapon and could have broken the ankle of a predator," said paleontologist Victoria Arbour of North Carolina State University and the North Carolina Museum of Natural Sciences, who led the study published this week in the *Journal of Anatomy*. "But in living animals today, weapons are also often used for battling members of your own species—consider the horns of bighorn sheep or the antlers of deer—so perhaps ankylosaurs did something similar."

—WILL DUNHAM, *KING OF CLUBS: INTRIGUING TALE OF THE "TANK" DINOSAUR'S TAIL*

THE ARMORED DINOSAURS

In the summer of 1906, American Museum paleontologists Barnum Brown and Peter Kaisen were in the badlands of the Upper Cretaceous Hell Creek Formation, near Gilbert Creek, in central Montana. The temperatures were in the scorching range, over 40°C (104°F) most of the day, and the heat reflected off the light-colored rock, making it even more unbearable. There is no water anywhere, so each of the scientists had to carry water or die of dehydration, and they carried food and water for their horses as well. They had been working the Hell Creek for several years, and previous field seasons had produced the specimens of *Tyrannosaurus rex* that had generated great publicity for the American Museum (see chapter 14). Day after day they continued their search. They saw plenty of fossils of broken turtle shells and armor of crocodilians (not worth collecting), and the distinctive parallelogram-shaped scales of garfish. Particularly common were the small, finger-shaped cylinders of rock that represented ossified tendons from the tails of duckbills. Dinosaur bone fragments were also common, but if loose on the ground and not very diagnostic, they were not worth picking up. However, if a few of them were together and not too broken, experienced collectors such as Brown and Kaisen knew to follow the trail of bone fragments like Hansel's breadcrumbs to see if the rest of the bone was eroding out of the slopes above. Sometimes they found nothing, but occasionally they got lucky.

On one particular summer day in 1906, Kaisen followed a trail of bone scraps and found a partial skeleton eroding out of the cliff (figure 21.1). He poked around to determine how much was there, then hiked back and brought Brown to see it. Soon they were both digging carefully around it, exposing the top and sides but leaving the specimen embedded in the soft siltstones to support it as they dug. The specimen just kept getting bigger and bigger, and going farther and farther into the cliff face. As they dug, they left each large bone on a pedestal of soft siltstone, giving it minimal

Figure 21.1 ▲

The Hell Creek badlands of central Montana, where Brown and Kaisen found the first *Anky-losaurus* specimen (just above the pick). (Courtesy of Wikimedia Commons)

support. Brown probably used a bit of dynamite to blow away the overburden, a risky practice that is rarely used today. After days of hard work, they had the fossil exposed and surrounded in a hard plaster jacket, ready to move. They pried the blocks off their pedestals, added a plaster jacket to the newly exposed bottom surface, and then brought the horse and buckboard wagon to the site to load each block and transport it to the nearest rail line. Eventually, all of the huge blocks were on their way to New York.

Once back at the American Museum, Kaisen began to prepare the specimen, cutting away the plaster jacket and then carefully cleaning off the rock and siltstone that still encased the fossil bone. After two years, the specimen was clean enough for Brown to publish a description. The specimen consisted of the skull, a few teeth, the shoulder girdle, a few vertebrae from the shoulders, back and tail, ribs, and the dense bony pieces of armor called osteoderms. The rest of the creature was unknown, so Brown reconstructed the limb proportions to match those of stegosaurs, with long hind limbs and an arched back. (Ankylosaurs were considered stegosaurs for a long time.) Brown's reconstruction (figure 21.2) made the creature look more like a stegosaur with the arched back and was criticized by Samuel Wendell Williston as being based on too little material for a reliable reconstruction. Instead, Williston thought *Ankylosaurus* should be reconstructed with a straight back and equal length forelimbs and hind limbs, as Baron Franz Nopcsa had done with *Polacanthus* a few years earlier. Williston also thought that Brown's dinosaur might be the same as *Stegopelta*, which Williston had named years earlier.

Brown gave it the name *Ankylosaurus magniventris*. *Ankylos* literally means "bent" or "crooked" in Greek, and it is the root word for the medical condition called "ankylosis" in which joints are fused and stiff. Brown

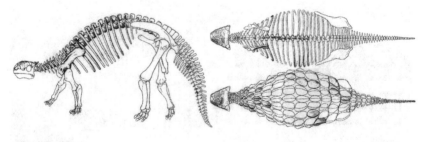

`Figure 21.2` ▲
Barnum Brown's 1908 reconstruction of the first *Ankylosaurus* specimen, with the hind legs too long so it has high hips and a back like a stegosaur. (Courtesy of Wikimedia Commons)

probably was thinking not of the original meaning but of the idea of "fused" armor plates. Today the meaning of the genus is usually translated as "fused lizard" or "stiff lizard" rather than "bent" or "crooked" lizard. The species name *magniventris* means "broad belly," which refers to the large flat shell on the back and the flat belly below.

It turned out that Brown had collected parts of this animal before. In 1900, while collecting a *Tyrannosaurus rex* in the Upper Cretaceous Lance Creek beds of eastern Wyoming, Brown had found about 77 of the bony osteoderms, but both Brown and Osborn thought they might belong to *Tyrannosaurus*. In fact, the dense, bony osteoderms were durable fossils that looked and weathered like small cannonballs, and they had been found many times by Native Americans in the past.

In 1910 Brown was on another American Museum expedition, working in the Late Cretaceous of the Red Deer River badlands in Alberta (see chapter 22). Collecting in the Scollard Formation, he found a much more complete specimen of *Ankylosaurus*, with a skull, jaws, ribs, vertebrae, limbs, armor, and the first and only specimen with the tail club preserved. This is the amazing specimen now on display in the American Museum. Together these specimens gave the most complete image of the kind of dinosaurs ankylosaurs were. In addition, the discovery of *Euoplocephalus* further improved our understanding of ankylosaurs (figure 21.3).

Figure 21.3 ▲

Skeleton of the ankylosaur *Euoplocephalus*. (Courtesy of Wikimedia Commons)

Their image became iconic when the American Museum mount led to life-sized sculptures, such as that created for the 1964 World's Fair in New York (figure 21.4). The turtle-like reconstruction of *Ankylosaurus* painted by artist Rudolph Zallinger for the famous 1947 mural *The Age of Reptiles* (still on the walls of the Peabody Museum of Natural History at Yale) has been the template for nearly every book and reconstruction since then. We know the groups as the ankylosaurs, and all the mystery fossils found earlier but poorly preserved and understood began to fit into an image.

Ankylosaurus turns out to be a relatively rare dinosaur, with only a handful of specimens found in more than a century, and most of them are fragmentary. Only three skulls are known (figure 21.5), but it was widespread in Upper Cretaceous rocks all over the Rocky Mountains, including the Hell Creek Formation of Montana, Lance and Ferris formations of Wyoming, and Scolland and Frenchman formations of Alberta (but not in New Mexico Upper Cretaceous rocks). The largest known specimens were about 10 meters (33 feet) long, 1.5 meters (5 feet) wide, although several were only 6 meters (20 feet) long; they weighed from 4.8 metric tonnes (5.2 tons)

Figure 21.4 ▲
Life-sized sculpture of *Ankylosaurus* displayed at the 1964 World's Fair in New York. (Courtesy of Wikimedia Commons)

up to 8 metric tonnes (9 tons). The head was covered with thick plates of armor on top, along with a pair of pyramid-shaped horn-like plates in the back corners of the skull pointing up and back, and their nostrils faced sideways (figure 21.5). The lower jaw was small and the bite force was also weak, with tiny leaf-shaped teeth similar to those of stegosaurs, so it could

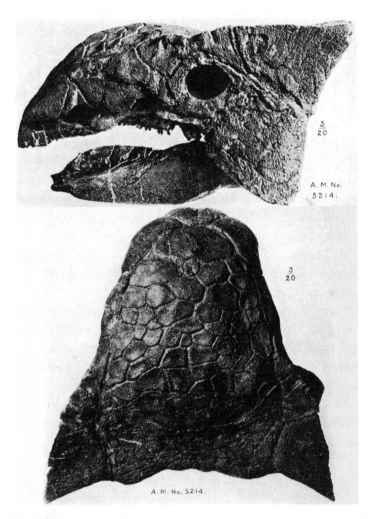

Figure 21.5 ▲

The best of the three known skulls of *Ankylosaurus*, showing the distinctive head armor, small eyes, weak lower jaws with small teeth, and pyramid-shaped horns on the back corners of the skull roof. (Courtesy of Wikimedia Commons)

only eat low vegetation such as shrubs and small bushes. *Ankylosaurus* had large sinuses and nasal chambers, possibly for water and heat balance or for sound amplification. Rings of bone covered the neck and protected it between the armored head and back. It is most famous for its semisolid shell of fused osteoderms on the back, composed of a "pavement" of small polygonal osteoderms, alternating with longitudinal and transverse rows of big osteoderms. Ankylosaur legs were relatively short and equal in length, so it did not have the high hips of stegosaurs and was built very low to the ground. The vertebrae of the back and hips were strongly fused together and partially fused to the shell above them. The tail vertebrae were highly flexible, so it could swing its tail club easily. All in all, *Ankylosaurus* had formidable defenses. It was capable of rapid maneuvering to keep predators like *T. rex* at bay, and it could bring its body around to break the shins of the predator with the bone-crushing club on its tail.

More than two dozen genera of the family Ankylosauridae are now known, and most were described in the past 30 years. They were restricted to the Cretaceous of North America and Asia, and they never appeared in Gondwana continents. They first appeared in the Early Cretaceous with fossils like *Gastonia* and *Cedarpelta* from the Lower Cretaceous Cedar Mountain Formation of Utah and *Liaoningosaurus* and *Chuanqilong* from the Early Cretaceous of China. Most of the ankylosaurids come from the Late Cretaceous, when they apparently pushed out the nodosaurids that had dominated the Late Jurassic and Early Cretaceous. One of the most interesting is *Crichtonpelta*, a middle Cretaceous ankylosaurid from China that was named in honor of Michael Crichton, the author of the *Jurassic Park* novels. Another nearly complete skeleton from the Hell Creek Formation was described by Victoria Arbour and David Evans in 2017 and given the name *Zuul* after the evil demi-god and Gatekeeper in the 1984 movie *Ghostbusters* (in the movie the character Zuul had a head shaped like this dinosaur).

"MR. BONES"

Barnum Brown was one of the greatest fossil collectors who ever lived, and he probably found more famous dinosaurs than anyone in his time or before (figure 21.6). A short list of these includes theropods *Albertosaurus*, *Struthiomimus*, *Dromaeosaurus*, and *Tyrannosaurus rex*; ankylosaurs *Ankylosaurus*,

Figure 21.6 ◄ ►
Barnum Brown: (A) class picture in the 1897 University of Kansas yearbook, showing the young college grad about to embark on a great career finding fossils; (B) a young apprentice Barnum Brown (*left*) and Henry Fairfield Osborn (*right*) at Como Bluff in 1897, excavating the first *Diplodocus* for the American Museum; (C) Brown in his later years, coating a fossil with a protective plaster and burlap jacket; (D) Brown (*right*) in 1941, with Erich M. Schlaikjer, and Roland T. Bird with the reconstructed skull of the gigantic Cretaceous crocodilian *Deinosuchus* from the Big Bend of Texas. Bird holds the skull of a modern Nile crocodile for scale. ([A, C] Courtesy of Wikimedia Commons; [B] image #17808, [D] image #318634, courtesy of the American Museum of Natural History Library)

Figure 21.6 ▲
(continued)

Euoplocephalus, and *Edmontonia*; duckbills *Corythosaurus, Kritosaurus, Hypacrosaurus, Saurolophus, Prosaurolophus,* and *Parasaurolophus*; many ceratopsians including *Chasmosaurus, Leptoceratops, Centrosaurus*; and, of course, *Triceratops*; and many Morrison sauropods. Even though most people haven't heard of him today, he is considered a giant in the field. As Dingus and Norell wrote:

> Among paleontologists, Barnum Brown is almost universally recognized as the greatest fossil collector of all time, and although his name is no longer widely recognized among the general public, the name of the most famous dinosaur he ever discovered—*Tyrannosaurus rex*—most certainly is. The fossils he unearthed across continents from the Americas to Asia vastly expanded our knowledge not only of the evolution of dinosaurs but also of most major groups of reptiles and mammals. Crucial to his success was a penchant for eccentricity and risk, both professionally and personally. His expeditions, many of which he undertook with little or no backup, even from his trusted companions, led him to several of our planet's most remote and dangerous regions. As if that weren't risky enough, Brown often used his guise as a bone digger to cover a second, clandestine role as an American intelligence agent, gathering strategic geologic and geographic data that would aid both the country's exploration for oil and the government's war efforts.

Brown's career started modestly. Born in Carbondale, Kansas, on the anniversary of Abraham Lincoln's and Charles Darwin's birthday (February 12, 1873), he was named after the famous showman and circus founder, P. T. Barnum. However, no one could foresee the amazing show that Brown's fossil discoveries would eventually become. As Gil Troy wrote:

> The parental impasse over what to name the young lad ended after a few days when P. T. Barnum's dazzling circus visited. "There must be something in a name," Brown, who had his own genius for self-promotion would say "for I have always been in the business of running a fossil menagerie."

His family lived on top the Carboniferous coal beds of eastern Kansas. His father not only maintained a ranch but spent much of the Civil War running supplies in a wagon from the railheads in Kansas to remote outposts all over the Plains. He also had a small-scale strip mining operation for coal on his land, which was much cheaper to collect locally than to ship via rail and wagons from remote coal mines. As the workers stripped away the rock over the coal seams, they found incredibly fossiliferous limestones, as well as beautiful fossil ferns and other plants in the coal itself. Brown wrote in his notebook: "They unearthed vast numbers and—varieties of seashells, crinoid stems and parts, corals . . . and other sea organisms. . . . I followed the plows and scrapers, and obtained such a large collection that it filled all of the bureau drawers and boxes until one could scarcely move. Finally Mother compelled me to move the collection into the laundry house."

His love of fossils then led him to enroll at the University of Kansas where he studied paleontology under Samuel Wendell Williston, as well as learning the geology of his state and of coal and other fossil fuels such as oil (figure 21.6A). Williston wrote that Brown was "the best man in the field that I ever had. He is very energetic, has great powers of endurance, walking thirty miles a day without fatigue, is very methodical in all his habits, and thoroughly honest." During summer breaks in the mid-1890s, he went on university trips to the badlands of South Dakota and the plains of Wyoming, where he discovered his aptitude for strenuous fieldwork and fossil collecting. In 1896, he was hired as part of a field crew for the American Museum working in the San Juan Basin of New Mexico and the Bighorn Basin of Wyoming. He was such a talented collector that he was hired right away, even though he had not yet graduated from the University of Kansas. (He finally got his degree 10 years later.) In 1897, he was an apprentice collector in the American Museum crew in Medicine Bow, Wyoming, where they found Morrison dinosaurs in abundance (figure 21.6B). As the 24-year-old Brown wrote in his notes, "I was . . . fortunate in discovering a partial skeleton of . . . *Diplodocus*. This was the first dinosaur excavated by any American Museum expedition, and here I introduced the use of plaster of Paris in excavating fossils." Soon he was sent anywhere the American Museum needed a hardy, sharp-eyed, self-sufficient field collector. As Dingus and Norell wrote:

> On a December morning in 1898, a 25-year-old paleontologist named Barnum Brown trudged through the snow-choked streets of New York City for what he thought would be a routine day at the American Museum of Natural History. But before he could remove his hat, his supervisor, Henry Fairfield Osborn, anxiously summoned him. "Brown," he said, "I want you to go to Patagonia today with the Princeton expedition . . . to represent the American Museum. The boat leaves at 11; will you go?" "This is short notice," Brown replied. "But

I'll be on that boat if . . . I [can] go home to pack my personal belongings and arrange for my absence." Although he had never been out of the country before, Brown did not return for almost a year and a half. In Patagonia he prospected for fossils largely on his own. On one excursion, he was shipwrecked off the Patagonian coast when a large wave slammed his small cutter; despite not knowing how to swim, Brown floated safely to shore grasping a barrel. By befriending the locals, he got his project back on track and eventually shipped 4.5 tons of fossil mammal bones back to the museum. Brown's gamble—to strike out by himself to search for fossils in a foreign land—established a precedent for his expeditions and launched his legendary career.

After his Patagonia trip, Brown spent nearly every field season for the next decade in the Hell Creek beds of Montana, where he found *T. rex, Ankylosaurus*, and many other dinosaurs already discussed. Between 1910 and 1915, he worked in the Red Deer River badlands of Alberta (see chapter 22), where he discovered Late Cretaceous dinosaur fossils from 70 to 75 million years ago, older than the latest Cretaceous Hell Creek fossils. In Canada, he competed with the Sternberg family, who were hired by the Royal Ontario Museum to collect their native fossils—but Brown brought back amazing specimens that are now on display or in storage at the American Museum. He finished working in Canada in 1916, then went to Cuba in 1918, and back to the Rockies in 1918 and 1919.

When travel restrictions ended after World War I, he went abroad and added to the American Museum's collections from many regions, including fossils from Ethiopia and Egypt in 1920–1921, Miocene mammals from India (and part of what became Pakistan) and Burma from 1922–1923, and Miocene mammals from the Greek island of Samos in 1923 and 1924. Although he did a lot of collecting on Samos by himself, he took advantage of the low costs of labor (thanks to the refugees of the Turkish war with Greece) to hire a team of 18 men and 6 women to haul dirt in baskets for "the attractive sum of 35 drachmas (70 cents) per day for men and 20 drachmas per day for the girls." When the work was done, he had 56 cases of beautiful bones, "representing three species of three-toed horses, rhinoceroses . . . many species of antelope and gazelle . . . birds, and a variety of carnivorous mammals," including the short-necked giraffe named *Samotherium*.

From 1931 to 1935, he took part in excavating a huge Morrison bone bed at Howe Ranch in Wyoming. In the late 1930s, Brown and American

Museum assistant Roland T. Bird (with a crew of CCC workers) excavated some huge trackways from the legendary track site at Glen Rose, Texas. Later he collected the skull of the enormous crocodilian *Deinosuchus* in the Big Bend region of Texas (figure 21.6D). He also spent much of the Depression years of the 1930s driving to, exploring, and collecting from important American fossil beds, such as the Early Permian of Texas, the Petrified Forest of Arizona, and elsewhere. The American Museum was bursting at the seams with too many crates of dinosaurs, and they needed other kinds of fossils to round out their collections. Also, funds for expensive foreign travel were scarce during the Depression when even the rich people who once funded their expeditions had lost money in the Stock Market Crash of 1929.

Brown was legendary for his fieldwork and collecting prowess, but his personal life was a bit more complicated. He was a real ladies' man, famous for his affairs with ranchers' daughters while he was in the field and seducing foreign ladies during his travels abroad. I vividly remember my first summer in the field with my graduate advisor Malcolm McKenna of the American Museum in 1977, who had met Brown in his old age. Malcolm had also visited many of Brown's old localities, and more than once heard the stories of Brown's romantic exploits from the local ranch ladies, who would tell him how they still remembered Barnum Brown—and not for his paleontology. As Dingus and Norell document in their biography of Brown, he often stayed abroad for many months during periods when his New York lawyer was settling lawsuits of jilted lovers, and he had to hastily exit some foreign countries for the same reason.

As Gil Troy wrote:

> Someone that obsessed, who charmed New York's elite, effete Upper West Siders when he returned from his remote, sun-baked dry beds yielding ancient mysteries under layers of rock and dirt, was bound to be complicated. His first wife died of scarlet fever shortly after childbirth. The distraught Brown left their daughter to be raised by her grandparents. Brown remarried but, even in those more discrete times, was known as a cad. His second wife Lilian Brown got the last laugh, with a passive-aggressive, neglected-wife-of-a-workaholic memoir: *I Married a Dinosaur*.

Brown's later career was documented by Lowell Dingus and Mark Norell in their excellent biography of him, and by his second wife Lilian

McLaughlin Brown (nicknamed "Pixie" by Barnum Brown), who wrote amusing books documenting their travails and triumphs. Her first was *I Married a Dinosaur* in 1950, about their fieldwork in the Siwalik Hills of India (now part of Pakistan) and side trips to Burma. Her other books included *Cleopatra Slept Here* in 1951, about Barnum's collecting in Ethiopia and Egypt, and *Bring 'Em Back Petrified* in 1956, about their fieldwork in Guatemala excavating Ice Age mammals from the bottoms of rivers.

As Dingus and Norell wrote:

The peripatetic Brown collected so many specimens on his travels that even today dozens of large boxes of his fossils have yet to be opened. Crates labeled "mammals from Samos" and "ornithomimid from Hell Creek" rest on sanitized racks in the storerooms of the American Museum of Natural History simply because there are not enough staff workers to unpack them, remove the plaster, and prepare them all. As Brown neared his 90th birthday in 1963, his successor at the museum, Edwin Colbert, marveled at how much of the museum was Brown's single-handed work: "There are, in our Tyrannosaur Hall, 36 North American dinosaurs on display. . . . You collected 27, an unsurpassed achievement." For all of Brown's success as a collector, his direct impact as a scientist was strangely modest, compromised by his time abroad and by his laxity in publishing his discoveries. The CV that he compiled for his memoirs—which he never finished—shows that there were only five years between 1897 and his retirement in 1942 in which he did not participate in a major expedition. Yet he penned few scientific papers, and those that he did publish are mostly short notes. On the other hand, his longer publications, like his monograph on *Protoceratops*, a Mongolian relative of the great horned dinosaurs, are classics, both for the freshness of the presentation and the quality of the analysis.

Brown was such a prolific collector and spent so little time writing up his discoveries that many finds were left to others. For example, in the 1930s he was working on the Crow Reservation in Canada and found a partial skeleton of a small, agile carnivorous dinosaur. He began to write up a scientific paper describing its bird-like characteristics and even gave it a tentative name (the drawer is labeled "Daptosaurus"), but he never finished or published that project. Brown would eventually show the specimens and discuss the project with one of Colbert's students at the American Museum, John Ostrom. This got Ostrom thinking about making his own discovery of the same bird-like dinosaur that he named *Deinonychus* (see chapter 16). Eventually, Ostrom combined his own finds on his Yale expeditions with the Brown American Museum specimens. Brown never got credit for being the first to find *Deinonychus* because he never finished or published his work.

Brown was a hard-working collector for the American Museum, but he had many commitments on the side, which Dingus and Norell described:

Brown's succession of exotic discoveries made him one of the greatest scientific celebrities of his day. The public nicknamed him "Mr. Bones," and one writer noted that "wherever Brown went on his expeditions in the American West . . . he was feted by the local populace. Droves of people would meet his train and vie for the honor of driving him from the station to town." Museum archives reveal that Brown's zest for geologic exploration inspired a second, clandestine life. From early on Brown's expeditions for fossils had served as a smoke screen for occasional sojourns as an intelligence agent and corporate spy. During both world wars he funneled geologic knowledge gleaned from his fossil-hunting expeditions to the United States government. He occasionally confided in museum colleagues, as in a 1921 letter to paleontologist W. D. Matthew stating that he had "an exciting time in Turkey, and secured much desired data for the State Department."

In 1941 the American government contacted museums to find out where their curators had done fieldwork in order to harvest information about remote and strategically vital areas. Although Brown was about to retire, he happily obliged, citing his travels to Canada, Cuba, Mexico, Patagonia, France, England, Turkey, Greece, Ethiopia, Egypt, Somalia, Arabia, India, and Burma. . . . Because of his passion for both geology and paleontology, Brown also forged close ties with oil and mining companies. Part of the industry's appeal was financial; he charged a consulting fee of $50 per day, almost $800 in today's currency. These contacts could also be lucrative sources of fossil-hunting cash. In 1934, with museum funds for fieldwork in short supply, Brown approached officials of the Sinclair Oil Corporation in search of financial support. The company's president became so enamored of Brown and his work that he personally backed Brown's expedition to northern Wyoming, where he discovered the famous Jurassic dinosaur graveyard named Howe Quarry. To this day the Sinclair logo features an image of Diplodocus in deference to this partnership.

Brown's oil company contacts were extensive. In 1920 he quite likely fled a jilted lover's lawsuit by skedaddling for Abyssinia (now Ethiopia) under the banner of the Anglo American Oil Company, an offshoot of John

D. Rockefeller's Standard Oil. His mentor, Henry Osborn, was supportive of such ventures because Brown kept his keen eyes peeled for fossils while prospecting for oil. In this way Osborn reaped the fruits of Brown's labors while the oil company paid for them. Brown's letters—in which he invariably addresses his supervisor as "my dear Professor Osborn"—maintain the careful guise of an eccentric fossil hunter, but they also contain carefully worded notes about both oil prospects and the activities of European diplomats, military personnel, and businessmen. Those notes also contain telling details about his targets' wives and daughters.

Barnum Brown officially retired from the American Museum in 1942 at age 69. During the war years of 1942–1945, he used his extensive knowledge of world geology to help the Office of Strategic Services (predecessor of the CIA), giving them strategic information about geology and natural resources and helping plan military strategy. He spent the late 1940s and 1950s, when he was in his 70s and early 80s, in semiretirement. He still came in to the American Museum to finish projects when he felt like it, helped supervise mounting of his specimens, and participated in occasional field expeditions as his energy and funds allowed. He mostly worked on consulting jobs for oil companies, as well as advising the people who were preparing dinosaur exhibits for the 1964 World's Fair in New York (see figure 21.4).

Barnum Brown died in 1963 at the ripe old age of 89. As Dingus and Norell wrote, "First and foremost he was the greatest dinosaur collector the world has ever known. Through Brown's efforts, dinosaurs gained a strong foothold in the psyche of both the scientific community and the general public." As Colbert noted above, Brown collected 24 of the 37 dinosaurs that were once in the Cretaceous Hall in the American Museum (57 of his specimens are currently on display in all the American Museum halls), so his finds were the public face of dinosaur paleontology for almost a century. He was as public a celebrity as any paleontologist was in those days, with weekly radio broadcasts, articles in newspapers, and many public appearances. Brown was also the main consultant on Walt Disney's pioneering movie *Fantasia*, in which the dinosaurs and their world are animated to Stravinsky's *The Rite of Spring*. This animation of dinosaurs was inspiring and iconic for many generations of future paleontologists (including me). As Dingus and Norell wrote: "Brown's real legacy lies with his discoveries themselves, which still serve as a foundation for numerous geologic and biological research projects in paleontology."

TANTALIZING MYSTERIES: THE NODOSAURS

Ankylosaurus itself is one of the more extreme examples of the ankylosaur family, with a relatively solid shell of armor, horns on its head, a tail club, and spikes along the edge of its shell. A number of ankylosaurs had been found prior to its discovery, but these fragmentary specimens gave little or no reliable indication of what they would have looked like in life. These represent the other main group of ankylosaurs, called nodosaurs. They did not have a tail club, and the armor in the skin of their backs and sides were made of numerous smaller unfused osteoderms rather than the solid bony shell of *Ankylosaurus, Gastonia, Crichtonpelta, Pinacosaurus*, or *Euoplocephalus*. Given how fragmentary most of the early specimens were, it is not surprising that no one was able to reconstruct them accurately until nearly complete articulated specimens were finally found. Today at least two dozen genera of nodosaurs are known, and they are found on almost every continent, including Antarctica. Yet most amateur dinosaur enthusiasts do not recognize a nodosaur, nor can they guess that it is a kind of ankylosaur.

The very first ankylosaur to be discovered (long before *Ankylosaurus* in 1906) was *Hylaeosaurus*, from the Lower Cretaceous Wealden beds of southeastern England. It was only the third dinosaur to be named (after *Megalosaurus* and *Iguanodon*) and was included among the genera in Owen's original definition of Dinosauria. Gideon Mantell obtained the first specimen on July 20, 1832 (figure 21.7A) from one of the Wealden sandstone quarries in the Tilgate Forest near Cuckfield in Sussex (the same one that produced *Iguanodon*). The workers dynamited a rock face to pry loose more slabs and exposed a boulder with dinosaur bones in it. The bones were shattered into about 50 pieces, but Mantell bought it from the workmen anyway. With much effort, he was able to piece it together into a single incomplete skeleton that was partially articulated. This was a big improvement over *Megalosaurus* and *Iguanodon*, which were all based on isolated bones. Mantell thought it might belong to his *Iguanodon*, but William Clift of the Royal College of Surgeons Museum looked at the plates and spikes that were pieces of body armor and suggested that it was something new. In November 1832, Mantell formally named it *Hylaeosaurus*, which means "forest lizard" in Greek. The name could refer to the Tilgate Forest where it was found or to the old Anglo-Saxon word "Weald," meaning "forest," which had given its name to the formation where the dinosaurs were found.

Figure 21.7 ▲

The British nodosaur *Hylaeosaurus*: (*A*) the original partial skeleton by Mantell; (*B*) Mantell's reconstruction as a giant armored lizard; (*C*) a modern reconstruction of the animal in a life pose. ([*A, B*] Courtesy of Wikimedia Commons; [*C*] courtesy of N. Tamura)

The preserved bones (figure 21.7A) include most of the front part of the animal, with the shoulder girdles, dermal armor, and shoulder spikes, but most of the skull, forelimbs, and the entire back half of the animal were missing. Mantell's specimen remained in the partially prepared block until recently. Modern acid etching techniques have managed to free the bones and allow scientists to see them from all sides for the first time. Additional specimens of *Hylaeosaurus* have been reported from the Isle of Wight, France, Germany, Spain, and Romania, although they may belong to some other type of nodosaur.

Despite the incompleteness of the skeleton, many attempts have been made to reconstruct how *Hylaeosaurus* might have looked. Richard Owen commissioned Waterhouse Hawkins to create a concrete sculpture for the Crystal Palace Exhibition that made it look like a giant iguana with some spikes along its back (figure 21.7B). More recent reconstructions make it look like other nodosaurs (figure 21.7C), although *Hylaeosaurus* is so incomplete that it is impossible to reconstruct most of the animal with confidence.

The next nodosaur discovered was *Polacanthus*, found by the Reverend William Fox in the Lower Cretaceous beds on the Isle of Wight in 1865 (some specimens had been discovered as early as 1843 but were never fully reported or described). Fox originally asked Alfred Lord Tennyson to name it, and he chose "Euacanthus vectianus." However, Fox reported on it during a lecture to the British Association in late 1865, and the text of a description appeared anonymously in the *Illustrated London News*. The author was probably Richard Owen, although some speculate it might have been Thomas Henry Huxley, or even Fox. Whoever wrote the article called it *Polacanthus foxii* (*poly* meaning "many" and *acanthus* meaning "thorn" in Greek), and the species name honors Fox—something that Fox could not do because the rules of zoological names prevent an author from naming something after himself.

The skeleton itself consists of some of the back vertebrae, the hip bones, parts of both hind limbs, tail vertebrae, ribs, and the armor of osteoderms and spikes—but no head, neck, forelimbs, or the rest of the front of the body. In 1881 the specimen was acquired by the British Museum from Fox. Fox had let the specimen deteriorate over the years, and the armor was falling apart. British Museum preparator Caleb Barlow glued it all together with a natural resin called Canadian balsam, miraculously restoring it to its original condition. The specimen was not fully studied and described

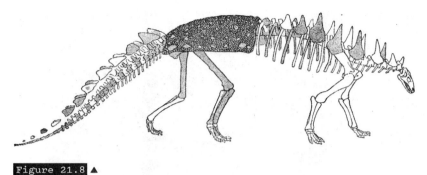

Figure 21.8 ▲
Franz Nopsca's 1905 reconstruction of *Polacanthus foxii*. (Courtesy of Wikimedia Commons)

until 1887 when John Whitaker Hulke took over the task. It was redescribed by Nopcsa in 1905, who published the first reconstruction of the animal (figure 21.8). As *Polacanthus* was one of the first relatively complete nodosaurs to be published, many other fragmentary European fossils have been referred to that genus, making it a taxonomic wastebasket for a long time.

Additional fragmentary skeletons of nodosaurs continued to turn up. *Struthiosaurus* (ostrich lizard) was described by Emmanuel Bunzel in 1871 from fragments found in a coal mine in Austria. It was the first nodosaur with a partial skull (a braincase), plus a forelimb, hind limb, hip and tail vertebrae, and some ribs. The armor consisted of knob-like osteoderms plus a number of spines and plates, although it is impossible to reconstruct where they fit in the skin of the animal. *Nodosaurus* (knobbed lizard) was found in the Upper Cretaceous Frontier Formation of Wyoming and described by O. C. Marsh in 1889. It was the first ankylosaur found in North America and was represented by a relatively complete skeleton that showed the dermal armor arranged in bands around the body, with narrow bands over the ribs alternating with wider plates between them. This was one of the first nearly complete nodosaurs found, and it established the basic body plan for the group. With so many partial skeletons with broken and missing pieces, *Nodosaurus* soon became the name for the entire group.

The best preserved of the early nodosaur discoveries was *Edmontonia*, discovered by Barnum Brown in the Red Deer River badlands of Alberta in 1915 (figure 21.9). When it was finally cleaned and prepared, it turned out to be the nearly complete articulated front end of the dinosaur, and it forms a dramatic display in the American Museum even today. It was first referred

Figure 21.9 ▲

(A) Barnum Brown's nearly complete fossil of the front end of the nodosaur *Edmontonia*, on display in the American Museum. (B) Reconstruction of *Edmontonia*. (Courtesy of Wikimedia Commons)

by William Diller Matthew in 1922 to Leidy's invalid genus *Palaeoscincus*, but another specimen found by Charles M. Sternberg in the same beds was named *Edmontonia* in 1928 (in reference to the Edmonton beds where it was found). The misnamed specimens were all eventually assigned to *Edmontonia*, making it among the best known of the nodosaurs. The articulated American Museum specimen (figure 21.9A) has an armored skull, neck armor, and a solid armored back fringed with large spikes on its shoulders and along the edge of the back shield. It was also among the largest nodosaurs, estimated at 6.6 meters (22 feet) long (figure 21.9B).

In the past few decades, the pace of nodosaur discoveries has accelerated, and many new genera were recovered from China and North America. The oldest known nodosaurs are *Gargoyleosaurus* and *Mymoorapelta* from the Upper Jurassic Morrison Formation. Nodosaurs become more diverse in the Early Cretaceous with *Hylaeosaurus* from England, *Polacanthus* from Austria, *Gastonia* from the Cedar Mountain Formation of Utah, and *Hoplitosaurus* from the Lakota Formation of South Dakota. By the late Early Cretaceous, nodosaurs began to diversify explosively with many new genera like *Nodosaurus, Sauropelta, Stegopelta, Animantarx, Peloroplites,* and *Tatankacephalus* from the Rocky Mountains of the United States; *Pawpawsaurus* and *Texasetes* from Texas; *Silvisaurus* from Kansas; teeth referred to *Priconodon* and *Propanoplosaurus* from the Maryland coast; *Europelta* from Spain; *Anoplosaurus* from England; *Hungarosaurus* from Hungary; and *Dongyangopelta* and *Zhejiangosaurus* from China. Found in the latest Cretaceous are the Alberta nodosaurs *Edmontonia* and *Panoplosaurus*, *Glyptodontopelta* from the very end of the Cretaceous of New Mexico, *Struthiosaurus* from eastern Europe, and a handful of teeth from James Ross Island on the Antarctic peninsula named *Antarctopelta*.

THE TAR SAND CARCASS

For a group as plagued by incomplete specimens as the nodosaurs, in 2017 it was amazing to hear news that a new nodosaur was known from an almost complete specimen in a death pose, with preservation of skin and scales between and around the bones, and even preservation of some of its color (figure 21.10). The discovery was made on March 21, 2011, when an excavator scooping up tar sands from the Suncor Energy strip mine near Fort McMurray, Alberta, hit a hard object in the midst of the sands. The

The tar sand nodosaur mummy, *Borealopelta*, with nearly all the armor of the head and front of the body preserved intact. (Courtesy of Wikimedia Commons)

excavator operator, Shawn Funk, immediately recognized that it was an unusual fossil, and together with his supervisor, Mike Gratton, they notified the Royal Tyrrell Museum in Drumheller, Alberta, which was responsible for paleontological salvage at the oil sand pit. Tyrrell paleontologists Donald Henderson and Darren Tanke rushed up to the mine expecting to find a plesiosaur in these marine sand deposits. Instead they discovered the carcass of a nodosaur that had apparently washed out to sea before being buried. The specimen was found upside down on the seabed, apparently weighted down by its armor. It had floated that way for weeks, and when the soft tissues rotted and the trapped gases escaped it sank to the bottom and was buried in the soft organic-rich sand. The upper part of the specimen was preserved (including the nearly complete head and back shield) because it was buried upside down when it hit the bottom. The belly and legs and tail, however, are mostly missing. They would have been exposed and must have rotted away or been scavenged. The inside of the shell was filled with sand, washed into the body cavity after the soft tissues were lost. Then the chemistry of the rotting tissue precipitated a concretion of siderite (iron carbonate) around it, which protected the specimen and preserved it not only against scavengers but also against deformation when it was buried by later deposits.

The specimen was embedded in a sandy cliff about 8 meters (26 feet) above the floor of the mine. Tyrrell and Suncor employees needed 14 days to recover all the pieces and jacket them in plaster for transport. It then took Tyrrell preparator Mark Mitchell five years to put all the pieces together and clean them properly, which is the reason it finally made a big splash on May 12, 2017. It debuted to the public during a big press event the same day that it was published as *Borealopelta markmitchelli*. *Borealopelta* means "northern shield" in Greek, and the species name honors the preparator who spent all those years putting it together.

The specimen (figure 21.10) preserves the head shield, neck rings, the pattern of bony armor on its back, and the spines protruding from its shoulders—all articulated as they were in life. There were none of the usual problems of guessing where to put the spines, as so often happens with other nodosaurs. The skin between the bones is well preserved, showing the detailed texture and pattern of the scales. Most remarkable of all, the pigment cells, or melanosomes, can be detected. It apparently was reddish-brown in life, with countershading along the sides and bottom.

Beginning with the puzzling fragments of the earliest ankylosaurs, we now have one of the most complete articulated dinosaurs every found. Ankylosaurs were certainly among the most bizarre and interesting of all the dinosaurs.

FOR FURTHER READING

Brinkman, Paul D. *The Second Jurassic Dinosaur Rush: Museums and Paleontology in America at the Turn of the Twentieth Century*. Chicago: University of Chicago Press, 2010.

Brown, Lilian. *Bring 'Em Back Petrified*. New York: Dodd, Mead, 1956.

——. *Cleopatra Slept Here*. New York: Dodd, Mead, 1941.

——. *I Married a Dinosaur*. New York: Dodd, Mead, 1940.

Carpenter, Kenneth. "Ankylosauria." In *Encyclopedia of Dinosaurs*, ed. Philip J. Currie and Kevin Padian, 16–20. San Diego: Academic Press, 1997.

——, ed. *Armored Dinosaurs*. Bloomington: Indiana University Press, 2001.

Colbert, Edwin. *Men and Dinosaurs: The Search in the Field and in the Laboratory*. New York: Dutton, 1968.

Dingus, Lowell, and Mark Norell. *Barnum Brown: The Man Who Discovered Tyrannosaurus rex*. Berkeley: University of California Press, 2010.

Farlow, James, and M. K. Brett-Surman. *The Complete Dinosaur*. Bloomington: Indiana University Press, 1999.

Fastovsky, David, and David Weishampel. *Dinosaurs: A Concise Natural History*, 3rd ed. Cambridge: Cambridge University Press, 2016.

Holtz, Thomas R., Jr. *Dinosaurs: The Most Complete, Up-to-Date Encyclopedia for Dinosaur Lovers of All Ages*. New York: Random House, 2011.

Naish, Darren. *The Great Dinosaur Discoveries*. Berkeley: University of California Press, 2009.

Naish, Darren, and Paul M. Barrett. *Dinosaurs: How They Lived and Evolved*. Washington D.C.: Smithsonian Books, 2016.

Spaulding, David A. E. *Dinosaur Hunters: Eccentric Amateurs and Obsessed Professionals*. Rocklin, Calif.: Prima, 1993.

Vickaryous, Matthew K., Teresa Maryańska, and David B. Weishampel. "Ankylosauria." In *The Dinosauria*, 2nd ed., ed. David B. Weishampel, Peter Dodson, and Halszka Osmólska, 363–392. Berkeley: University of California Press, 2004.

CORYTHOSAURUS

The proper outfit for a collecting expedition consists of a good team of ponies or small mules, a light lumber wagon, cover, wall-tent, camp-stove or "Dutch oven," knives and forks, tin plates and cups, and other cooking-utensils. Each member of the party should be provided with a rubber blanket and coat, and a couple of pairs of woolen blankets; besides these but little extra baggage should be taken; a good pair of woolen shirts are valuable. The tools should consist of several small hand-picks, miner's-picks, with one point made into a duck-bill with sharp edge; butcher-knives, shovels and collecting-bags—made after the pattern of mail-carrier's bags, of heavy ducking, with two apartments—one for cotton, paper and string, and the other for fossils. There should always be kept on hand a supply of burlap sacks, old newspapers, cotton, manila paper and hop-needles; boxes and barrels for shipping.

—CHARLES H. STERNBERG, 1884

THE CANADIAN BONE RUSH

After years of collecting latest Cretaceous dinosaurs from the Hell Creek beds of Montana, Barnum Brown and Henry Fairfield Osborn of the American Museum of Natural History felt that they needed to find another underexploited prospect. As early as 1884, a surveying party led by Joseph B. Tyrrell (for whom the Tyrrell Museum in Alberta is named) had found dinosaur bones in the Red Deer River badlands of Alberta. Another party was led by Thomas C. Weston in 1888. In 1897 and 1898, the first major expedition to the region by the Canadian Geological Survey, led by pale-ontologist Lawrence Lambe (see chapter 23), produced more discoveries, some of which he shared with Osborn.

In 1909, a Canadian rancher from the area visited the American Museum, saw their exhibits, and told Osborn that he had dinosaur bones littering the canyon walls on his ranch. He even invited them to come collect on his ranch, giving the Americans the rancher's permission to work in Canada (although this was not official permission from the Canadian government). Brown and Osborn both agreed that there was tremendous potential in the Red Deer badlands, and also that these slightly older Cretaceous beds would help fill the gap between their Upper Jurassic Morrison Formation dinosaurs and the latest Cretaceous Hell Creek fossils.

In the spring of 1910, Brown and Peter Kaisen visited the region, collected a few bones, and saw that the rancher was right—the region was littered with dinosaur bones weathering out everywhere. They began to lay out plans for future field seasons. The biggest problem was that the dinosaur-bearing beds were eroded into badlands, which plunged steeply down from the settled plateau to the bottom of the Red Deer River valley, so it was hard work getting in and out. More to the point, there were few roads or towns in most of the wild, unsettled region, so getting a horse and wagon down to the fossils below in the canyon, then hauling them out, would be impossible. In a few places the roads reached low spots in the canyon, and ferries were used at those spots to get in and out. Brown took inspiration from the earlier Canadian surveys by Tyrrell, Weston, and Lambe, who had taken boats or rafts down the river to get the best look at the wild badlands without excessive riding or climbing.

By August 1910, Brown, Kaisen, and two other American Museum staffers were rafting down the river on a scow they build of local lumber (figure 22.1). They recovered several specimens of the theropod *Albertosaurus*, the ostrich dinosaur *Struthiomimus*, several ceratopsians, and lots of hadrosaur bones. They would take the scow a short distance each day, tie up to one bank or another, then prospect the badlands for bones weathering down from the cliffs. Once they found something, they would climb up to it, and if it was worth collecting, they excavated it and jacketed it, then lowered it carefully down with ropes and block and tackle until they could load it onto the raft. The conditions for rafting down the river plucking dinosaurs from the cliffs were not nearly as easy as some people think. The journals and photographs show the crew completely covered in thick clothing with mosquito nets draped over their hats because the biting bugs were so thick.

Brown and the crew had another productive season in summer 1911, but by then the Canadians were getting upset. All these incredible dinosaurs were

Figure 22.1 ▲

Barnum Brown, Peter Kaisen, and American Museum crew rafting down the Red Deer River on a scow in 1912, which provided a floating base for their exploration. The mosquito netting over their hats testifies to the bug problem. (Image #5418, courtesy of the American Museum of Natural History Library)

being collected by Americans and going to the museum in New York—none were remaining in Canada. As Colbert wrote, "cries were beginning to be heard in protest against the invasion of western Canada by Yankee bone hunters, who had been robbing the Dominion of its paleontological treasures." The Canadian Geological Survey needed experienced fossil collectors to bring more of those fossils to Canadian museums. Ironically, they hired American fossil collectors, the Sternberg family from Kansas, to do the job for Canada because no Canadian had more experience or know-how to collect dinosaurs.

THE STERNBERGS: THE FIRST PROFESSIONAL DINOSAUR HUNTERS

The Sternberg family is legendary in the lore of paleontology. The father, Charles Hazelius Sternberg (figure 22.2), and his three sons, George F.

Figure 22.2 ▲

The family patriarch, Charles Hazelius Sternberg, in his late career collecting Ice Age horses in tar pits near the oil fields of McKittrick, California, with a portrait of him inset in the upper left. (Courtesy of Wikimedia Commons)

Sternberg, Charles Mortram Sternberg, and Levi Sternberg, were the first significant professional dinosaur hunters in the United States. Cope, Marsh, and Leidy had collected for academic institutions, hiring men to work with them, but these collectors were not independent for most of their careers and were paid a salary by others. Charles H. Sternberg received his training from Cope, then he and his sons became collectors for hire, selling specimens to make a living. They were very conscientious, principled collectors, and most of their important specimens went to scientific collections and were available for study. Such specimens should not adorn the mansions of rich people as happens today with the commercial collectors and the black market in poached dinosaurs.

The father of the clan, Charles Hazelius Sternberg, was born in 1850 in Otsego County, New York, but grew up in Iowa, and then came to live on a ranch near Ellsworth, Kansas. He was a hardy Kansas farm boy like Barnum Brown. As a teenager, he made his first collections of the beautiful fossil leaves from the Lower Cretaceous Dakota Sandstone in central Kansas, which he eventually sold to the Smithsonian. Charles wrote, "At the age of seventeen, therefore, I made up my mind what part I should play in life, and determined that whatever it might cost me in privation, danger, and solitude, I would make it my business to collect facts from the crust of the earth." His pastor father was against this frivolous career choice, but Charles persisted, and in 1875 he enrolled at Kansas State Agricultural College in Manhattan, Kansas (now Kansas State University). There he studied under paleontologist Benjamin Mudge, who had collected for Marsh and still worked for him, taking groups of students out in the summer of 1876 to find more fossils that would go to Yale. But there was no spot in Mudge's crew for Sternberg, so he wrote to Cope. As Sternberg recounted in his autobiography:

> I put my soul into the letter I wrote him, for this was my last chance. I told him of my love for science, and of my earnest longing to enter the chalk of western Kansas and make a collection of its wonderful fossils, no matter what it might cost me in discomfort and danger. I said, however, that I was too poor to go at my own expense, and asked him to send me three hundred dollars to buy a team of ponies, a wagon, and a camp outfit, and to hire a cook and driver. I sent no recommendation from well-known men as to my honesty or executive ability, mentioning only my work in the Dakota Group. I was in a terrible state of suspense when I had despatched the letter, but fortunately the Professor responded promptly and when I opened the envelope, a draft for three hundred dollars fell at my feet. That letter bound me to Cope for four long years.

With that money, Sternberg got his supplies, a wagon, and a cook and driver and headed out to the Cretaceous chalk beds of the Smokey Hill

River in western Kansas. Basing his expedition out of Buffalo, Kansas (the only place with clean well water), every two weeks Sternberg and the Mudge crew worked in the same area, finding incredible marine reptiles such as mosasaurs and plesiosaurs. These reptiles used to swim in the shallow chalky seas that covered the mid-continent in the Western Interior Seaway, a shallow body of water that stretched from the Arctic to the Gulf of Mexico and cut North America in half. Later that summer, Cope came out to join him, and they went up to the Judith River badlands of Montana to find more dinosaurs (hostile tribes were wandering this area at that time as it was just after the Battle of the Little Bighorn that wiped out George Armstrong Custer and his men). In 1877, Sternberg and Cope collected not only in Kansas but also in the John Day beds of central Oregon, becoming one of the first to discover important fossils from these critical middle and late Oligocene deposits full of fossil mammals.

After four years of service to Cope, Charles then began to work independently with his three sons as his crew. In 1882, he began four years of collecting the Lower Permian beds of Texas, finding the finback protomammal *Dimetrodon* and the huge amphibian *Eryops* for various museums. From 1895 to 1897, he was collecting for Cope again. The family heard the news that Cope had died while they were in the field.

As the three sons grew more independent, each set up his own fossil collecting business. George remained in western Kansas, where most of his collection ended up at the Sternberg Museum in Hays, Kansas, which is still an amazing display of incredible fossils from the region. Charles Mortram Sternberg, the second son, eventually began working in Canada full time and was employed by the Geological Survey of Canada. Levi also ended up spending most his career in Canada, and he was employed by the Royal Ontario Museum in Toronto for most his career.

In the summer of 1908, the entire family was working in the Upper Cretaceous Lance Formation in eastern Wyoming, trying to find *Triceratops* for the British Museum. After weeks of finding very little, both Charles H. and Charles M. set out for Lusk, Wyoming, to restock supplies as they only had potatoes to eat. They left George and Levi in camp to continue hunting for fossils. As Ned Colbert wrote:

> Day after day hoping against hope we struggled bravely on. Every night the boys gave answer to my anxious inquiry, "What have you found? Nothing."

End of August, Sternberg finally discovered a fossil horn weathering out of the rock; subsequent excavation unearthed a 198 cm long Triceratops skull. Soon after, George, the oldest son, found bones sticking out of the rock while prospecting new territory with Levi, the youngest son. Levi discovered additional bones nearby, apparently belonging to the same skeleton. At this point of time, the group has departed 65 miles (105 km) from its base camp, and food reserves became scarce. Sternberg instructed George and Levi to carefully remove the sandstone above the skeleton, and himself set of for Lusk together with Charles Jr., in order to purchase new supplies and to initiate the shipping of the *Triceratops* skull to the British Museum. At the third day after their father's departure, George and Levi already had found out their find was an apparently complete skeleton lying on its back. When removing a large piece of sandstone from the chest region of the specimen, George discovered, to his surprise, a perfectly preserved skin impression. In 1930, George remembered: "Imagine the feeling that crept over me when I realized that here for the first time a skeleton of a dinosaur had been discovered wrapped in its skin. That was a sleepless night for me."

The Sternbergs had made one of their most remarkable discoveries, a mummified duckbill skeleton found in complete articulation, with the skin still preserved around the bones (figure 22.3). It showed all the limbs in natural articulation, and it was the first complete look at dinosaur skin (previously known only from fragments); it even preserved the soft skin between the fingers on the hands of the specimen, which held the fingers together in

Figure 22.3 ▲

The famous duckbill dinosaur mummy, collected by the Sternbergs and now in the American Museum of Natural History. (Image #330491, courtesy of the American Museum of Natural History Library)

a hoof-like arrangement. The specimen was first called by Leidy's invalid name *Trachodon*, but it is now considered to be *Edmontosaurus*. The Sternbergs had promised to find *Triceratops* for the British Museum, so they had a potential obligation to send this specimen across the Pond. Osborn got word of the amazing find, and sent Albert Thomson of the American Museum to purchase it. He appealed to the patriotism of the Sternberg family, promised to put it on permanent exhibit, and paid cash on the spot, so the specimen ended up in the American Museum where it is still on display. Meanwhile, William J. Holland of the Carnegie Museum had also arrived at Lusk to buy it, so Osborn upped his bid to make sure the Sternbergs sold it to him.

In the spring of 1912, the Sternbergs happened to be in Ottawa when the Canadian Geological Survey hired them to go west and get dinosaurs for Canada under the supervision of Lawrence Lambe. That summer they were working out of Drumheller (where the Tyrrell Museum is today), using rowboats as well as wagons to collect fossils, especially lots of duck-billed dinosaurs. Brown and the American Museum crew were in their third field season, working about 75 miles further south down river from the Sternbergs. That summer Brown and the American Museum crew collected an amazing complete articulated skeleton of *Corythosaurus* (figure 22.4), including skin impressions, along with the horned dinosaur *Monoclonius*. It was such a rich area that Brown worked it again in the summer of 1913, and so did the Sternbergs. By this point, they had copied Brown's idea of working from a large wooden scow, and both had purchased motorboats to pull and steer the raft around the river and make it easier to maneuver. The Sternbergs had their share of adventures too. As father Charles wrote in his biography:

> We were early astir, and Charlie hauled us in mid-stream. A strong east wind blew in our faces, it was disagreeable, because we had to lower our tents to the deck, and they acted as sails, and the power of the wind on them was [so] strong that the current and the five-horsepower motor would have driven us up stream. The choppy wave beat constantly against the front and side of our scow curling over the deck itself. The wind howled in the few cottonwoods along the shore and on the islands that we passed. . . . About nine o'clock we reached the fifth ferry below Drumheller. The ferry man had stretched a barbed wire across the river; Charlie saw it as he drove his motor under it and shouted to us, Jack rushed for the rear guiding oar and I for the front one, they

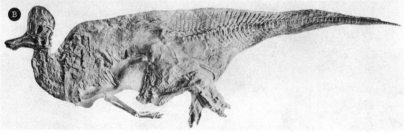

Figure 22.4 ▲

(A) Barnum Brown working on a complete articulated skeleton of *Corythosaurus* in the Red Deer badlands. (B) The specimen has all the bones in position, and the ossified tendons in the tail are visible; much of the specimen still has skin impressions. ([A] Image #131482, courtesy of the American Museum of Natural History Library; [B] courtesy of Wikimedia Commons)

were both stuck several feet up in the air, and if the wire had caught one, it would have swamped us. Jack had his back to the wire and when he released the oar and stood up, it caught his hat and threw it in to the river. If the wire had been six inches lower, or the river six inches higher, it would have cut his head off as easily and thrown it into the river.

The Sternbergs escaped this close call and moored not far from Brown's camp. Both groups worked the badlands around them steadily, and luckily there were enough fossils to keep them both busy for the duration. Unlike Cope and Marsh, who were hypercompetitive and destroyed each other in their feuds, the Brown American Museum crew and the Sternbergs maintained a friendly competition without creating any bad feelings. The Sternbergs found two amazing skeletons of *Corythosaurus*, several other duck-billed dinosaurs, the theropod *Gorgosaurus*, the horned dinosaur *Chasmosaurus* (with skin impressions), and the spiky *Styracosaurus*.

The Canadian bone rush between Brown and the Sternbergs (figure 22.5) continued for the next several years, even as World War I raged in Europe. Both groups were active in 1914 and 1915, after which Brown was finished in the region. In 1916, the elder Sternberg resigned from the Canadian Geological Survey, having gotten his fill of supervision by Lawrence Lambe, and he and Levi struck out on their own. Charles M. Sternberg and George Sternberg remained with the survey. The two family parties were now independent of each other and working in nearby areas for different employers:

`Figure 22.5` ▲

The Sternberg field party on July 21, 2017, in Little Sandhill Creek, Alberta, working for the University of Alberta. Standing (*left to right*), assistants Gustaf Lindblad and Ralph Rutherford, ornithologist Percy Tavenner, Charles M. Sternberg (*holding the flagstaff*), Tavenner's assistant C. H. Young, Charles H. Sternberg, and John A. Allan, a geologist from University of Alberta. Seated (*left to right*): an unnamed cook, another assistant, and Bruce McKee, both assistants for Charles H. Sternberg. Only Levi Sternberg is missing from the picture. (Courtesy of Wikimedia Commons)

Charles H. and Levi Sternberg for the British Museum and Charles M. and George for the Canadian Geological Survey.

The eldest and youngest Sternberg had great luck and found two complete *Corythosaurus* skeletons to ship to the British Museum. They carefully covered them in plaster jackets, crated them up, and shipped them by rail, where they were loaded on a British ship, the SS *Mount Temple*. The waters of the North Atlantic were a war zone at that time, and the Germans tried to stop ships out of North America from supplying the British forces. On December 6, 1916, the SS *Mount Temple* was attacked by a German warship and sank to the bottom of the Atlantic, along with its precious cargo of dinosaur skeletons. The Sternbergs and the British Museum lost the fossils forever, and the Sternbergs went to all that work and expense without getting paid, something commercial collectors cannot afford.

In 1917, Charles H. and Levi Sternberg found three fine dinosaurs, a duckbill that was sold to the San Diego Museum of Natural History and a *Gorgosaurus* and an ankylosaur bought by the American Museum. These specimens were first offered to the British Museum, but they were leery of buying them during a war when their precious cargo might once again be sunk to the bottom of the Atlantic.

After 1917, the Canadian bone rush came to an end. From then on, each of the Sternbergs began to pursue their own independent careers. Charles M. and Levi eventually worked for Canadian museums, and George returned to Kansas. Their father worked on other projects, such as collecting Ice Age mammals in the McKittrick tar pits near Taft, California (see figure 22.2). Each of the Sternberg brothers returned to the badlands of Alberta off and on over the years, collecting amazing fossils for Canadian museums. Today the Royal Tyrrell Museum of Paleontology in Drumheller is the center for all paleontological research in the Alberta badlands, with many more discoveries brought about by their active collecting program and incredible displays that make it one of the most amazing dinosaur museums in the world. The badlands are nearly all protected as part of Dinosaur Provincial Park, so no poachers are allowed.

CRESTED HADROSAURS

The first dinosaur found in North America, the duck-billed genus *Hadrosaurus*, was found in the New Jersey Cretaceous marl pits (see chapter 4).

However, the specimen was so fragmentary that it provided only a glimpse of what duckbills were like. As with the rest of American dinosaur paleontology, it wasn't until the discoveries of Cope and Marsh, and especially the large expeditions by the American Museum to the Rockies and by the Sternbergs and Barnum Brown to the Alberta badlands, that scientists began to get complete articulated skeletons of a wide diversity of hadrosaurs, with unusual features that beggared the imagination when they were found and described.

The early finds of duckbills in the late 1800s were mostly fragments (especially prisms from their densely packed battery of teeth), and they acquired names like *Trachodon* (Leidy, 1856) or *Diclonius* (Cope, 1876). Other names based on nondiagnostic fragments include Cope's isolated vertebra named *Hypsibema crassicauda* and another called *Cionodon arctatus*, Marsh's *Claosaurus annectens*, Cope's name *Ornithotarsus immanis* for their specimens of Leidy's New Jersey *Hadrosaurus*, and *Pteroplex grallipes* for a headless skeleton that might be *Corythosaurus*. Nearly all of these names have been abandoned because the fossils are too poor to tell to which of several possible duckbill genera they might belong. It was not until the Canadian bone rush in the early 1900s that good specimens with skulls attached were found in enough abundance to get a clearer picture of what kinds of duckbills were there.

Today the family Hadrosauridae (figure 22.6) is classified into two subfamilies: the Saurolophinae (such as *Saurolophus, Prosaurolophus, Gryposaurus, Kritosaurus, Edmontosaurus, Maiasaura*, and half a dozen other genera), which have solid crests or lack crests; and the Lambeosaurinae, which have hollow crests, plus a number of primitive duckbills that are related to these subfamilies. *Corythosaurus* (see figure 22.4) is one of best known and the most interesting of all the crested hadrosaurs, which come mostly from the Alberta badlands and were nearly all collected by either Brown's crew or the Sternbergs. More than 20 skulls are now known from this dinosaur, as well as several complete articulated skeletons, so we know it very well compared to most dinosaurs. It was one of the larger hadrosaurs, reaching 8.1–9.4 meters (27–31 feet) in length, and weighing about 3.1 to 3.8 metric tonnes (3.4–3.8 tons). Brown's first complete skeleton described in 1914 is the type of the genus, *Corythosaurus casuarius*. *Corythos* means "helmet" in Greek (it most resembles the ancient Greek Corinthian helmet), so it is the "helmet lizard"; the species name refers to its similarity to the crest

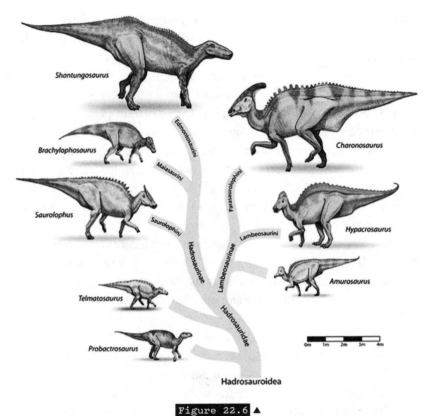

Shantungosaurus

Brachylophosaurus

Saurolophus

Telmatosaurus

Probactrosaurus

Charonosaurus

Hypacrosaurus

Amurosaurus

Edmontosaurini

Maiasaurini

Parasaurolophini

Saurolophini

Lambeosaurini

Hadrosaurinae

Lambeosaurinae

Hadrosauridae

Hadrosauroidea

0m   1m   2m   3m   4m

Figure 22.6 ▲
Family tree of the duckbill dinosaurs. (Redrawn from several sources)

of the cassowary bird of Australia. The crest rises up from the nasal region of the skull to above the eyes, and so do the internal nasal passages, which open above the eyes; there are pseudonares in the snout. There are large eye sockets, but the beak is rather narrow at the base and broadens out to the typical duck-like bill at the tip.

The type specimen and several other specimens are well preserved enough that skin impressions are also preserved (see figure 22.4B). Most of the body was covered by a mosaic of polygonal scales that were separated by rows of scales shaped like a shield. The entire back, hips, and tail are held rigid with a trusswork of ossified tendons, which held the tail out rigid when it walked (unlike the old restorations showing specimens dragged their tails).

As many as seven species were originally named for *Corythosaurus*, but in 1975 Peter Dodson studied all the skulls closely and concluded that the species differences in the crests were mostly due to the gender and age of the animal, with less prominent crests appearing in juvenile specimens and in presumed females. In addition, a number of specimens appear to be juvenile *Corythosaurus*, although it is hard to assign them to a species because their crests are not developed. Many of the old invalid names—*Procheneosaurus* and *Tetragonosaurus*, for example—are actually just juveniles of *Corythosaurus*. Their crests did not begin to develop until juveniles reached about half the body size of adults, whereas *Parasaurolophus* and other lambeosaurs developed their crests when they were only a quarter of adult body size.

The lifestyle of the hadrosaurs has long been a controversial topic. In the early days of study, it was presumed that they were slow, sluggish, lizard-like creatures like sauropods, dragging their tails and living in swamps or bodies of water, feeding on water plants. Thus they were restored with webbed feet.

Nearly all of these scenarios have since been falsified. The "webbing" on the feet of the mummified specimens turns out to be normal fleshy footpads that have flattened out in the mummification process; their feet were not webbed after all. They were active walkers and runners, with tails held straight out by a trusswork of tendons. Trackways show no signs of tail drag marks, and most of their trackways (and the environments of the rocks where they lived) are not from swamps but from dry floodplains with seasonal rains and droughts.

In addition, there is positive proof that they did not eat water plants at all. Several *Corythosaurus* specimens have been found with preserved stomach contents, and these included conifer needles, seeds, twigs, and fruits. Their relatively narrow beaks enabled them to be selective feeders, picking out delicate fruits and shoots among the vegetation. Their tough dental batteries were suited to grinding up very coarse vegetation, not the soft plants found in the water.

DINOSAUR TROMBONES

The argument over the purpose of dinosaur crests has been long and complicated. For a long time, their crests were interpreted as snorkels or as devices for holding more air in their heads as they dived for water plants. Others suggested that the crest might be an air trap to keep water out of the lungs or that it was attached to a mobile proboscis and aided in feeding or that it was a weapon for combat with other members of the species. Less outrageous were the ideas that the crest housed olfactory tissue for improving sense of smell, or possible were salt glands.

Most telling, none of their crests were built like snorkels. The nasal opening is near the bottom of the snout, not at the top where it would allow them to keep their head below water. Nor is the volume of airspace inside their nasal cavities enough to give them any advantage in holding their breath under water. In fact, if they had dived deeper than about 3 meters (10 feet), the water pressure on their bodies would have been too great to allow them to inflate their lungs. More to the point, good evidence shows that hadrosaurs were mostly land dwellers and spent very little time in the water.

After all the dust has settled, good evidence exists for only a few possibilities. The likeliest is that the crest served to advertise their sex and rank, let others in their herd know who they were, and to help in recognition of other unrelated hadrosaurs. This would explain why the crests grew as they did, and why there seems to be a significant difference between male and female crests, and between adults and juveniles.

The second line of evidence came from looking at the detailed internal anatomy of the nasal passages (figure 22.7). In *Corythosaurus* and *Lambeosaurus*, they are wide and roughly S-shaped. The long curving crest of *Parasaurolophus* is even more remarkable, curving back from the nasals to the tip of the crest in the back, then returning along the lower part of

Figure 22.7 ▲

Bones. *From left to right: Corythosaurus, Parasaurolophus, Lambeosaurus.* (Redrawn from several sources)

the crest until it reached the throat region. Such a long curving nasal passage linking the throat to the nasals makes the most sense if it was some kind of amplification chamber for sounds generated in the vocal cords. In fact, the U-shaped loop of the crest in *Parasaurolophus* resembles the long tubes of a trombone. David Weishampel of Johns Hopkins University tested this idea by building a scale model of the crest out of PVC tubes that was able to generate a deep hooting sound much like a trombone. As he pointed out, these low-frequency sounds (below 400 Hertz) are excellent for communication among similar species because low-frequency sounds travel much farther than high-frequency sounds. Not only do these sounds travel farther, but their direction is more difficult to locate. The hoots and honks of hadrosaurs could communicate that a predator has been spotted without the hadrosaur giving its own position away. Living elephants also use low-frequency sounds (most are below the range of human hearing) to communicate within their own herd and among herds that are long distances away.

DUCKBILLS AROUND THE WORLD

Hadrosaurs came to dominate Late Cretaceous herbivorous niches in most of the northern continents of North America and Eurasia, but they rarely reached any of the Gondwana continents. The earliest hadrosaurs include *Eotrachodon* from the Mooreville Chalk of Alabama, about 86 million years old, and *Hadrosaurus* itself from beds about 80 million years old in New Jersey (see chapter 4). Slightly more primitive is *Tethyshadros*, a small form from the Early Cretaceous in the region near Trieste between Italy and Croatia. The most primitive forms are *Telmatosaurus*, named by Baron Franz Nopcsa from specimens from the Late Cretaceous of Romania, and *Jintasaurus* from the Early Cretaceous of China (see figure 22.6). Thus it appears that hadrosaurs originated in Eurasia before spreading to North America by the middle part of the Cretaceous.

Almost 50 genera of hadrosaurs are now known, not counting all the outdated or invalid genera based on nondiagnostic scraps that have been mostly forgotten (like *Trachodon*). New genera are described almost every year now. Their family relationships have been controversial but are more or less worked out today (see figure 22.6). Their geographic distribution is also very interesting. Some, like *Corythosaurus*, is only known from Alberta despite the large number of specimens, and *Parasaurolophus* is found in both Alberta and New Mexico. But the closest relative of *Parasaurolophus*, called *Charonosaurus*, comes from the Amur region of Siberia. The close relative of *Corythosaurus*, the hypacrosaurs, largely come from the American west, but *Olorotitan* and *Amurosaurus* comes from the Amur region of Siberia, and *Blasisaurus* from Spain. Primitive lambeosaurines include *Aralosaurus* from the Aral Sea region of Russia, *Canardia* from France, *Jaxartosaurus* from Kazakstan and China, *Tsintaosaurus* from China, and *Pararhabdodon* from Spain.

Among saurolophines, *Edmontosaurus*, *Prosaurolophus*, and *Saurolophus* come from the American Rockies and Alberta, but their relative *Kerberosaurus* is from the Amur region of Siberia, and *Shantungosaurus* is from China. *Kritosaurus* is found in New Mexico and Texas, and the kritosaurs *Secernosaurus* and *Willinakaqe* are from the latest Cretaceous of Argentina but not further north. Yet the kritosaur *Wulagasaurus* comes from China.

Hadrosaurs seemed to be very wide ranging and mobile, switching back and forth between Eurasia and North America freely during the Late Cretaceous. Most remained in the northern Laurasian continents with the exception of the two kritosaurs, *Secernosaurus* and *Willinakaqe*, that show up in the latest Cretaceous Argentina, one of the few Cretaceous groups that reached Argentina from North America.

FOR FURTHER READING

Brinkman, Paul D. *The Second Jurassic Dinosaur Rush: Museums and Paleontology in America at the Turn of the Twentieth Century*. Chicago: University of Chicago Press, 2010.

Colbert, Edwin. *Men and Dinosaurs: The Search in the Field and in the Laboratory*. New York: Dutton, 1968.

Dingus, Lowell, and Mark Norell. *Barnum Brown: The Man Who Discovered Tyrannosaurus rex*. Berkeley: University of California Press, 2010.

Eberth, David, and David Evans, eds. *Hadrosaurs*. Bloomington: Indiana University Press, 2014.

Farlow, James, and M. K. Brett-Surman. *The Complete Dinosaur*. Bloomington: Indiana University Press, 1999.

Fastovsky, David, and David Weishampel. *Dinosaurs: A Concise Natural History*, 3rd ed. Cambridge: Cambridge University Press, 2016.

Holtz, Thomas R., Jr. *Dinosaurs: The Most Complete, Up-to-Date Encyclopedia for Dinosaur Lovers of All Ages.* New York: Random House, 2011.

Naish, Darren. *The Great Dinosaur Discoveries.* Berkeley: University of California Press, 2009.

Naish, Darren, and Paul M. Barrett. *Dinosaurs: How They Lived and Evolved.* Washington, D.C.: Smithsonian Books, 2016.

Rogers, Katherine. *A Dinosaur Dynasty: The Sternberg Fossil Hunters.* Missoula, Mont.: Mountain Press, 1991.

Spaulding, David A. E. *Dinosaur Hunters: Eccentric Amateurs and Obsessed Professionals.* Rocklin, Calif.: Prima, 1993.

Sternberg, Charles H. *Hunting Dinosaurs in the Badlands of the Red Deer River, Alberta.* Edmonton, Canada: NeWest Press, 1917.

——. *The Life of a Fossil Hunter.* New York: Holt, 1909.

Weishampel, David B., and Jack R. Horner. "Hadrosauridae." In *The Dinosauria*, 1st ed., ed. David B. Weishampel, Peter Dodson, and Halszka Osmólska, 534–561. Berkeley: University of California Press, 1990.

STEGOCERAS

The frontal and parietal have become so firmly coalesced that there is no remaining indication of a sutural contact between them. The development of these two bones into a very much enlarged dome-like mass is the most striking single feature of the skull. So extreme is the vaulting in this area that more than 6 inches of solid bone lie above the portion of the endocranial cavity that was occupied by the olfactory stock of the brain and 9 inches above the region of the cerebellum. In no other reptile has the skull roof become so thickened. . . . Most of the mass is composed of a compact fibrous structure made up of many small columns of bone which radiate out from a thin dense ventral zone and terminate in an equally dense outer zone. The outer surface is without ornamentation and presents numerous perforations leading to canals which penetrate into the fibrous zone.

—BARNUM BROWN AND ERICH M. SCHLAIKJER, "A STUDY OF THE TROÖDONT DINOSAURS WITH THE DESCRIPTION OF A NEW GENUS AND FOUR NEW SPECIES," 1943

LAWRENCE LAMBE AND THE "UNICORN DINOSAUR"

The early work in western Canada by Americans such as Barnum Brown and the Sternbergs has been discussed, but the "Father of Canadian Dinosaur Paleontology" was Lawrence Lambe. Born in 1863, he was just slightly younger than the pioneering generation of Cope and Marsh, but he was close in age to the "next generation" of paleontologists such as Osborn, Scott, Matthew, and Granger. Unlike Tyrrell and Weston, the first men into the Red Deer River badlands who were geologists doing a survey and mapping project, Lambe had some training in paleontology.

Born in Montreal, Lambe originally planned a military career, and he graduated from the Royal Military College in Kingston, Ontario, ready to become an Army officer. As he waited for his first commission, he was assigned to be an assistant construction engineer on the Canadian Pacific Railroad, which was blazing a trail across the Rocky Mountains and giving Canada its first transcontinental railroad. While working out there, he contracted typhoid fever, which was typically fatal in those days. He recovered but had to give up on an Army career because his health was impaired. He was already a talented artist, and he took a position in the Geological Survey of Canada in 1885, drawing fossils under the supervision of paleontologist J. F. Whiteaves. Soon he was not only doing drawings but also doing research in fossil corals, picking up the fundamentals of paleontology as he worked. Nearly all of his generation of paleontologists were self-taught because there were no college courses in paleontology. Most got a background in either anatomy or geology, then they learn paleontology by doing it.

Lambe had an opportunity to return to western Canada in 1897. The government was conducting test boring in northern Alberta, and they needed a geologist from the Geological Survey of Canada on the site. Soon he was running his own boat trip down the Red Deer River, following the tracks of Weston. He did a quick reconnaissance of the badlands over the course of a month, drifting down to the Saskatchewan Landing on the South Saskatchewan River. The next year he brought a crew with horse and wagon all the way from Medicine Hat to the site of Steveville, where he focused on collecting in the Berry Creek area (figure 23.1) because his camp was not down on the river level in the bottom of the canyon. In the 1898 and 1899 field seasons, he collected theropods, duckbills, and horned dinosaurs, as well as lots of turtles and crocodiles. However, he was just learning to collect in the field and did not know about the careful excavation techniques, the use of hardeners to hold specimens together, or plaster jackets to protect them during shipping that American paleontologists were pioneering, so most of what he brought home was pretty fragmentary.

After two field seasons, he had a lot of specimens that he needed to compare to other known fossils. He spent a number of weeks at the American Museum as a guest of Henry Fairfield Osborn, who took and interest in his work and offered to collaborate. Here Lambe learned of the latest in dinosaur paleontology, saw the best specimens in the United States, and quickly

Figure 23.1 ▲
Lawrence Lambe and his assistant in camp near the Red Deer River, Alberta, 1901.
(Courtesy of Wikimedia Commons)

made up for his lack of training in field collection and preparation methods. Lambe returned the favor by helping appoint Osborn the Honorary Vertebrate Palaeontologist for the Geological Survey of Canada. Lambe mounted one more collecting season in 1901 to explore the badlands downstream from Berry Creek, but his best collection was made in his last weeks on the Red Deer River.

In 1902, Osborn and Lambe published the first paleontological monograph on the specimens from the Red Deer badlands, *On Vertebrata of the Mid-Cretaceous of the North West Territory*. It was beautifully illustrated by Lambe himself and laid the foundation for all the future dinosaur studies in the region. In this volume, Osborn established that these fossils and the beds that produced them were definitely earlier in the Cretaceous than the specimens from the Hell Creek and Lance formations in Wyoming and Montana. Unfortunately, many of Lambe's specimens were fragmentary, and it was often hard to interpret what they were. Later work by Brown and the Sternbergs recovered much better fossils, which made Lambe's broken type specimens hard to use.

During the 1898 field season, Lambe had collected two strange lumps of solid bone that were very puzzling. There was a smooth dome on the top of the specimen and spongy bone all around the broken sides of the skull

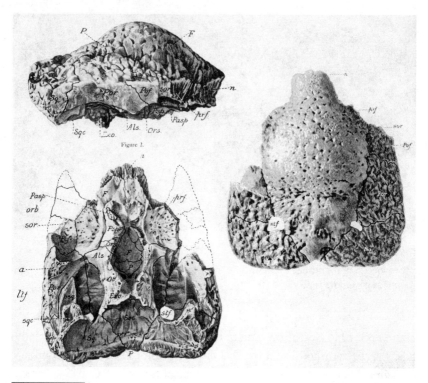

Figure 23.2 ▲

Lambe's original illustration of the bony dome of *Stegoceras*, showing the broken spongy bone around the edges and the top of the braincase exposed on the bottom of the bony dome. (From Osborn and Lambe, 1902)

(figure 23.2), but it was hard to tell what kind of animal had produced this fossil because no other parts were preserved. Nevertheless, in a short paper in 1902, he named the fossils *Stegoceras validum*. *Stegos* means "roof" and *ceras* means "horn" in Greek, so it was the "horn-roofed" dinosaur; the species name *validum* means "strong," in reference to the thickness of the skull roof. Lambe did the best he could with interpretations of the specimens, but it wasn't even clear what part of the skull the small domes had come from. He thought it came from the snout of a large dinosaur, like the short horns on the nose of *Triceratops*. In 1903, Baron Franz Nopcsa published a different interpretation, arguing that it was a blunt horn on the forehead of the dinosaur, possibly between the eyes. In 1903, Lambe endorsed Nopcsa's suggestion, calling it a "unicorn dinosaur."

Paleontologists were puzzled not only about what the complete skull of this fossil might have looked like but also whether this fossil was a weird kind of ceratopsian, or an even weirder kind of stegosaur. In 1907, the ceratopsian expert John Bell Hatcher (see chapter 25) weighed in on the fossils. He doubted whether the two specimens were from the same species, and whether they were even from dinosaurs at all. However, he did correctly suggest that they came from the top of the skull, from the frontal bones above the eyes to the parietal bones in the back of the skull. In 1918, Lambe added another dome fossil to the collection of *Stegoceras* and decided it was a member of the group that included stegosaurs and ankylosaurs, which he called the Psalisauridae (now called the Thyreophora).

This was his last paper on the mystery fossils; Lambe died at the relatively young age of 56 in 1919. The Father of Canadian Dinosaur Paleontology had named not only *Stegoceras* but also a slew of important Cretaceous dinosaurs, including the ceratopsians *Centrosaurus, Chasmosaurus, Styracosaurus,* and *Eoceratops;* the theropod *Gorgosaurus;* the ankylosaurs *Euoplocephalus* and *Panoplosaurus;* the duckbills *Gryposaurus* and *Edmontosaurus;* as well as crocodilians, such as *Leidysuchus,* and several turtles. After his death, Canadian paleontologist W. A. Parks named the crested hadrosaur *Lambeosaurus* in his honor.

The breakthrough in understanding *Stegoceras* came in 1924 when a complete, unbroken skull and jaws and partial skeleton was found by George F. Sternberg and acquired by the University of Alberta. It was described by Smithsonian dinosaur expert Charles W. Gilmore (figure 23.3). The complete skull (figure 23.3A) not only proved that Lambe's bony dome sat on top of the head from the eyes to the back of the skull but also showed many other interesting features. It had large eyes, roofed by a shelf of bone protruding above them and below the dome of the skull. This ridge of bone continued to the back of the skull, producing a short "frill" over the neck that is found in all pachycephalosaurs and ceratopsians. This is one of many features that demonstrate that these two groups are close relatives, now known as the Marginocephalia. The entire frill around the back and side of the skull was covered with bumps and ridges of bones. The skull went from broad in the back to a short, narrow, pointed snout, with large forward-facing nostrils. CAT scans of the large olfactory bulbs of the brain showed that these dinosaurs had a good sense of smell. The jaw contained small leaf-shaped teeth, with a gap between the front teeth and the cheek tooth row.

PLATE XV.

Figure 23.3 ▲

Stegoceras: (*A*) complete unbroken skull and jaws, found by George F. Sternberg for the University of Alberta and published in 1924 by C. W. Gilmore; (*B*) Gilmore's reconstruction of the entire animal, based on the skeletal elements found with the skull. (Courtesy of Wikimedia Commons)

The front teeth were conical nipping teeth with a small set of ridges and cusps on them, and the cheek teeth were triangular in cross section.

The teeth of *Stegoceras* caused confusion for a long time. The front teeth were very similar to the teeth that Leidy had named *Troodon* in 1856, so initially Lambe's *Stegoceras validus* was renamed *Troodon validus*. Later studies showed that the shape of the teeth was deceptive and that *Troodon* teeth better matched a group of small predatory dinosaurs. For a long time, all the early pachycephalosaurs were referred to Leidy's genus, which confuses people who only associate the name with the small predator.

The partial skeleton of *Stegoceras* (figure 23.3B) is like that of many other small bipedal ornithischians, with no specialized armor or other features typical of the advanced groups such as stegosaurs, ceratopsians, or duckbills. Gilmore reconstructed some of the fossils he found as belly ribs, or gastralia, but they are now known to be the ossified trusswork of intermuscular bones in the tail. Like most other dinosaurs, *Stegoceras* held its tail straight out in the back. Because *Stegoceras* is known from a partial skeleton, it is one of the most completely known pachycephalosaurs because most others are known only from skulls. *Stegoceras* was about 2.0–2.5 meters (6.6–8.2 feet) long counting the long tail and weighed about 10–40 kilograms (22–88 pounds), about size of a goat. The front limbs were quite small, so the dinosaur was completely bipedal, unlike many larger ornithischians. The animal must have run in a bird-like fashion, with its tail stuck straight out behind it and the body balanced horizontally over the long hind limbs.

"THICK-HEADED LIZARD"

Stegoceras remained the only known pachycephalosaur until 1931 when a much larger skull with a thicker dome was found in the uppermost Cretaceous Lance Formation of Wyoming. Charles Gilmore described this larger genus as *Troodon wyomingensis*, thinking it was a larger version of *Stegoceras* (then mistakenly called *Troodon*). In 1943, Barnum Brown of the American Museum and Erich M. Schlaikjer from Harvard were collecting in the Hell Creek beds of the Ekalaka Hills in the southeast corner of Montana and found a beautiful skull of Gilmore's "*Troodon wyomingensis*" (figure 23.4). They gave it the name *Pachycephalosaurus*, from the Greek words *pachy* meaning "thick," *cephalos* meaning "head," plus *sauros* for "lizard." In 1945, Charles M. Sternberg realized that Leidy's "*Troodon*" was based on teeth

Figure 23.4 ▲

(A) The beautiful complete skull of the large genus *Pachycephalosaurus*, now on display at the American Museum of Natural History. (B) The skeleton of the *Pachycephalosaurus* nicknamed "Sandy," now on display in the National Museum of Science and Nature in Tokyo. (Courtesy of Wikimedia Commons)

that mostly belonged to a small theropod and removed that name from the pachycephalosaurs. Since they were no longer in the Troodontidae, they were placed in a new family, the Pachycephalosauridae (although properly they should have been named the Stegoceratidae after the earliest valid genus in the group).

Although no skeleton was known from the original specimens of *Pachy-cephalosaurus*, the skull is very impressive. The bulging dome on the skull was 25 centimeters (10 inches) thick and cushioned a tiny brain (the brain cavity is also preserved). All around the rear and sides of the dome are bony knobs, and there are bony spikes on the nose and snout. Like *Stegoceras*, it had large eyes covered by a rim of bone above the eye sockets. The snout was short, with a pointed beak. *Pachycephalosaurus* had tiny leaf-shaped teeth similar to those of others in the group.

The rest of the skeleton was unknown from the original skull fossils, but a partial skeleton has since been found in the Hell Creek Formation in South Dakota. Found by commercial collectors, it was sold to the National Museum of Science and Nature in Tokyo, where it was described by Taka Tsuihiji. Nicknamed "Sandy," it was about 1.5 meters (5 feet) tall and 3 meters (10 feet) long. Its partial skull includes most of the back, side, and snout region, but not the dome. In addition, the fossil includes the hind limbs and hips, plus some neck and back vertebrae and ribs (figure 23.4B). Based on this specimen and other related pachycephalosaurs known from skeletons, *Pachycephalosaurus* was a medium-sized bipedal animal that weighed about 450 kilograms (990 pounds), had a fairly short, thick, S-shaped neck, very short forelimbs, a bulky body with a large gut for fermenting and digesting plants, long thick hind legs, and a heavy tail held out straight behind it by ossified tendons.

After 1943, the only known genera in the group for decades were *Stegoceras* and *Pachycephalosaurus*. Then, in the 1970s, the pace of discovery accelerated dramatically, and now more than two dozen genera are known. The first major discoveries came from the Polish-Mongolian expeditions of the 1960s and 1970s (see chapter 16). Polish paleontologists Teresa Maryańska and Halszka Osmólska published a major review of their finds in Mongolia. They propose a suborder Pachycephalosauria, and they added several new genera based on remarkable fossils found during the years of the expeditions. One of these was *Homalocephale*, which had a flat skull roof rather than the dome seen in other genera. Another genus was *Prenocephale*, which had a bulging dome and ridges of bone sloping down to a narrow snout, very much like *Stegoceras* but larger. Based on the proportions of the preserved bones, it was about 2.4 meters (8 feet) long. Some scientists argue that *Homalocephale* is just a juvenile version of *Prenocephale* that has not yet acquired the bulging dome-shaped forehead, but new finds have discredited this idea. A third genus is *Tylocephale*, which is known from a fragmentary skull that doesn't preserve the dome or snout, represents an animal that may have been 1.4 meters (4.6 feet) long. Finally, Mongolian paleontologist Antangerel Perle and Maryańska and Osmólska published the genus *Goyocephale* in 1982. It consists of a partial skull and parts of the skeleton. The skull is much narrower than other genera, with a relatively flat top and no large dome, and it has lots of bony bumps on the back margin and all over the surface of the top of the skull roof.

I cannot mention all the other genera that have been described since 1974 here, but they have come from many different Cretaceous localities in North America and Asia. The most important to mention is the very primitive *Wannanosaurus*, from the early Late Cretaceous of China. It consists of only a partial skeleton, but it is more primitive than that found in any other pachycephalosaur. More important, the flat skull roof has almost no dome, and it still has the large openings in the sides and back of the skull found in most dinosaurs. Advanced pachycephalosaurs developed their thick bony domes and closed up these holes in the skull. Not only is it older and more primitive than the rest of the pachycephalosaurs, but it is one of the smallest dinosaurs known, with an estimated body length of only 60 centimeters (2 feet). This transitional fossil shows how the weird pachycephalosaurs evolved from much more primitive ancestors.

DINOSAUR RAMS?

The first people to describe *Stegoceras* and *Pachycephalosaurus* did not speculate much about how they behaved or the way the thick dome of bone was used. In 1955, Edwin Colbert was the first to suggest that the pachycephalosaurs were like dinosaurian rams, head butting with their heavily armored skulls. In addition to the solid helmet of bone, the shape of the neck suggested that they had strong neck muscles, with an S-shaped curve to absorb the shock of each blow. Others have suggested that they used their bony helmets to head butt the flanks of other members of their herd, giving a less lethal glancing blow. They had wide trunks and bellies, which would protect the internal organs from a head blow. One genus, *Stygimoloch*, had horns on the side of its face, which would have been even more effective in flank butting.

In 2004, Mark Goodwin and Jack Horner argued that pachycephalosaurs could not have endured direct head butting because the bone structure allegedly could not have absorbed such stresses. In their opinion, the dome was for species recognition only. This has been disputed by numerous analyses since then, which established that the spongy bone of the skull supporting the solid bone of the dome is indeed capable of absorbing head-to-head collisions. Their bone structure is much like that of rams and muskoxen, which also engage in head-to-head impacts. In addition, the domes do not appear to differ much among adults. Such differences between males and females would be expected if they were use for species recognition or mate recognition.

The battering ram model was further supported by a 2013 study of the pathologies of the specimens that had been injured. About 22 percent of the domes had damage or lesions consistent with osteomyelitis, a bone infection caused by penetrative trauma. The flat-skulled pachycephalosaurs show no such rate of injury, suggesting that they did not engage in head-to-head combat. This would make sense if they were females or juveniles who did not have to compete to become masters of their herds.

HORNY YOUNG DINOSAURS?

The variation in dome size also may be due to growth through time. There are enough different *Stegoceras* skulls known now to construct a growth series from juveniles to adults. Not only do the skulls get larger, but their domes go from flat to a slight bump to a large dome structure over the course of growth (figure 23.5).

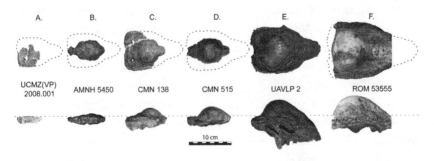

Figure 23.5 ▲

Different specimens of *Stegoceras* interpreted as a growth series, with the small flat skulls considered to be juveniles, which grew into large domed forms as adults. (Courtesy of Wikimedia Commons)

The weird horned genus *Stygimoloch spinifer* was a great puzzle when it was first described. It got its name from the River Styx, the river at the entrance to Hades; Moloch was a Canaanite god who demanded child sacrifice; and "spinifer" means "spiny." Then there was the even smaller horned genus *Dracorex hogwartsi* (dragon king of Hogwarts), which was described in 2006 as another small, horned adult pachycephalosaur. But several paleontologists have looked closely at these fossils and concluded that they were probably juveniles of the larger forms like *Pachycephalosaurus* that had not yet developed their dome but had relatively large spines (figure 23.6). The horns of all pachycephalosaurs seem to show a lot of developmental plasticity, with their shape and number and orientation changing within individuals of the same population as well as during their presumed growth. All of these dinosaurs came from the same Upper Cretaceous beds of Montana and Wyoming (except *Stegoceras* from Alberta, which has its own growth series). It makes sense that they might all represent growth stages of the same few genera that lived in that place and time. Of course, it's impossible to know for sure from such a small sample of specimens, most of them fragmentary, which have no living descendants; so their behavior, growth features, and sexual differences cannot yet be determined.

PACHYCEPHALOSAURS THROUGH SPACE AND TIME

Pachycephalosaurs, like many groups we have just seen (especially tyrannosaurs, ceratopsians, most duckbills, and ankylosaurs), were a strictly Laurasian group from the Late Cretaceous and were never found anywhere else. (Specimens of *Majungatholus* from Madagascar and *Yaverlandia* from England are no longer considered pachycephalosaurs.) The oldest member of the group is *Wannanosaurus* from China, and it was the starting point of the first wave of migration in the early Campanian from Asia to North America, including such descendants as *Stygimoloch, Stegoceras, Tylocephale, Prenocephale,* and *Pachycephalosaurus.* The second migration from Asia occurred in the late Campanian, producing the lineage that led to *Prenocephale* and *Tylocephale.*

But where did pachycephalosaurs and other marginocephalans come from? In 2004, the most amazing fossil in this sequence was discovered with the description of *Yinlong* (figure 23.7) from much earlier Upper Jurassic

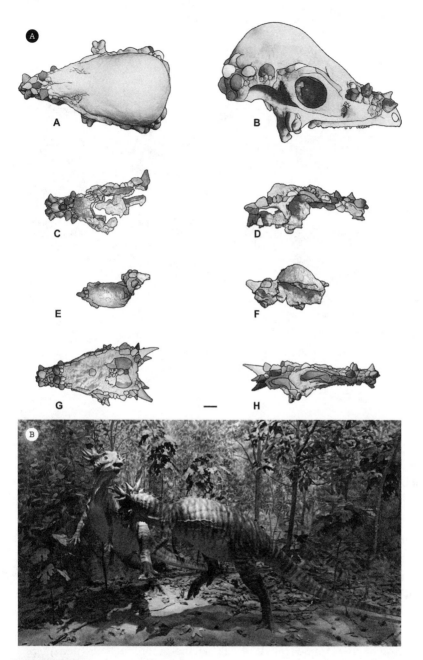

(A) The small, spiky specimens of *Stygimoloch* and *Dracorex* were first described as different genera, but now some paleontologists view them as juvenile stages of *Pachycephalosaurus*. (B) Reconstruction of battling *Stygimoloch* in head-butting poses. ([A] Courtesy of Wikimedia Commons; [B] photograph by the author)

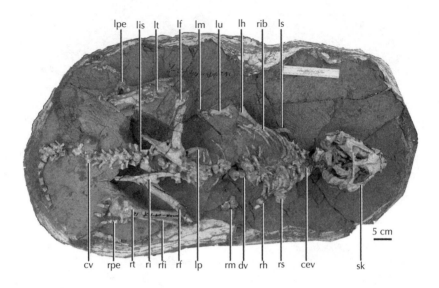

Figure 23.7 ▲

The complete skeleton of *Yinlong*, a fossil from the Late Jurassic of China that appears to be the common ancestor of pachycephalosaurs and ceratopsians. It has the frill on the back of the head like both groups, and features such as the rostral bone are found in ceratopsians. (Courtesy of J. Clark)

beds of China. Its name means "hidden dragon" in Mandarin, a reference to the popular movie *Crouching Tiger, Hidden Dragon*, part of which was filmed close to the locality where the fossil was found. *Yinlong* consists of a beautifully preserved skeleton of a bipedal dinosaur not too different in proportions from the primitive ceratopsian *Psittacosaurus* (see chapter 24). *Yinlong* has the rostral bone, a feature unique to ceratopsians, in its upper beak. However, its skull roof has a unique configuration of bones found in the pachycephalosaurs, which are famous for having a thick dome of bone in their skulls protecting their tiny brains. Paleontologists have long thought that ceratopsians and pachycephalosaurs are closest relatives, based on the fact that they both have a frill of bone around the back of the skull (hence their name, "Marginocephalia"). Like all marginocephalians (pachycephalosaurs plus ceratopsians), there is a frill in the back of the skull of *Yinlong*. But *Yinlong* shows features of both ceratopsians and pachycephalosaurs before their lineage split into the two families that every kid recognizes. Thus it forms a transition between more primitive ornithischians in the

Jurassic and the most primitive pachycephalosaurs such as *Wannanosaurus* and the earliest ceratopsians such as *Psittacosaurus*.

FOR FURTHER READING

Brown, Barnum, and Erich M. Schlaikjer. "A Study of the Troödont Dinosaurs with the Description of a New Genus and Four New Species." *Bulletin of the American Museum of Natural History* 82, no. 5 (1943): 115–150.

Colbert, Edwin. *Men and Dinosaurs: The Search in the Field and in the Laboratory.* New York: Dutton, 1968.

Farlow, James, and M. K. Brett-Surman. *The Complete Dinosaur.* Bloomington: Indiana University Press, 1999.

Fastovsky, David, and David Weishampel. *Dinosaurs: A Concise Natural History,* 3rd ed. Cambridge: Cambridge University Press, 2016.

Holtz, Thomas R., Jr. *Dinosaurs: The Most Complete, Up-to-Date Encyclopedia for Dinosaur Lovers of All Ages.* New York: Random House, 2011.

Kielan-Jaworowska, Zofia. *Hunting for Dinosaurs.* Cambridge, Mass.: MIT Press, 1969.

——. *In Pursuit of Early Mammals.* Bloomington: Indiana University Press, 2013.

Maryańska, Teresa, Ralph E. Chapman, and David B. Weishampel. "Pachycephalosauria." In *The Dinosauria,* 2nd ed., ed. David B. Weishampel, Peter Dodson, and Halszka Osmólska, 464–477. Berkeley: University of California Press, 2004.

Naish, Darren. *The Great Dinosaur Discoveries.* Berkeley: University of California Press, 2009.

Naish, Darren, and Paul M. Barrett. *Dinosaurs: How They Lived and Evolved.* Washington, D.C.: Smithsonian Books, 2016.

Russell, Loris Shano. *Dinosaur Hunting in Western Canada.* Toronto, Canada: Life Sciences, Royal Ontario Museum, 1966.

PROTOCERATOPS

This is one of the most picturesque spots that I have ever seen. From our tents, we looked down into a vast pink basin, studded with giant buttes like strange beasts, carved from sandstone. One of them we named the "dinosaur" for it resembled a huge Brontosaurus sitting on its haunches. There appear to be medieval castles with spires and turrets, brick-red in the evening light, colossal gateways, walls and ramparts. Caverns run deep into the rock and a labyrinth of ravines and gorges studded with fossil bones make a paradise for the palaeontologist. One great sculptured wall we named the "Flaming Cliffs" for when seen in early morning or late afternoon sunlight it seemed to be a mass of glowing fire. On the floor of the basin, to the north, is an area of old dead sand-dunes covered by a miniature forest of stunted trees, which we first supposed were tamarisk. . . . To the south, the rolling plain sweeps back to the Gurbun Saikhan, the last of the prominent uplifts of the Altai system.

—ROY CHAPMAN ANDREWS, *THE NEW CONQUEST OF CENTRAL ASIA*

THE MONGOLIAN GRIFFIN

Most creatures of ancient mythology are clearly not real or even that similar to real animals. There is some evidence that legends about the Indian rhinoceros became distorted into the myth of the unicorn, with its horse-like body and straight horn sticking out of its forehead, as the Indian rhino has a single horn on its skull. Many other mythical beasts have some combination of real animals in them, although they don't make much biological sense and are not based on anything that actually lived.

One of these mythical creatures is known as the griffin, a beast with the body, tail, and back legs of a lion; the head and wings of an eagle; and eagle talons for its front feet (figure 24.1A). It was a combination of the "king of the beasts" (the lion) and the "king of the birds" (the eagle), so it was doubly royal. The earliest representations of the griffin can be seen in ancient Iranian and ancient Egyptian art dating to 3000 BCE. There are griffin sculptures and images in many of the Bronze Age cultures of the Middle East, especially in Mesopotamia and Persia, as well as griffins in the throne room of the Minoan Palace of Knossos in Crete. The image remained common in Greek and Roman mythology and art, and was carried on through the Middle Ages as part of the fantastic bestiary in which ancient and medieval cultures believed. They appear in Dante's *Divine Comedy*, Milton's *Paradise Lost*, and many other works of literature, right through the Harry Potter universe, where they are the namesake of Gryffindor ("golden griffins" in French) House in Hogwarts. Griffins also were incorporated into architecture in the Middle Ages and later, so they are common images throughout buildings in Europe.

The late Roman author Flavius Philostratus wrote this about griffins around 244 CE:

> As to the gold which the griffins dig up, there are rocks which are spotted with drops of gold as with sparks, which this creature can quarry because of the strength of its beak. "For these animals do exist in India" he said, "and are held in veneration as being sacred to the Sun; and the Indian artists, when they represent the Sun, yoke four of them abreast to draw the images; and in size and strength they resemble lions, but having this advantage over them that they have wings, they will attack them, and they get the better of elephants and of dragons. But they have no great power of flying, not more than have birds of short flight; for they are not winged as is proper with birds, but the palms of their feet are webbed with red membranes, such that they are able to revolve them, and make a flight and fight in the air; and the tiger alone is beyond their powers of attack, because in swiftness it rivals the winds."

The griffin is well established in mythology, but was there any basis for it in reality? Folklorist and science historian Adrienne Mayor of Stanford University has argued that the griffin myth has been amplified by ancient accounts of *Protoceratops*. Reading ancient stories, she found that the *I Ching* (around 1000 BCE) talks about the "dragons encountered in the fields." Even today fossil bones from Mongolia and other parts of China are called "dragon bones" and sold in Chinese "medicine" as a ground-up powder that can cure most anything. Later accounts from the thirteenth century in

Figure 24.1 ▲

(A) Bronze figure of a griffin from the Roman period, between 50 and 270 CE. (B) *Protoceratops* skull eroding out of the Mongolian badlands, which emphasizes the beak and eyes; the arch of the frill suggests wings. (Courtesy of Wikimedia Commons)

China described how travelers feared the "field of white bones" and "heaps of bright white stones like bones" in the areas near the Flaming Cliffs.

About 600 BCE there are extensive accounts and art objects from the Scythian culture, a nomadic culture that originated in what is now Iran. The Scythians were among the first cultures to master mounted warfare, and they were excellent horsemen. They also had trade routes that ran all the way to China (long before the famous "Silk Road" routes followed by Marco Polo). At their peak, their empire stretched from the Carpathian Mountains in Poland to the edge of China, so they traveled through modern Mongolia on a regular basis, bringing gold mined from the Altai, Tien Shan, and Hindu Kush mountains. Their legends say that griffins guarded these gold deposits.

From the American Museum accounts and photos, Mayor noticed that when *Protoceratops* skulls were first exposed to weathering, they have the prominent beak, the "eagle eyes," and a large arch of bone behind the head that could be interpreted as "wings" when only partially exposed (figure 24.1B). Some of the comparisons to known specimens are very striking. Given the abundant mention of griffins in Scythian lore, it seems possible that Scythian traders were influenced by seeing the ghostly bones of a *Protoceratops* weathering out of the rock. Griffins also laid their nests of eggs in the ground, and the abundance of nests full of eggs in the Dja-dokhta Formation would have fed this part of the legend.

Greek accounts of griffins date to about 675 BCE, which was when they first made contact with Scythian nomads, so it seems likely that they shaped the way Greeks and later cultures viewed the griffin. This does not address why even older cultures have similar concepts of the griffin; there is no clear evidence that pre-Scythian trade routes extended to Mongolia from the Middle East. It is possible that *Protoceratops* influenced the post-675 BCE versions of the griffin myth, but some version of that myth was already established as early as 3000 BCE.

TO THE FLAMING CLIFFS

The American Museum Central Asiatic Expeditions of the 1920s to Mongolia and China were led by Roy Chapman Andrews and paleontologist Walter Granger (see chapter 16). After spending time in the Mongolian capital Ulaanbaatar, on September 1, 1922, they headed southwest with their entire party in Dodge touring cars, followed by the slower caravan of camels bringing up the supplies. Eventually they reached a place called Shabarakh Usu (now known as Bayn Dzak), where they spotted the amazing sculpted Cretaceous sandstones of the Djadokhta Formation that they named the "Flaming Cliffs" (figure 24.2). As Andrews described it in *The New Conquest of Central Asia*:

> My car was far in advance of the others and I asked Shackelford [the official photographer of the expedition] to stop the fleet while I ran over to the yurts for a conference with the inmates. During the time that I was gone he wandered off a few hundred yards to inspect some peculiar blocks of earth which

Figure 24.2 ▲

The American Museum Central Asiatic Expedition in 1923. *Middle row, left to right:* Walter Granger, Henry Fairfield Osborn (*in pith helmet*), Roy Chapman Andrews, geologist Frederick K. Morris, and Peter Kaisen, an American Museum preparator who had worked with Barnum Brown in Canada. *Back row, third from left:* Albert Johnson, another preparator who worked with Brown; *third from right,* George Olsen, an American Museum preparator. In addition, there are American cooks and mechanics and numerous Mongolian crew members. (Image #251731, courtesy of the American Museum of Natural History Library).

attracted his attention north of the trail. From them he walked a little farther and soon found that he was standing on the edge of a vast basin, looking down upon a chaos of ravines and gullies cut deep into the red sandstone. He made his way down the steep slope with the thought that he would spend ten minutes searching for fossils, and if none were found, return to the trail. Almost as though led by an invisible hand he walked straight to a small pinnacle of rock on top of which rested a white fossil bone. Below it the soft sandstone had weathered away, leaving it balanced ready to be plucked off. Shackelford picked the "fruit" and returned to the cars, just as I arrived. Granger examined the specimen with keen interest. It was a skull, obviously reptilian, but unlike any with which he was familiar. All of us were puzzled. Granger and Gregory named it *Protoceratops andrewsi* in 1923. Shackelford reported that he had seen other bones, and it was evident that we must investigate the deposit. . . . Quantities of white bone were exposed in the red sandstone, and at dark we had a sizable collection. However, Shackelford's skull still remains the best specimen, with the possible exception of the skull and jaws of a small reptile found by Berkey [one of the expedition geologists]. Granger brought in, among other things, a part of an eggshell which we supposed was that of a fossil bird, but which subsequently was recognized as dinosaurian. It was evident that the formation was Cretaceous and very rich in fossils, but at that time we could do no more than mark it as one of the localities for future work. We could hardly suspect that we should later consider it the most important deposit in Asia, if not the entire world.

As Andrews mentions, the very first reconnaissance visit to the Flaming Cliffs produced both *Protoceratops* and eggshell fragments. During the expedition of 1923, they spent much more time at the locality and collected dozens of skulls and skeletons of *Protoceratops*, ranging from juveniles to adults. Walter Granger and William King Gregory of the American Museum wrote up the first descriptions of this dinosaur in 1923. The name *Protoceratops* means "first horned face" in Greek, and the American Museum scientists could tell from the skull that it was "the long-sought ancestor of *Triceratops*" and the other horned dinosaurs, complete with the large frill over the back of the skull, but without any horns on the nose, face, or frill like more advanced ceratopsians.

Even more important, *Protoceratops* fossils were so abundant that the American Museum retrieved dozens of skulls and skeletons, some of which they swapped with other museums for fossils they didn't have and needed. The fossils were so well preserved that even the delicate ring of bones around the eyeball (sclerotic ring) was fossilized. In addition, they collected a complete growth series from the smallest juveniles to large adults, one of the very few dinosaurs known from more than a handful of juvenile specimens. These were mentioned by Granger and Gregory in 1923 and formed a striking exhibit in the American Museum that is still there today (figure 24.3). In 1976, Peter Dodson restudied these juvenile specimens and applied modern methods of analyzing growth and development to them. Over the years, more and more *Protoceratops* have been found by other expeditions that have visited the Flaming Cliffs, including the Soviet expeditions of the 1950s, the Polish-Mongolian expeditions of the 1960s and 1970s, and the return of the American Museum expeditions collaborating with the Mongolian Academy of Sciences in the 1990s.

Protoceratops was not a big dinosaur, reaching about 1.8 meters (6 feet) long and 0.6 meters (2 feet) high at the shoulder (figure 24.4). It may have weighed about 180 kilograms (400 pounds), about the size of a large sheep. Like other ceratopsians, *Protoceratops* had a sharp beak on its snout made of the rostral bone, a feature unique to the ceratopsians and some of their ancestors such as *Yinlong*. The large frill over the back of the head and neck were perforated with large holes or "windows" (fenestrae in anatomical terms), which made them lighter and may have added attachment points for jaw muscles. The development of this frill from a tiny edge of bone to the broad flaring structure of adults is one of the most striking features of

Figure 24.3 ▲

Growth series of *Protoceratops* skulls on display at the American Museum. (Photograph by the author)

Figure 24.4 ▲

(A) George Olsen and Andrews in Mongolia in 1925 excavating a nest of dinosaur eggs. (B) Two mounted skeletons of *Protoceratops*, with a nest of eggs once thought to belong to this dinosaur, on display at the American Museum. ([A] Image #410760, [B] image #324205, courtesy of the American Museum of Natural History Library)

their development. Initially it was thought that the frill was mainly to protect the neck, but more recent analysis has argued that it wasn't very effective as neck protection. It is more likely that it served as a display structure to communicate with its own herd and advertise its age and status. The fact that the frill grew dramatically from juveniles to adults is consistent with its ability to advertise the age and strength of adults, comparable to the way larger horns or antlers in adult antelopes and deer show who is boss.

There is good evidence that *Protoceratops* had a powerful bite and was able to chew tough vegetation, especially with its dental battery packed with small, wedge-shaped teeth (a feature distinctive to ceratopsians as well). It had large eyes, consistent with the idea that it could see well in the dark and may have been nocturnal. We know that it coped with another night dweller, the turkey-sized dromaeosaur *Velociraptor*, because of the famous fossil of a *Protoceratops* with a *Velociraptor* attacking it, then both dinosaurs dying as they fought, and buried in sand (see figure 17.4B).

DINOSAUR EGGS—AND THE SLANDEROUS NAME

In the 1922 season, the American Museum found a few eggshell fragments but left the Flaming Gorge after only a brief visit. When they came back in 1923, they spent a long time collecting there, and they found not only lots of *Protoceratops* but several nests full of eggs (figure 24.4). Dinosaur eggs from the Cretaceous of the Provence region in France had been collected and described by a Catholic priest, Father Jean-Jacques Pouech, in 1859, but he thought they were just fossil bird eggs. They were not recognized as dinosaurian until much later.

Andrews and the American Museum staff did not know about the Provence discovery, and they were justifiably excited to find intact nests full of oblong eggs, all in the position in which they had been laid.

On July 13, George Olsen [American Museum preparator and field collector] reported at tiffin that he had found some fossil eggs. Inasmuch as the deposit was obviously Cretaceous and too early for large birds, we did not take his story very seriously. We felt quite certain that his so-called eggs would prove to be sandstone concretions or some other geological phenomena. Nevertheless, we were all curious enough to go with him to inspect his find. We saw a small sandstone ledge, beside which were lying three eggs partly broken.

The brown striated shell was so egglike that there could be no mistake. Granger finally said, "No dinosaur eggs ever have been found, but the reptiles probably did lay eggs. These must be dinosaur eggs. They can't be anything else." The prospect was thrilling but we would not let ourselves think of it too seriously and continued to criticize the supposition from every possible standpoint. But finally we had to admit that "eggs are eggs" and that we could make them out to be nothing else. It was evident that dinosaurs did lay eggs and that we had discovered the first specimens known to science. The eggs which had broken out of the sandstone block are eight inches long by seven inches in circumference. They are red-brown in color and are rather more elongate and flattened than those of modern reptiles; they differ greatly in shape from the eggs of any known birds, living or fossil. The outer surface of the shell is striated, with broken, longitudinal rugosities, but the inner surface is smooth; the shell is about one millimeter thick. (Andrews, *The New Conquest of Central Asia*, 205)

Granger collected several nests of eggs and brought them home to the American Museum, where Osborn made the biggest publicity splash he could to promote the Museum and its discoveries and to help raise more funds for future expeditions. In the winter of 1923–1924, the American Museum held an auction for one of the eggs as a fundraiser. The highest bidder, Col. Austin Colgate, donated it to the Colgate College Museum (where it still resides). Unfortunately, the Mongolians later saw this as a sign that the American Museum had plundered their country and planned to sell their finds at a profit. After the 1930 expedition, the Mongolian government refused to let the American Museum return to the Gobi Desert, and the Museum did not get another chance to return until the 1990s when the political situation had changed many times.

The eggs themselves led to much further research on dinosaur eggs, and many new discoveries were made, as well as recognition for what the French dinosaur eggs really were. Dinosaur eggs have been found in many different places, such as Jack Horner's famous "Egg Mountain" localities in Montana or Luis Chiappe's Auca Mahuevo sites in Argentina. The study of dinosaur eggs is now very well established and has its own specialists.

The nests that had been recovered by the American Museum were put on display and are still on exhibit (figure 24.4B). Because *Protoceratops* was the most abundant dinosaur in the Djadokhta Formation, Osborn and

Granger assumed that it must have been the egg layer. But in the process of excavating one of the nests, the American Museum expedition made another discovery.

> In the ledge beside which the eggs lay we could see many bits of shell imbedded in the rock and it was obvious that other specimens might be enclosed in the sandstone matrix. When Olsen brushed away the loose sediment on top of the ledge, he exposed the fragmentary skeleton of a small dinosaur. It proved to represent a toothless type, and Professor Osborn subsequently named it *Oviraptor philoceratops*. In referring to its habits he remarks, "The generic and specific names of this animal, *Oviraptor*, signifying the 'egg-seizer', *philoceratops*, signifying 'fondness for ceratopsian eggs', may entirely mislead us as to its feeding habits and bely its character. The names are given because the type skull was found lying directly over a nest of dinosaur eggs, the one photographed being actually separated from the eggs by only four inches of matrix. This immediately put the animal under suspicion of having been overtaken by a sandstorm in the very act of robbing the dinosaur egg nest." (Andrews, *The New Conquest of Central Asia*, 209)

The dinosaur they recovered, *Oviraptor philoceratops*, was fragmentary, so not much could be said about its broken skull. It was a weird looking dinosaur, with a toothless beak and a very bird-like skeleton, but Osborn couldn't make much sense of it based on the broken parts he had in 1924. In the 1980s, Mongolian paleontologist Rinchen Barsbold found a number of much more complete specimens, and the appearance of *Oviraptor* turned out to be even weirder than anyone had imagined (figure 24.5A). It was built like a large ground-dwelling bird, almost like an ostrich, with a high crest on top of its head and an oddly shaped beak that resembled a broad scoop.

In the 1990s, the American Museum crews, under the leadership of Mike Novacek and Mark Norell in collaboration with Mongolian paleontologists, got another chance to return to Mongolia and make new collections. One of the most sensational discoveries was another nest of the *"Protoceratops"*

Figure 24.5 ▶

(A) The skeleton of the oviraptorid *Hagryphus*, with the weird crest on its head and broad toothless beak. (B) An intact nest of eggs covered by a female oviraptorid *Citipati* in brooding position, discovered by the American Museum crews in the 1990s. These showed that the eggs were not from *Protoceratops* but from the oviraptorid parents. ([A] Photograph by the author; [B] courtesy of Wikimedia Commons)

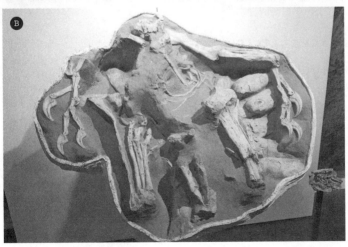

eggs—but this nest was covered by an oviraptorid in brooding position (figure 24.5B)! The new oviraptorid was described in 2001 by Jim Clark, Mark Norell, and Rinchen Barsbold and named *Citipati osmolskae*. *Citipati* is Hindi for "funeral pyre lord." According to Tibetan Buddhist folklore, the Citipati were two monks who were beheaded by a thief while in a meditative trance, and they became a pair of dancing skeletons hovering over a circle of flame (hence the comparison to the specimen, hovering over the ring of the nest). The species name honors Halszka Osmólska for her many years of work on the dinosaurs of Mongolia.

If this oviraptorid was really brooding the eggs, then who was their real mother? Several eggs were X-rayed and then opened, and inside were the embryonic bones of tiny baby oviraptors. So the answer was clear: *Oviraptor* and *Citipati* were not stealing the eggs—they were the mothers of the eggs, protecting them against real predators! Thus the name *Oviraptor* slanders the bearer of the name because it was not the "egg thief" but the "good mother lizard." Unfortunately, the rules of zoological names are that you cannot change a name, no matter how inappropriate it becomes.

If the abundant eggs that Osborn and Granger and Andrews incorrectly assigned to *Protoceratops* were in fact from oviraptors, what did the real eggs of *Protoceratops* look like? A nest of actual eggs and young *Protoceratops* was discovered in 2011, and this time the babies and eggs seemed to match the true parents. These young dinosaurs appeared to be too weak to have fended for themselves; they were found in the nest, having grown since hatching, suggesting that the parents cared for them (as Jack Horner found with *Maiasaura*, the "good mother lizard").

THE "PARROT LIZARD"

Protoceratops gave us a glimpse of an early primitive horned dinosaur with a broad frill but no horns on its nose or forehead or spikes on its frill. But from what kind of dinosaur did it evolve?

Ironically, the American Museum Central Asiatic Expeditions had found that dinosaur too. Among the more common fossils from the Lower Cretaceous (Aptian-Albian) Ashile Formation in the Artsa Bogdo basin (about 100–125 million years old) was a smaller bipedal dinosaur described by Osborn in 1923 as *Psittacosaurus mongoliensis*. *Psittacos* means "parrot" and *sauros* means "lizard" in Greek, so this was the "Mongolian parrot lizard." Osborn noticed how the short, narrow head with large eyes and the prominent deep beak resembled a parrot's head, and this was the inspiration for the name. Osborn had abundant material of this genus, including a nearly complete articulated type skeleton (figure 24.6). This specimen suggested an animal about 2 meters (6.5 feet) in length, and weighing about 20 kilograms (44 pounds). Although he gave a brief description of it in 1923, in no place does Osborn mention its similarities to the skulls of

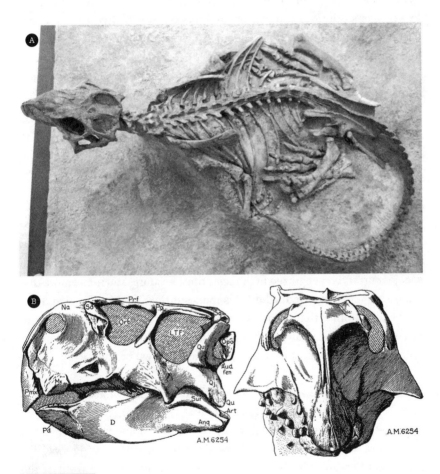

Figure 24.6 ▲

Psittacosaurus mongoliensis: (A) articulated type skeleton in the American Museum; (B) Osborn's diagram of the weird-looking skull, with the broad flaring cheekbones. (Courtesy of Wikimedia Commons)

ceratopsians, including the *Protoceratops* from the same expedition. It is not clear from reading the paper that he realized it was a very primitive relative of horned dinosaurs.

Altogether, more than 75 specimens of this dinosaur and 20 complete skeletons were found in the same area of Mongolia, and hundreds representing numerous species are now known from all over Asia, including China, Siberia, and possibly Thailand as well as Mongolia. One of these skeletons was found near the original type specimen of *Psittacosaurus* in 1923. Osborn originally named it *Protiguanodon mongoliense* because it looked like a primitive ornithopod, and its skull was badly weathered and virtually uninterpretable. Subsequent study has shown that it is merely a juvenile of *Psittacosaurus mongoliensis*.

(On a personal note, I got to know these fossils well. When I was a student at the American Museum in the late 1970s, my officemate Daniel J. Chure was working on psittacosaurs and *Protoceratops* for his thesis, and he had most of the best specimens in our office so he could study them. Trying to get around these huge flat trays with specimens without damaging the delicate bones collected in 1922 was a constant nuisance. Dan has recently retired after spending his career as the paleontologist for Dinosaur National Monument.)

The large number of *Psittacosaurus* fossils from many different Lower Cretaceous beds in Mongolia, China, and elsewhere in Asia has led to lots of arguments about how diverse they were. There are about 11 different species currently recognized in the scientific literature, and many more that are no longer considered valid. This makes it the most species-rich dinosaur known (other than birds). Other scientists consider *Psittacosaurus* to be a highly variable genus, with many differences due to gender and developmental stage, so most of these species are not real biological species, in their opinion.

Because there are so many specimens from so many localities, a lot is known about the biology of *Psittacosaurus*. The distinctive skull (figure 24.6B) with the parrot-like beak made of the rostral and predentary bone is the most striking feature. The beak also had a horny covering to it, like many reptilian and bird beaks do over their jaw bones. The front of the beak was toothless, but small cheek teeth were self-sharpening and had a prominent crest across the top, presumably for chopping up plants, seeds, and nuts for its herbivorous diet. The type skeleton even preserved a pile of small stones in the gut cavity that were probably gastroliths, grinding stones for use in its stomach or gizzard.

Another striking feature is the flared cheekbones, which stick out almost like small horns on the side of the face. The eyes were rather large, with a ring of bone (sclerotic ring) protecting the large eyeball, suggesting that it might have been nocturnal. There was no large ceratopsian frill, but the back of the skull still has a significant shelf of bone that is true of all the Marginocephala, including pachycephalosaurs. It was long thought to have a small brain, but a 2007 study showed the brain was larger and more advanced than most other herbivorous dinosaurs, with a brain/body size ratio close to that of *Tyrannosaurus rex*.

Psittacosaurus was a bipedal dinosaur with small forelimbs, less than 60 percent the length of the long hind limbs. Not only were the arms too short for quadrupedal locomotion but it could not rotate its hand to bring the palms flat on the ground, so it was almost certainly bipedal as an adult. However, the limb sizes and bone structure in a study of juvenile specimens indicated a quadrupedal posture in young animals that gradually became fully bipedal as the legs grew faster than the arms. The arms were too short to reach the mouth, so they were only good for two-handed grasping of near objects and for scratching and fighting. *Psittacosaurus* had only four fingers on the hand, compared to five in most other ornithischians, and four toes on the feet like most other dinosaurs.

There is one complete specimen (figure 24.7) from the Yixian Formation in the Liaoning Province of China in which all the soft tissues and even the melanosomes indicating its color are preserved. The body was mostly covered by scales rather than feathers, but along the back were an array of bristles that glow like feathers under UV light. The melanosomes preserved on this specimen indicate that it was countershaded, with dark on the top and light on its belly, so it was easier for it to hide in the shade of the forests in which it lived. There were patches of color on the face for display as well as around the cloaca and on the membranes of its hind legs.

A number of juvenile *Psittacosaurus* fossils are known, including a hatchling that is only 11–13 centimeters (4–5 inches) long with a skull barely 2.5 centimeters (1 inch) long. Looking at the histology of their bones, three-year-olds weighed only 1 kilogram (2.2 pounds), whereas the nine-year-olds reached 20 kilograms (44 pounds). This is rapid growth for a reptile, but it is slower than that of birds or mammals. Based on the

0.25 m

(*A*) The complete articulated specimen of *Psittacosaurus* from the Liaoning beds, showing the melanosomes with different color patches and the spine-like feathers along the backbone and tail. (*B*) Reconstruction of *Psittacosaurus*. ([*A*] Courtesy of Wikimedia Commons; [*B*] courtesy of N. Tamura)

growth rings of the bones of the largest adults, most lived only 9–11 years. There are numerous specimens of juveniles found together, including a block with six young individuals from two distinct age groups, which were buried together in a volcanic mudflow. Even the youngest hatchlings have worn teeth, suggesting that they could fend for themselves once they hatched and needed little parental care (like most modern reptiles). Another specimen had 34 articulated juvenile skeletons of *Psittacosaurus* closely bunched together, with a larger skull of a six-year-old individual found on top of them. The six-year-old is too young to breed, suggesting that it was babysitting the 34 tiny juveniles before they all died when their burrow collapsed on them.

Finally, we always think of dinosaurs as the rulers of the Mesozoic, and mammals as tiny shrew-sized creatures that hid in the darkness to avoid being eaten by their reptilian overlords. In one instance, however, an opossum-sized Mesozoic mammal called *Repenomamus* was found in China with chunks of *Psittacosaurus* babies in its stomach contents. Mammals were not always the victims during the 130 million years of the Late Triassic through the end of the Cretaceous when they shared the world with dinosaurs. Now and then they were the predators rather than the prey!

Along with *Protoceratops* and *Psittacosaurus*, a number of primitive Asian dinosaurs (*Xuanhuaceratops, Chaoyangsaurus, Liaoceratops, Auroraceratops, Yamaceratops, Helioceratops, Archaeoceratops, Koreaceratops,* and *Leptoceratops*) now provide a bridge between *Yinlong* and the pachycephalosaurs, on one hand, and the advanced horned dinosaurs, on the other. The final chapter explores the enormous radiation of the horned dinosaurs.

FOR FURTHER READING

Andrews, Roy Chapman. *The New Conquest of Central Asia: A Narrative of the Explorations of the Central Asiatic Expeditions in Mongolia and China, 1921–1930.* New York: American Museum of Natural History, 1932.

Colbert, Edwin. *Men and Dinosaurs: The Search in the Field and in the Laboratory.* New York: Dutton, 1968.

Dodson, Peter. *The Horned Dinosaurs.* Princeton, N.J.: Princeton University Press, 1996.

Farlow, James, and M. K. Brett-Surman. *The Complete Dinosaur.* Bloomington: Indiana University Press, 1999.

Fastovsky, David, and David Weishampel. *Dinosaurs: A Concise Natural History*, 3rd ed. Cambridge: Cambridge University Press, 2016.

Holtz, Thomas R., Jr. *Dinosaurs: The Most Complete, Up-to-Date Encyclopedia for Dinosaur Lovers of All Ages*. New York: Random House, 2011.

Kielan-Jaworowska, Zofia. *Hunting for Dinosaurs*. Cambridge, Mass.: MIT Press, 1969.

Mayor, Adrienne. *The First Fossil Hunters: Paleontology in Greek and Roman Times*. Princeton, N.J.: Princeton University Press, 2000.

Naish, Darren. *The Great Dinosaur Discoveries*. Berkeley: University of California Press, 2009.

Naish, Darren, and Paul M. Barrett. *Dinosaurs: How They Lived and Evolved*. Washington, D.C.: Smithsonian Books, 2016.

Spaulding, David A. E. *Dinosaur Hunters: Eccentric Amateurs and Obsessed Professionals*. Rocklin, Calif.: Prima, 1993.

You, Hailu, and Peter Dodson. "Basal Ceratopsia." In *The Dinosauria*, 2nd ed., ed. David B. Weishampel, Peter Dodson, and Halszka Osmólska, 478–494. Berkeley: University of California Press, 2004.

TRICERATOPS

A group of blind men heard that a strange animal, called an elephant,
had been brought to the town, but none of them were aware of its shape
and form. Out of curiosity, they said: "We must inspect and know it by
touch, of which we are capable." So, they sought it out, and when they
found it they groped about it. In the case of the first person, whose
hand landed on the trunk, said "This being is like a thick snake." For
another one whose hand reached its ear, it seemed like a kind of fan. As
for another person, whose hand was upon its leg, said, the elephant is a
pillar like a tree-trunk. The blind man who placed his hand upon its side
said, "Elephant is a wall." Another who felt its tail, described it as a
rope. The last felt its tusk, stating the elephant is that which is hard,
smooth and like a spear.

—*THE BLIND MEN AND THE ELEPHANT*, A HINDU PARABLE

THE BLIND MAN AND THE ELEPHANT

As with the history of so many early dinosaur discoveries, the story of *Triceratops* begins even before Edward Drinker Cope and O. C. Marsh and the Bone Wars. The first ceratopsian specimens were found in 1872, by Fielding Bradford Meek. Born in 1817 in Madison, Indiana, across the Ohio River from Louisville, as a boy Meek collected the abundant Paleozoic corals from the Falls of the Ohio between Louisville and Madison, and other invertebrate fossils all around southern Indiana. After failing in business, he tried his luck at science. He was hired to work for the U.S. Geological and Geographical Survey of the Territories in Iowa, and later for surveys in Wisconsin and Minnesota. Between 1852 and 1858 he was in New York, learning from pioneering paleontologist James Hall, who was the first to map the geology of New York and describe its fossils. On a break from that job, he joined Ferdinand Vandeveer Hayden's Geological Survey of the Territories on his famous 1853 expedition to the Big Badlands of South Dakota and other areas. This expedition eventually provided lots of early mammal fossils for Joseph Leidy. In 1858, Meek moved to Washington, D.C. and worked full time compiling the reports and documents of the various early geological surveys of which he had been a part. In the 1860s, he was in California at the end of the Gold Rush, collecting fossils and publishing the first reports on the paleontology of California.

In 1872 Meek was in the Upper Cretaceous Laramie Formation in the Black Butte and Bitter Creek area of southwestern Wyoming, working with Henry Martyn Bannister, looking for fossil shells. These beds of terrestrial rocks were rich in dinosaurs (and a few freshwater mollusks), so they could not help but stumble upon dinosaur bone fragments. They were still working for the Hayden Survey, so they notified Cope of their finds. Cope got to Black Butte as soon as he could and found Meek's site, which was full of huge bones protruding from the ground near a coal vein, with fossil sticks and leaves packed around it. He quickly rushed a short publication into print about the bones. He named it *Agathaumas silvestris*, and based the new genus on just 16 vertebrae from the back, hips, and tail, plus a partial pelvis and some ribs (figure 25.1). *Agan* means "much" and *thauma* means "wonder" in Greek, and *sylvestris* means "of the forest" in Latin, so it was the "great wonder of the forest."

In 1873, Cope rhapsodized about the fossil as "the wreck of one of the princes among giants" because it was the largest land animal known anywhere in the world at that time. The huge stegosaurs and sauropods did not began to emerge from places in Colorado and Wyoming for several years, and the British dinosaurs *Cetiosaurus* and *Pelorosaurus* were still considered marine reptiles then (see chapter 3). As more fossils were found over the years, Cope added six additional species to *Agathaumas*; none are considered valid now because they were based on partial scrappy specimens with no diagnostic features that can be used today. Cope could not really tell what kind of dinosaur these bones came from, but he assumed it was from a hadrosaur as those were the only dinosaurs described in North America at that time.

One of these fossils was eventually reassigned to *Monoclonius*, which was found and named by Cope in 1876 when he and young Charles H. Sternberg were in the Judith River badlands of Montana looking for Late

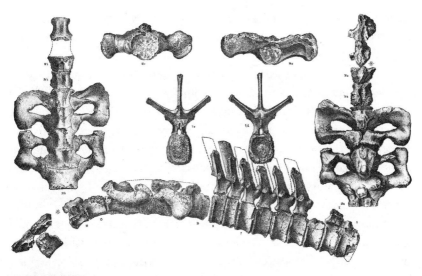

Figure 25.1 ▲
Cope's original fragmentary fossils of *Agathaumas silvestris*, a genus based on a few vertebrae and hip bones. (From J. B. Hatcher, O. C. Marsh, and R. S. Lull, *The Ceratopsia*, 1907)

Cretaceous dinosaurs. They hastily collected the fossil (a short straight nose horn and part of the rounded frill) because the Battle of the Little Bighorn had recently ended about 150 kilometers (100 miles) to the east, and hostile warriors roamed the entire area. Assuming the hostiles would stay together for fear of counterattack from the U.S. Army, Cope persisted with his expedition. At one point, he was only a day away from Sitting Bull's camp, and some of his local men deserted him. However, the only natives Cope's crew saw were Crows, who were the enemies of the hostile Lakota (Sioux) and Cheyenne warriors found at Little Bighorn. They were impressed with Cope's trick of removing his false teeth and reinserting them.

As more and more specimens of ceratopsians were found, it was clear that most of Cope's *Agathaumas* fossils were nondiagnostic. With only the back end of the skeleton in the original specimen (figure 25.1), and without the skull and especially the horns, you can't tell most ceratopsian dinosaurs apart. As much better specimens with skulls of *Centrosaurus* and *Styracosaurus* and eventually *Triceratops* were found, it was impossible to know where Cope's fossils of *Agathaumas* belonged. Most paleontologists dropped the name, and it is nearly forgotten. To the end of his life, however,

Cope refused to use Marsh's names such as *Triceratops*, always insisting that his 1872 name *Agathaumas* had priority over the names of ceratopsians that were proposed later.

Unfortunately, Marsh's first attempts at describing ceratopsian fossils were not much better than Cope's. George M. Cannon and his crew were working for the U.S. Geological Survey in the Denver area in 1887, and they found fragmentary bones and a partial skull roof with two horns more than 60 centimeters (2 feet) long (figure 25.2). They came from the Laramie Formation, which we now know (and they also thought) were Cretaceous rocks. By this point, Marsh was the official paleontologist of the U.S. Geological Survey, so the fossils were sent to him to identify. Marsh thought they looked like enormous bison horn cores (figure 25.2), so he overruled the field men and even his own assistants and asserted that the fossils must have come from the younger Pliocene or Pleistocene beds in the area. Rushing to print, he named the fossils *Bison alticornis* (tall horned bison), and he wrote that it was "one of the largest of the American bovines, and one differing widely from those already described." Like Cope, he had no idea that there was an entire family of horned dinosaurs because nothing like that had yet been found complete. Cope and Marsh were like the blind men and the elephant in the famous parable: each had different parts of the animal and jumped to false conclusions based on what was known of dinosaurs at the time.

Within a few months of his "*Bison alticornis*" paper, Marsh received new fossils from John Bell Hatcher who was collecting in Cope's old area in the Judith River badlands near Cow Creek, Montana, in the summer of 1888. These fossils consisted of a pair of horn cores and the back of the skull, but they were from known Cretaceous beds. Marsh suddenly realized that his "*Bison alticornis*" was a Cretaceous dinosaur after all. He called the horn cores *Ceratops montanus*, meaning "horned face from Montana." From this name comes the name for the entire group, the Ceratopsia.

The name *Ceratops* hasn't fared too well either because the broken horn cores were not diagnostic enough to really know what genus of horned dinosaur they represent. In 1906, Marsh's successor at Yale, Richard Swann Lull, realized that the name *Ceratops* had already been used for a bird in 1815, so it was not available to be used on another animal. Lull wanted to rename the fossil *Proceratops*, but this was unnecessary. No one can tell where the fossil belongs, so *Ceratops* (like *Agathaumas*) is a *nomen dubium*, or "doubtful name."

On the way home to Nebraska from collecting in Montana in late 1888, Hatcher made a stop in the Lance Formation north of Lusk in eastern

Figure 25.2 ▲
Marsh's original fossils of "*Bison alticornis*," based on horn cores attached to part of a skull.
(From J. B. Hatcher, O. C. Marsh, and R. S. Lull, *The Ceratopsia*, 1907)

Wyoming. He ran into the local rancher, Charles Gurney, whose cowboys told him they'd found a skull "with horns as long as a hoe handle and eye holes as big as your hat." They had seen it sticking out of a bank just out of reach, lassoed it, and dislodged the horn and part of the skull, which then broke up as it fell into the wash. It was a nearly complete skull, with the paired horns over the eyes and the short nose horn and rostral beak in front. Hatcher told Marsh when he sent the broken horn that it had been attached to a massive skull embedded in a concretion and stuck in the bank of the creek. Marsh told Hatcher to go back and get the rest of the skull, which weighed more than a ton when finally collected. In May 1889, he wrote to Marsh, "The big skull is ours." When Marsh received the specimen and described it in late 1889 (figure 25.3A), he named it *Triceratops horridus* (horrible three-horned face). Other specimens that he had been calling *Ceratops* were gradually reassigned to *Triceratops*.

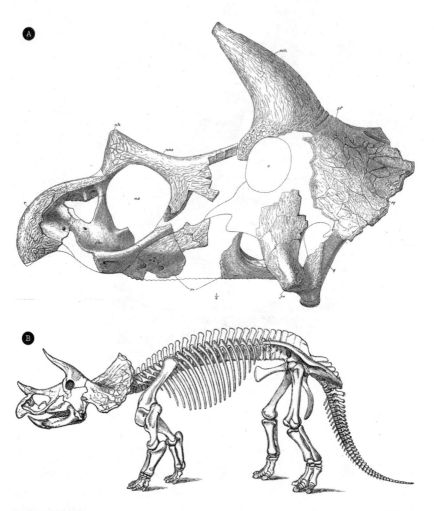

Figure 25.3 ▲ ▶

(A) Marsh's original type specimen of *Triceratops horridus*, the first partial skull retrieved by Hatcher. (B) Marsh's reconstruction of *Triceratops prorsus*, based on the specimens known at the time. (C) The mounted skeleton of *Triceratops horridus* found by Barnum Brown in the Hell Creek beds of Montana, which is still on display today at the American Museum with its front limbs splayed out to the side like those of a sprawling alligator or a lizard. (D) A more modern reconstruction of *Triceratops*, with the front limbs held almost vertically beneath the body. ([A–B] From J. B. Hatcher, O. C. Marsh, and R. S. Lull, *The Cera-topsia*, 1907; [C] courtesy of Wikimedia Commons; [D] photograph by the author)

Figure 25.3 ▲
(*continued*)

Hatcher spent the next four years (1889–1892) collecting mostly in the Lance Formation where *Triceratops* were everywhere. In that short time, he shipped more than 50 individual skulls and many partial skeletons to Marsh at Yale, and 33 of those were nearly perfect and unbroken. Some of these huge skulls weighed more than 3.5 metric tonnes (3.86 tons). By 1892, *Triceratops* was the best-known dinosaur of the time, with the largest sample of skulls found for any dinosaur. Even better more complete specimens were found in other Late Cretaceous localities, such as the Hell Creek Formation by Barnum Brown in 1902–1908; *Triceratops* was mounted and exhibited in the American Museum and other institutions (figure 25.3C). By the early twentieth century *Triceratops* was almost as familiar as *Tyrannosaurus rex*, and today it is an iconic dinosaur that everyone knows when they see it.

THE GAMBLER

The enormous quantity of *Triceratops* fossils found by Hatcher was fairly typical of his prowess as a collector and as a paleontologist.

> Determined, confident, hard-working and often very temperamental, John Bell Hatcher was the most talented and respected field paleontologist of his generation, but he struggled mightily to find his own level in the hierarchical world of America vertebrate paleontology. (Brinkman, *The Second Jurassic Dinosaur Rush*, 123)

Hatcher was born a few months after the beginning of the Civil War, in October 1861, in Cooperstown, Illinois (figure 25.4). When he was very young, his family moved to Cooper, Iowa. There he became fascinated by the fossils he found lying on the ground all around him. He started his higher education in 1880 at Grinnell College in central Iowa. To earn money, he worked in the local coal mines and became intrigued with the beautifully preserved fossil plants he found. After just one term, he transferred to Yale's Sheffield Scientific School, where he graduated in 1884. Just before graduating, he showed his fossil collection to the school's mineralogist, George Jarvis Brush. Impressed with the young man's enthusiasm, Brush introduced the 23-year-old college graduate to Marsh.

Figure 25.4 ▲
Portrait of John Bell Hatcher. (Courtesy of Wikimedia Commons)

Marsh needed a new young eager field assistant, so he sent Hatcher to Kansas to be trained by Charles H. Sternberg. (Cope was now broke and could not pay him, so Sternberg was now working for Marsh.) Sternberg wasn't exactly excited to be training the young greenhorn, but he tried to be positive. He wrote that he "gave promise of a future even then by his perfect understanding of the work in hand and the thoughtful care which he devoted to it."

The college-educated Hatcher, however, looked down on Charles H. Sternberg, especially his slapdash methods of getting the fossil out of the ground. Already trained in geology, Hatcher was a perfectionist, and he was interested in the details of the sedimentary rocks in which the fossils were found as a clue to their ancient environments, and how the fossil had died and become buried and fossilized (something most commercial collectors ignore even today). Hatcher wrote this about one of his finds:

> On one side of a bone, the matrix will be made up entirely of sand, while on the opposite side the stem and leaves of plants have been dropped.... This ... shows the direction of the current to have been from that side containing only sand, and towards the side containing the plants. So shallow were the waters, the bone itself became an obstacle sufficient to produce an eddy on the lower side, in which the leaves and other vegetable materials accumulated and sank to the bottom.

Sternberg helped Hatcher begin his training by collecting the hippo-like Miocene rhinoceros *Teleoceras* from Long Island Rhino Quarry in northern Kansas in 1885. Soon Hatcher was independent of Sternberg and reporting directly to Marsh. He roamed over much of the Rocky Mountain region collecting fossil mammals that he sent to Yale. In 1887, he married the sister of paleontologist Olaf A. Peterson and settled in Nebraska. That summer he revisited some of Cope's localities in Montana and sent a ton of bones back to Marsh. It was on the way home to his new bride that he made the discovery of the first good *Triceratops* skull roped by the cowboys in Lance Creek (figure 25.3A). That was the focus of his fieldwork for the next four years until 1892. In addition to all the *Triceratops* specimens, 1891 was highlighted by the discovery of the skull of *Torosaurus*, the longest skull that had ever been found in a land animal.

In 1889, Hatcher made the first-ever discovery of the teeth of Cretaceous mammals, which Marsh found extremely exciting. Hatcher stumbled on a technique used by many paleontologists since then—look at the anthills. In strata where there is nothing but muds and silts and fine sands and no gravel, the harvester ants will collect tiny, hard, gravel-sized objects like fossil teeth, creating an armored layer over the anthill to prevent it from blowing away. For more than a century, paleontologists have been abusing poor harvester ants by picking over the tops of their anthills, or scooping them away and running them through fine screen mesh to retrieve all the precious fossil teeth the ants have collected.

By this time, Hatcher was fed up with Marsh and his abusive ways. Marsh paid his collectors poorly, and he paid them infrequently as well. He insisted that Hatcher labor through the blizzards of the hard western winters rather than work in the museum on his finds of the previous year. More important, Marsh hogged all the credit for doing the work. He would not even give his assistants coauthorship on papers that they had written, putting his name over their writing. This had already driven Samuel Williston (who had started with Marsh in 1874) to leave in 1890 and start over in Kansas, becoming a professor at the University of Kansas.

Hatcher finally resigned in 1893 while he was still in the field. Osborn was aware of the discontent Marsh has fostered in his staff, and Osborn tried to recruit Hatcher for the American Museum paleontology program he was trying to build. Hatcher backed out, however, when he realized that Osborn himself didn't have his own job secured yet. Hatcher got a job as a collector at Scott's and Osborn's alma mater, Princeton University, in 1893, through the invitation of Osborn's friend William Berryman Scott. Osborn soon forgave him and helped him get established, even arranging for an honorary degree.

The 32-year-old Hatcher was suffering from rheumatism and assorted injuries caused by years of rough fieldwork, and he could barely get on his horse. Nevertheless he conceived of the idea of leading an expedition to Patagonia to collect fossil mammals where Darwin had once been. By this time, the self-taught Argentinian paleontologists, the brothers Florentino and Carlos Ameghino, were making amazing and puzzling discoveries of weird mammals that completely challenged what people thought about mammal evolution. Despite his infirmities, Hatcher planned an audacious trip to South America, and he received funding from millionaire J. Pierpont Morgan for the first Princeton University Expeditions to Patagonia in 1896.

During the month-long voyage to South America in March 1896, Hatcher became famous as a first-rate card shark and gambler. He used the long boring periods at sea to play poker with the other passengers, and he cleaned them out. He continued to do so any chance he got while in Argentina and raised enough money to fund his Patagonian expeditions for three more years (1896–1899) in this way. In the first two years, the Princeton group was just Hatcher and his brother-in-law, Olaf A. Peterson. The final season, in 1899, included Barnum Brown of the American Museum, who had just been hired and was given just a few hours' notice by Osborn to join the expedition (see chapter 21).

When Hatcher returned from the 1899 expedition, he was upset with the way Princeton was treating him. Despite the concerns of Osborn and

Scott, he was ready to jump ship again. Meanwhile W. J. Holland was trying to get the Carnegie Museum paleontology program back on track after Jacob Wortman was forced to resign. To make up for the loss, Holland hired Olaf Peterson and then Hatcher to replenish their staff of paleontologists. Hatcher was soon working on the Morrison localities that the Carnegie Museum was collecting in Wyoming, and he led the excavations of a *Diplodocus* skeleton at Sheep Creek Quarry, near Como Bluff and Medicine Bow (see chapter 8). His monograph on *Diplodocus*, published in 1901, is still considered a masterpiece of descriptive paleontology.

By the spring of 1904, Hatcher was done with the Morrison dinosaurs and ready to return to Lance Creek for more *Triceratops*. He also sent his brother-in-law Peterson to collect more fossil mammals in Nebraska. However, Marsh's death in 1899 left a huge project unfinished, the monograph on all the *Triceratops* Hatcher had once collected for Yale. Osborn had replaced Marsh as the official paleontologist of the U.S. Geological Survey, and he was legally in charge of all the specimens at the Smithsonian that Marsh had accumulated. Osborn believed Hatcher would be the best person to complete the task. In addition to his job at the Carnegie Museum, Hatcher made frequent trips to see all the *Triceratops* specimens at Yale and at the Smithsonian, so he was overworked.

In late June, Hatcher was feeling "somewhat run down, having exerted himself rather strenuously." Holland urged Hatcher to go home and get some rest, but the workaholic Hatcher refused. He returned to his office, locked the door, and worked late into the night, despite a high fever. "He joked about his illness and ignored his doctor's orders." Holland had to order him out and had him put in ambulance to the hospital, where he was diagnosed with typhoid fever. Given Hatcher's legendary stamina in the field, the doctors thought he would be over it in a week. Hatcher did not recover, and he died suddenly on July 3, 1904, at the young age of 42.

Holland sent Peterson a letter in the field in Agate Springs, Nebraska, notifying him of his brother-in-law's death. But Holland demanded that Peterson stay and continue the work because Hatcher would already be buried by the time the letter reached him. Osborn, on the other hand, sent a much more sensitive letter to the grieving Peterson and his sister, Hatcher's widow:

> I hasten to write you of my deep sorrow and sympathy in learning of the death of Hatcher.... I was greatly shocked, for when I saw him last he was looking

very well and was full of bright plans for the future. This is a great personal loss for I greatly admired Hatcher's scientific ability and enthusiasm, and always felt a fresh inspiration from talking to him. It is a hard blow to American paleontology, to which Mr. Hatcher was making such splendid contributions following his many years of magnificent work in the field—he was certainly our greatest collector. It is especially sad to think of his dying at the beginning of what promised to be the brightest and most satisfactory period of his life— when people could see his work and recognize his ability. . . . I return early in September to take up my work and I shall always miss Hatcher.

The entire paleontological world mourned Hatcher and praised his enormous contributions over such a short life. His death meant that the big monograph on *Triceratops*, for which all the illustrations were complete but that he had just begun to write up, outlived both Marsh and Hatcher. Marsh's successor at Yale, Richard Swann Lull, took over the task, and it was finally published in 1907 with Hatcher as first author and Marsh as second author. Lull put himself last, even though he wrote nearly all of it.

"THREE-HORNED FACE"

Triceratops is now known from hundreds of skulls and many partial skeletons, although no complete skeleton of a single individual has ever been found. Barnum Brown claimed he had seen as many as 500 skulls in the field (in various states of completeness), and Bruce Ericson of the Science Museum of Minnesota reported over 200. John Scannella commented that "it is hard to walk out into the Hell Creek Formation and not stumble upon a *Triceratops* weathering out of a hillside."

The dinosaur is now so familiar to us that it's difficult to appreciate how startling its appearance is (figure 25.3C). Most of the largest specimens were about 9 meters (30 feet) in length, 3 meters (10 feet) high at the shoulder, and are thought to have weighed between 6 and 12 metric tonnes (6.5–13 tons). The skulls of it and most ceratopsians are enormous; they are the largest skulls known for any group of land animals.

In addition to the famous combination of three horns on its face, *Triceratops* had many other anatomical peculiarities. The nose horn was sometimes compressed and narrow, but in other skulls it was more rounded and conical in cross section, located above a remarkably large opening for the nostrils. Like all ceratopsians, *Triceratops* had a prominent upper beak made of the rostral bone, which occluded with a lower beak made of the predentary bone found in all ornithischians. It also had dozens of small teeth arranged in stacked rows, called a dental battery (figure 25.5). However, their teeth looked very different from the dental battery of hadrosaurs, which were built of closely packed tall polygonal prisms (see figure 4.3). In *Triceratops*, each battery was made of 35–40 tooth columns, and each column was built of 3–5 stacked teeth, which shed teeth off the top as they wore out. Each battery occluded against the inclined surface of the battery in the opposite jaw. Altogether they had between 432 and 800 teeth in their mouth at once (hadrosaurs had more than 1,000 tiny prisms in their mouth). Their teeth sheared in a vertical plane and were able to chop up even the toughest vegetation. Grasses were not yet common, so the likeliest food would have been palms and cycads and ferns. *Triceratops* could not reach its head high enough to eat tall shrubs or trees.

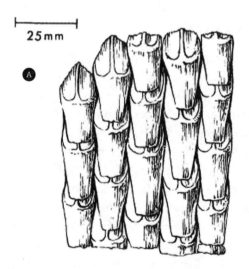

Figure 25.5 ▲ ▶

Dental battery and jaw mechanics of *Triceratops*: (A) external view of the teeth, which overlap like shingles and are shed as they move up and wear out; (B) cross section of the dental battery, showing how the teeth nest one inside another as they are pushed up and out as they wear down; (C) cross section of the upper and lower jaws, showing how the external surface of the lower dental battery slices and occludes against the inner surface of the upper dental battery. (Modified from Ostrom 1966)

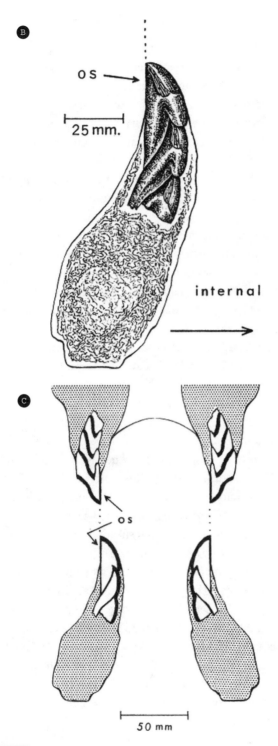

os →

25 mm.

internal →

os

Figure 25.5 ▲
(continued)

50 mm

The function of the frill has long been a source of speculation among paleontologists. The conventional story was that the frill served to protect its neck against the bite of *Tyrannosaurus rex*, an idea first proposed by Charles H. Sternberg in 1917 and revived many times. There is some evidence of tyrannosaur bite marks on the brow horns and cheek bones of *Triceratops*, but not on the frill. In some cases, horns were bitten off and regrew during the life of *Triceratops*. But there is no direct evidence that tyrannosaurs bit the frill in an effort to reach down and bite the neck of their prey. Many *Triceratops* carcasses do show bite marks and even shed teeth of tyrannosaurs, so they were definitely scavenged.

Many paleontologists believe the frill was more important for display and dominance within the species. This would explain the wide variation in shapes and sizes of frills, as well as the dramatic change in size and shape of the frill as *Triceratops* grew up. Juvenile *Triceratops* even had a small frill before they reached the age of sexual reproduction, so the frill helped aid in communication visually and in species recognition.

Juveniles and subadult *Triceratops* fossils are less common than adults but have been studied. The smallest babies known were only 38 centimeters (15 inches) long. As they grew up, the bones around the skull remodeled dramatically, and the horns enlarged, changed shape, and became more hollow. In 2009, at the Society of Vertebrate Paleontology meeting in Bristol, England, John Scannella and Jack Horner discussed Marsh's 1881 genus *Torosaurus* (also found by Hatcher in the Lance Formation), which has the longest frill of all (figure 25.6). It also had large holes in the frill, a feature found in many other adult ceratopsians. Scanella and Horner argued that *Torosaurus* was a very mature example of *Triceratops* and that the rest of the skulls referred to *Triceratops* were less mature adults and subadults. If their interpretation is true, *Torosaurus* would just be a junior synonym of *Triceratops*.

The media got this claim completely backward, and many reported that the name *Triceratops* was no longer valid, which caused unnecessary panic in the public arena. To be clear, if they are the same animal, then *Torosaurus* is no longer valid because it was named later. However, a study by Nick Longrich and Daniel Field looked at 35 skulls of *Triceratops* and *Torosaurus* and found very old skulls that still had the *Triceratops* frill, and less mature skulls that were clearly *Torosaurus*. Currently, the evidence seems to be against the idea that the two genera are the same animal representing different growth stages.

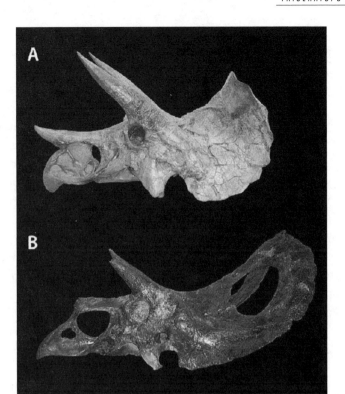

Figure 25.6 ▲

Comparison of the skull of *Triceratops* and *Torosaurus*. (Courtesy of Wikimedia Commons)

Triceratops was never found in large herd assemblages like *Centrosaurus*. Although they were among the most common dinosaurs of the latest Cretaceous, they apparently lived in small family groups or were solitary. There is evidence of combat between adult *Triceratops*. A study of skulls showed that about 14 percent had some damage from intraspecific combat, although this is low compared to other ceratopsians. There is little evidence, however, that *Triceratops* engaged in direct head-to-head jousting. We find, at most, one or two wounds caused by the horns puncturing the face of their opponents. Based on injuries the skull and face described by Andy Farke, it is more likely that they engaged in head-to-head wrestling.

Triceratops was clearly quadrupedal, with four sturdy legs tipped with short hooves. However, most of the early reconstructions were based on the "sprawling lizard" way of visualizing dinosaurs with their front limbs bent in a crouch (see figure 25.3C) like a crocodile or lizard, and with their tails dragging on the ground. More recent reconstructions show the limbs more upright beneath the body and only slightly flexed, and their tails held straight out (figure 25.3D). This is confirmed by trackways, which show their footprints much closer together than they would be if their front legs splayed to the side, and no tail drag marks have been found. Their hands did not have the ability to rotate to allow their fingers to face forward, unlike other quadrupeds such as stegosaurs and sauropods. Instead, *Triceratops* walked with most of their fingers pointing outward to the side and front. Their hands had three large functional fingers, with only vestigial remnants of the ring finger and pinky.

Triceratops has been one of the most well-known and popular dinosaurs since skeletons were first mounted in the American Museum in the early twentieth century (figure 25.3C). They were first reconstructed by legendary paleoartist Charles R. Knight, who made them look more lizard-like with sprawling forelimbs and dragging tail, consistent with the understanding at the time. Another famous 1942 mural by Knight showed *Triceratops* in combat with *T. rex*. These images influenced the portrayal of *Triceratops* in almost every medium for decades, especially in children's dinosaur books and movies and TV shows, such as Walt Disney's 1940 movie *Fantasia*. The heroine of the *Land Before Time* cartoon series for kids is the stubborn baby *Triceratops* named "Cera." *Triceratops* has even made an appearance in recent blockbuster dinosaur media: *Walking with Dinosaurs*, *Jurassic Park* (both the novel and the 1993 movie), *Jurassic Park: The Lost World*, *Jurassic Park III*, and *Jurassic World*.

THE HORNED DINOSAUR RADIATION

Triceratops and *Protoceratops* and *Psittacosaurus* were part of a huge Late Jurassic through Late Cretaceous radiation of horned dinosaurs. Currently, more than 77 genera are recognized in the group (figures 25.7 and 25.8). Not only were they very diverse, but many of them were extremely abundant. A high percentage of the specimens from the Lower Cretaceous of China and Mongolia are *Psittacosaurus*, whereas *Protoceratops* dominated the Djadokhta Formation, and *Triceratops* makes up five-sixths of the dinosaurs recovered from the Lance and Hell Creek formations. Some, like *Centrosaurus*, were almost certainly herding animals, but others, like *Triceratops*, were apparently loners. As the most common large herbivores through much of the Cretaceous, they must have been important prey for large predators such as the tyrannosaurs.

Beginning in the Late Jurassic with *Yinlong, Chaoyangsaurus*, and *Xuanhuaceratops* from China, ceratopsians became more common and diverse in the Early Cretaceous with *Psittacosaurus* and numerous other genera (chapter 24), almost all restricted to Asia until the very end of the Early Cretaceous. In the early Late Cretaceous, Protoceratopsidae were in Asia, and *Zuniceratops* and the Leptoceratopsidae were typical of North America. By the middle Late Cretaceous, ceratopsians apparently vanished from their original homeland in Asia, but they became increasingly more diverse and common in North America. During this last great evolutionary flowering of the group, bizarre and elaborate combinations of horns, frills, and spikes increased on various parts of the skull (figures 25.7 and 25.8). One group, the Centrosaurinae, include taxa with a single nose horn, no brow horns, and a broad frill with various ornamentations on its edges, like the spiked frill of *Styracosaurus*, or the pair of "devil horns" on *Diablocerátops*, or the blunt thick boss of bone instead of a nose horn on *Pachyrhinosaurus* (star of the CG movie *Walking with Dinosaurs*) and *Einiosaurus*. The other group, the Chasmosaurinae, include genera with prominent brow horns and long frills with few or no spikes or other ornamentation on the edge of the frill. These include not only the familiar genera *Triceratops* and *Torosaurus* but also the triangular-frilled *Pentaceratops* and the bizarre-looking *Medusaceratops* and *Mojoceratops*.

Although the ceratopsians were primarily Asian and North American, a handful are from Europe, including *Ajkaceratops* from Hungary, *Craspedodon* teeth from Belgium, and a possible leptoceratopsid from Sweden. Claims for possible ceratopsians have been made on other continents as well. The Australian genus *Serendipaceratops* is known only from an isolated lower arm bone; it is not certain that it is from a true ceratopsian, and if so, to what ceratopsian group it belonged. *Notoceratops* from Argentina was based on a single toothless jaw that has since been lost.

At the peak of their success, *Triceratops* and a few other genera were among the last nonbird dinosaurs still around when the Cretaceous came to an end 66 million years ago. *Triceratops* bones can be tracked to within a meter or so of sediment before the boundary itself. The media gives the

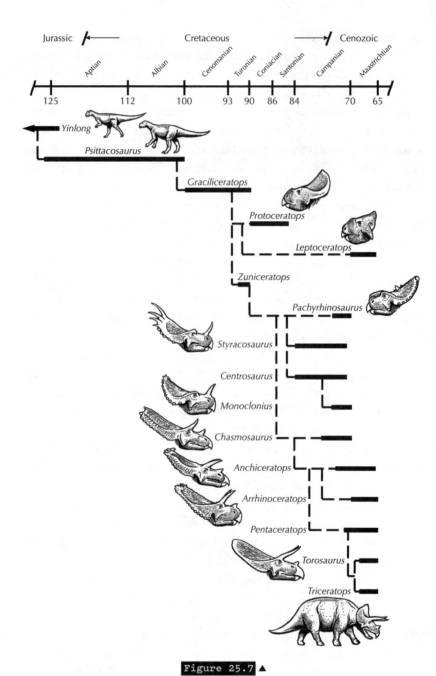

Figure 25.7 ▲

Family tree of ceratopsians. (Drawing by Carl Buell, based on several sources)

Figure 25.8 ▲

The "wall of skulls" family tree of ceratopsians at the Utah Museum of Natural History in Salt Lake City. The right hand branch (numbers 08–12) shows the chasmosaurines. The large skull on the upper right (number 11) is *Coahuilaceratops magnacuerna*. The two large skulls above it and to the left are (08) *Anchiceratops ornatus* and (09) *Triceratops horridus*. Number 10 is *Kosmoceratops richardsoni*, and 12 is *Chasmosaurus belli*. The branch on the left shows centrosaurines. In the top row on the left are (*left to right*) *Einiosaurus procurvicornis, Pachyrhinosaurus lakustai*, and *Achelousaurus horneri*. In the bottom row on the left (04–07) are (*left to right*) *Styracosaurus albertensis, Centrosaurus apertus, Nasutoceratops titusi*, and *Diabloceratops eatoni*. In the lower left corner are basal ceratopsians *Protoceratops* and *Zuniceratops*. (Photograph by the author)

oversimplified and incorrect impression that the extinction of the nonbird dinosaurs was simply due to the impact of an asteroid in Yucatán. However, for the past 39 years, strong evidence suggests that huge volcanic eruptions, the Deccan lavas of Pakistan and western India, were changing the climate well before the rock from space hit Earth. Among paleontologists (especially vertebrate paleontologists), the asteroid impact extinction hypothesis is not very popular. Most view the complex biological signals (such as the near total survival of crocodilians, frogs, salamanders, and many turtles through the extinction) as evidence that the event was much more complex than the media and the general public think it was. To learn more about this evidence, I encourage you to read my book, *The Story of the Earth in 25 Rocks* (especially chapter 20).

FOR FURTHER READING

Brinkman, Paul D. *The Second Jurassic Dinosaur Rush: Museums and Paleontology at the Turn of the Twentieth Century*. Chicago: University of Chicago Press, 2010.

Colbert, Edwin. *Men and Dinosaurs: The Search in the Field and in the Laboratory*. New York: Dutton, 1968.

Dingus, Lowell. *King of the Dinosaur Hunters: The Life of John Bell Hatcher and the Discoveries that Shaped Paleontology*. New York: Pegasus, 2018.

Dingus, Lowell, and Mark A. Norell. *Barnum Brown: The Man Who Discovered Tyrannosaurus Rex*. Berkeley: University of California Press, 2010.

Dodson, Peter. *The Horned Dinosaurs*. Princeton, N.J.: Princeton University Press, 1996.

Dodson, Peter, Catherine A. Forster, and Scott D. Sampson. "Ceratopsidae." In *The Dinosauria*, 2nd ed., ed. David B. Weishampel, Peter Dodson, and Halszka Osmólska, 494–515. Berkeley: University of California Press, 2004.

Farlow, James, and M. K. Brett-Surman. *The Complete Dinosaur*. Bloomington: Indiana University Press, 1999.

Fastovsky, David, and David Weishampel. *Dinosaurs: A Concise Natural History*, 3rd ed. Cambridge: Cambridge University Press, 2016.

Hatcher, John Bell, Othniel Charles Marsh, and Richard Swann Lull. *The Ceratopsia*. Washington, D.C.: Government Printing Office, 1907.

Holtz, Thomas R., Jr. *Dinosaurs: The Most Complete, Up-to-Date Encyclopedia for Dinosaur Lovers of All Ages*. New York: Random House, 2011.

Lull, Richard Swann. "A Revision of the Ceratopsia or Horned Dinosaurs." *Memoirs of the Peabody Museum of Natural History* 3, no. 3 (1933): 1–175.

Naish, Darren. *The Great Dinosaur Discoveries*. Berkeley: University of California Press, 2009.

Naish, Darren, and Paul M. Barrett. *Dinosaurs: How They Lived and Evolved*. Washington, D.C.: Smithsonian Books, 2016.

Ostrom, John H. "Functional Morphology and Evolution of the Ceratopsian Dinosaurs." *Evolution* 20, no. 3 (1966): 290–308.

Simpson, George Gaylord. *Discoverers of the Lost World*. New Haven, Conn.: Yale University Press, 1984.

Spaulding, David A. E. *Dinosaur Hunters: Eccentric Amateurs and Obsessed Professionals*. Rocklin, Calif.: Prima, 1993.

INDEX

American Rockies, 389
Amphicoelias, 109, 161–162
Amur region, Siberia, 389
Amurosaurus, 389
Anchiceratops, 202, 444–445
Anchiornis, 307
Andes Mountains, 239
Andesaurus, 71, 163–166
Andrews, Charles W., 276–277
Andrews, Roy Chapman, 223, 272–278, 406–414
Anglo American Oil Company, 362–363
Animantarx, 369
Ankylosauridae, 350–360
Ankylosaurus, 348, 350–354, 359, 364
Anning, Mary, 8, 10, 20, 21
Anoplosaurus, 369
Antarctica, 167, 175, 187–199, 364, 369
Antarctopelta, 201, 369
Antarctosaurus, 150, 165
antelopes, 3
Anthodon, 333
antorbital fenestra, 65–67
Apatosaurus, 44, 95, 106–120, 141–149, 161, 276
Arabia, 362
Aral Sea, 389
Aralosaurus, 389
Araucaria, 184, 200
Arbour, Victoria, 348, 354
Archaeoceratops, 423
Archaeopteryx, 39, 40, 86, 294–304, 307, 315
"Archaeoraptor," 304–306
archosaurs, 65–67, 298
Archueleta family (Ghost Ranch), 176

Arctic, 200–203
Arctic Ocean, 378
Argentina, 68, 70–73, 140, 146, 150, 153–168, 182, 236–253, 346, 389, 416, 443
Argentinosaurus, 71, 156–160, 165, 238
Argyrosauridae, 167
Argyrosaurus, 161, 165
Arizona, U.S., 173–179
Army Corps of Engineers, U.S., 100
Artiodactyla, 39
Artsa Bogdo Basin, 418
Ashdown Sand, 17–19
Ashile Formation, 418
Ashmolean Museum, Oxford, 5
Asiatic Zoological Expeditions, 274
Asilisaurus, 70
Assyria, 133
asteroid impact extinction hypothesis, 445
Atlascopcosaurus, 201
Attenborough, David, 153
Auca Mahuevo, Argentina, 164–165, 247, 416
Aucasaurus, 247, 250, 252
Audubon, John James, 48
Augustana College, 187, 194
Auroraceratops, 423
Austin, Texas, 17
Australia, 167, 197, 200, 201, 245, 443
Australian Museum, Sydney, 240
Australovenator, 245
Austria, 367
Austro-Hungarian Empire, 334–336
Avemetatarsalia, 67–68
Aves, 64